ISBN 978-0-266-72769-9
PIBN 10976628

1 MONTH OF
FREE
READING

at

www.ForgottenBooks.com

By purchasing this book you are eligible for one month membership to ForgottenBooks.com, giving you unlimited access to our entire collection of over 1,000,000 titles via our web site and mobile apps.

To claim your free month visit: www.forgottenbooks.com/free976628

English
Français
Deutsche
Italiano
Español
Português

www.forgottenbooks.com

Mythology Photography **Fiction**
Fishing Christianity **Art** Cooking
Essays Buddhism Freemasonry
Medicine **Biology** Music **Ancient**
Egypt Evolution Carpentry Physics
Dance Geology **Mathematics** Fitness
Shakespeare **Folklore** Yoga Marketing
Confidence Immortality Biographies
Poetry **Psychology** Witchcraft
Electronics Chemistry History **Law**
Accounting **Philosophy** Anthropology
Alchemy Drama Quantum Mechanics
Atheism Sexual Health **Ancient History**
Entrepreneurship Languages Sport
Paleontology Needlework Islam
Metaphysics Investment Archaeology
Parenting Statistics Criminology
Motivational

THE

ASSURANCE MAGAZINE,

AND

JOURNAL

OF THE

INSTITUTE OF ACTUARIES.

VOL. IX.

LONDON:

CHARLES & EDWIN LAYTON, 150, FLEET STREET.

EDINBURGH: KENNEDY. NEW YORK: H. BAILLIÈRE, 290, BROADWAY.
PARIS: J. B. BAILLIÈRE, LIBRAIRE, RUE HAUTEFEUILLE.
HAMBURG: PERTHES, BESSER, & MAUKE.

1861.

LONDON:
PRINTED BY CHARLES AND EDWIN LAYTON,
150, FLEET STREET.

CONTENTS OF VOL. IX.

A 2

171086

Contents of Vol. IX.

ERRATA.

Page 220, line 3,   for a_a read a_2.

,,   line 10, for $\dfrac{9}{9}$ read $\dfrac{9}{7}$.

Page 234, line 17, for 10,000 read 100,000.

THE

ASSURANCE MAGAZINE,

AND

JOURNAL

OF THE

INSTITUTE OF ACTUARIES.

New Method for Calculating the Value of an Assurance to the Survivor Nominated, &c. By M. Réboul, lately of the Paris Observatory, and now Consulting Actuary of the "Impériale" Assurance Company.

[Read before the Institute the 2nd January, 1860.]

THE following letter from M. Réboul will serve to introduce the several subjects upon which he has addressed the Institute:—

"Gentlemen,—The Institute of Actuaries being the most competent tribunal in matters of assurance, it is to you, as to my natural judges, that I have conceived it proper to submit my first essays in that science.

"The manuscript, which I have the honour to present to you, contains:—

"1st. Five new tables of life annuities.

"2nd. The introduction and plan of a work upon assurance.

"3rd. Note upon a new formula for calculating the assurance of a capital sum to the survivor nominated.

"The first table contains the annuities on a single life, calculated at 5 per cent. from Deparcieux's table of mortality, perfected by the Bureau des Longitudes. For the values of these annuities, are taken the figures common to the results obtained by two completely different methods; this process has permitted the rectifying of some errors, which had slipped in at the end of the table published by Baron Mazères, and which have since been reproduced by every writer quoting that table.

"The four tables which follow contain annuities on two lives, calculated at $4\frac{1}{2}$ per cent., by the same methods and from the same data for mortality, and are designed to complete the repertory of Mazères, and to facilitate interpolations.

" The work, of which I present to you the introduction and plan suc-
cinctly, is designed to arouse in France a taste for life assurance, which
seems now to slumber; the author is animated by a strong confidence in
the usefulness of his undertaking, the success of which appears to him certain,
if he have the good fortune to merit your approbation.

" In fine, gentlemen, I particularly call your attention to the method
which has conducted me to the formula that I propose for the calculation of
a capital sum payable to the survivor nominated, and the exposition of
which is more simple than has been the investigation.

" I was connected with the Observatory of Paris during four years; I
have applied to the study of astronomy under the auspices of Arago, when
I was invited to occupy myself with questions pertaining to assurance.
I have found an immense field to explore; and this first memorandum is
not, in anywise, other than a specimen of labours which I have undertaken,
and which I shall crave permission to submit, should this attempt merit
some encouragement at your hands.

" Be pleased to accept, gentlemen, the very respectful expression of my
most distinguished esteem.

                                                      " E. Réboul.

" *To the Members of the*
     " *Institute of Actuaries of London.*"

The annuity tables accompanying M. Réboul's communication
are sufficiently described by that gentleman himself. The rate of
mortality, and the rates of interest upon which they are based, will,
no doubt, be considered by our readers as circumstances affording
sufficient reason for not reproducing them here. They appear to
have been constructed with great care, and are very neatly executed.
The introduction and plan of a work upon assurance consists of a
very warm eulogy on the system of assurance in general, and on
that of life assurance in particular, and a table of its proposed
contents is also appended. A portion of it, which the author has,
it seems, already published or sought to publish, and which is, as
he says, intended to awaken in France a taste for life assurance,
is as follows :—

                    "Des Assurances sur la Vie.

       " A qui les assurances sur la vie sont-elles applicables ?—A tous.
       " A qui sont-elles appliquées ?—A personne.

" Les assurances sur la vie, dont les bienfaits ne font pas question pour
quiconque a étudié cette branche de l'économie politique, sont passées dans
les mœurs en Angleterre, en Allemagne, en Autriche, aux Etats-Unis—c'est-
à-dire, dans les trois-quarts du monde civilisé.

" En France elles sont complètement négligées. Le mot seul d'assurance soulève une idée de défiance. Enfin—le dirai-je?—la matière assurée
n'est pas la dix-millième partie de la matière assurable. Ce qui ne nous
empêche pas de nous croire à la tête de la civilisation.

" Voici une grande vérité: tant pis si elle a l'air d'un paradoxe!

" Il existe, même pour celui qui n'a d'autres ressources que son travail,
un moyen sûr, infaillible, à l'abri de toute restriction légale ou fiscale, de

créer un patrimoine, de constituer sans propriétés des nues-propriétés, des legs, des dotations, des usufruits, des rentes viagères, &c.

" Ce moyen, c'est les assurances sur la vie, dont nous venons d'indiquer les principales applications, et dont on ne connaît guère en France que le nom; car leurs principes, leurs fonctions sociales, leur portée économique —y sont complètement méconnus; j'ai presque dit—inconnus. Quant à leurs lois mathématiques, je n'en parle pas; pour tout le monde, c'est de l'algèbre.

" En Angleterre, où il existe 250 Compagnies d'Assurances, dont quelques-unes sont plus puissantes que la Banque de France, le crédit tout entier repose sur l'assurance. Il ne s'y fait pas un emprunt, une commandite, une transaction quelconque, sans contrat d'assurance. Il n'est même pas rare qu'un négociant ait plusieurs polices sur sa tête.

" Ces précautions sont tout-à-fait inusitées dans notre pays. D'où vient que nous sommes ainsi arriérés? Est-ce la faute du public? est-celle des Compagnies? (car il existe des Compagnies en France.)

" Avant de discuter cette grave question, remarquons bien que la solution, quelle qu'elle soit, ne peut infirmer le principe de l'assurance. Comme toutes les vérités, il est impérissable. Et si la faute était aux Compagnies, nous dirions volontiers, Périssent les Compagnies plutôt qu'un principe!

" Mais, hâtons-nous de la dire; les Compagnies anonymes sérieuses, les Compagnies à primes fixes, comme on les appelle, présentent des garanties surabondantes, et offrent au public des gages de sécurité plus que suffisants. Nous établirons qu'une Compagnie bien organisée, et qui fonctionne régulièrement, est d'une solidité inébranlable, et que ses opérations sont d'une moralité qui défie toute critique.

" Malheureusement, 'le corps politique,' dit Bacon, 'ainsi que le corps humain, a ses charlatans qui se mêlent aussi de le traiter.'

" Au début des assurances, des gens sont venus, qui étaient parfaitement étrangers aux lois des probabilités; obligés de ramper, parcequ'ils ne savaient pas marcher, dépensant cent fois plus d'habileté pour aller de travers qu'il n'en faut pour aller droit, ils ont tout gâté. Sous couleur de bien public ils n'ont fait que des dupes, et sous le nom d'assurances ils ont répandu des chimères et semé partout des déceptions.

" Grâce à leurs déplorables essais, les assurances sur la vie, qui n'ont avec eux rien de commun que le nom, ont été longtemps discréditées en France. C'est ainsi que les pires abus s'attachent aux meilleurs choses— *Corruptio optimi pessima.*

" En présence de tels désordres, nous conclurons comme ce juif, qui, témoin des scandales de la cour d'Alexandre VI., se convertit au catholicisme, avouant qu'il fallait bien que cette religion reposât sur des principes vrais et que Dieu la soutînt, puisqu'elle restait debout au-milieu de la dissolution générale.

" Aujourd'hui, un grand mouvement se produit. Une révolution s'opère dans les assurances sur la vie. Les questions de démographie et de statistique, qui leur servent de base, sont élucidées. De nouvelles Compagnies se fondent, les affaires se multiplient, les vrais principes triomphent, et les combinaisons fécondes se dégagent des mauvaises, comme, dans le travail de la fusion, le métal pur se sépare des scories.

" Dans quelques années, si ce mouvement continue, le nombre des contrats aura décuplé; la méfiance injuste, qui entrave le progrés des assurances sur la vie, sera dissipée; et cette institution, si utile, sera définitivement naturalisée dans notre pays.

"Il y a dans cette régénération des assurances un travail immense à faire. La matière est inépuisable. Le champ est vaste et inculte. Plus il sera cultivé, plus il produira. Nous ne craignons pas de convier à cette œuvre féconde tous les esprits positifs, qui aiment les affaires sérieuses, morales, satisfaisantes.

"Les bonnes affaires ont en effet ce caractère remarquable d'être bonnes pour tout le monde. Les assurances sont de ce nombre; elles sont bonnes pour les Compagnies et bonnes pour le public.

"Elles sont utiles à l'individu, à qui elles offrent le moyen le plus sûr, en même temps que le plus moral, de faire fructifier ses économies dans un but de prévoyance.

"Elles sont utiles à la société, qu'elles sauvegardent en favorisant l'ordre, qu'elles enrichissent en diminuant le nombre des pauvres. Rappelons-nous, en effet, cette admirable définition de la richesse donnée par G. Varennes, et qu' Achille Guillard appelle avec raison une sanglante critique du pédantisme économique,—'La nation la plus riche est celle qui a le moins de pauvres.'

"Elles sont utiles à l'état, qu'elles aident puissamment dans sa principale, pour ne pas dire dans son unique fonction, qui est, de garantir la sécurité. Car, en arrachant l'homme à l'incertitude du sort, en l'empêchant de devenir la proie du hasard, l'assurance augmente dans une proportion énorme la sécurité, qui est le premier besoin d'une société, puisque sans elle il ne peut y avoir ni épargne, ni travail producteur.

"Ne cherchons pas d'autre criterium. Une institution qui satisfait à cette triple condition—d'être utile à l'individu, à la société elle-même, et à l'état—est éminemment morale; elle est inattaquable.

"'Parmi les établissements fondés sur les probabilités de la vie humaine, les meilleurs sont ceux dans lesquels, au moyen d'un léger sacrifice de son revenu, on assure son existence et celle de sa famille pour un temps où l'on doit craindre de ne plus suffire à ses besoins. Autant le jeu est immoral autant ces établissements sont avantageux aux mœurs, en favorisant les plus doux penchants de la nature. Le Gouvernement doit donc les encourager et les respecter dans les vicissitudes de la fortune publique, car les espérances qu'ils représentent, portant sur un avenir éloigné, ils ne peuvent prospérer qu' à l'abri de toute inquiétude sur leur durée.'

"Qui a dit cela? Laplace, à l'Ecole Normale, in 1795. A voir où nous en sommes, on croirait volontiers que c'est lundi dernier à la séance de l'Institut.

"Pour nous, économiquement parlant, l'humanité se partage en deux classes; ceux qui possèdent, ceux qui ne possèdent pas. Les uns généralement amis de l'ordre, les autres plutôt partisans de la liberté—qui serait bien la plus belle chose du monde, si l'ordre n'était encore plus beau.

"Or, les assurances sur la vie, en procurant la sécurité, en consolidant l'épargne, en constituant des patrimoines, des dotations, des retraites, &c., créent des valeurs nouvelles, multiplient les capitaux et les propriétaires, suivant le vœu émis par Napoléon I., au Conseil d'Etat. En un mot, elles font passer ceux qui ne possèdent pas dans la classe de ceux qui possèdent. Donc elles enrichissent l'Etat; donc elles accroissent l'ordre, ce qui est toujours le vœu du Gouvernement, qui, en retour, leur doit toute sa sollicitude, comme le remarque Laplace.   Et nous n'oublions pas que sollicitude, en même temps que protection, implique surveillance.

"Quelques-uns ont voulu voir dans l'assurance une spéculation, d'autres y ont cherché une spéculation.   Disons le très-haut, l'assurance n'est pas, ne peut, et ne doit jamais être une spéculation.   Laplace a dit, 'L'assurance par laquelle on échange l'incertain contre le certain est le contraire du jeu.'   Je dis plus: l'assurance est *l'antidote* de la spéculation; j'entends, la spéculation parasite—l'agiotage, enfin, puisqu'il faut l'appeler par son nom.

"La spéculation fomente l'inquiétude, bouleverse les fortunes, déclasse les individus, ruine les familles.

"L'assurance rétablit le calme, maintient la propriété, moralise l'individu en augmentant son bien-être, enrichit la famille.   La spéculation est immorale, subversive, spoliatrice.   L'assurance est morale, conservatrice, féconde.   L'une est aveugle comme le hasard; l'autre est prévoyante comme la providence.

"Les majorats et les substitutions sont abolies; nous n'avons plus guère de biens inaliénables en France.   Qui sait!   C'est peut-être à la précaution qu'auront prise leurs parents de les placer sous la tutelle de l'assurance, que les rejetons de quelques nobles familles devront un jour de ne pas se mésallier, et de ne pas se galvauder dans des entreprises équivoques.

"Assurez-vous! vous laisserez à vos héritiers un trésor et un bon exemple.

"Disciple de l'illustre vulgarisateur de l'astronomie, nous nous proposons une tâche moins difficile; celle de vulgariser la science des assurances, convaincus que lorsque la vérité sur ce point se sera fait jour, le succès dans l'application ne se fera pas attendre.

• "Cette science ne s'improvise pas.   Elle repose sur la théorie des intérêts, et sur le calcul des probabilités, qui a conduit Laplace à tant de belles découvertes et auquel nous devons un chef-d'œuvre—son *Essai Philosophique.*

"'L'esprit a ses illusions, comme le sens de la vue; et de même que le toucher corrige celles-ci, la reflexion et le calcul corrigent les premières.' (Laplace, *Essai Phil.*, p. 197.)

"Ces paroles de l'illustre géomètre doivent toujours être présentes à l'esprit de ceux qui s'occupent d'assurances.   Ils ne sauraient pas plus se passer de l'analyse que l'astronome ne saurait se passer de la géométrie, ou que le médecin ne saurait se passer de l'anatomie.

"Rappelons aux Compagnies ce que le célèbre Docteur Price disait à ce sujet en 1762.

"'Il est d'une grande importance pour la sûreté d'un établissement de cette nature que ses opérations soient contrôlées par d'habiles mathématiciens.   Une déplorable expérience a montré qu'eux seuls peuvent avec sécurité fonder et conduire ces établissements.'

"Un dernier mot sur les assurances:—

"Quelle est la clef de voûte de cet édifice auquel nous nous efforçons d'apporter notre pierre?   La confiance publique.

"Nous nous proposons d'exposer le plan d'un système d'assurances qui, sans passer par l'Etat, commande la confiance et défie tout mauvais vouloir.

"Nous exposerons quelques combinaisons, ou, pour mieux dire, quelques applications nouvelles.   L'assurance en garantie de créance ou de recouvre-

ment, l'assurance de solvabilité, l'amortissement viager; en un mot—
l'organisation du crédit par l'assurance; enfin, nous donnerons une théorie
générale des transactions viagères.

"Notre but est suffisamment indiquée; quant à notre route, elle est
toute tracée—la vérité avant tout.　En moral comme en physique le plus
court chemin est la ligne droite.　C'est celle que nous suivrons.　C'est
aussi celle que suit la lumière.

"E. Réboul.

"2 *Décembre*, 1858."

We now come to M. Réboul's "New formula for calculating
the assurance of a capital sum to the survivor nominated"; and,
that there may be no room for any misunderstanding arising out of
an erroneous translation, we give as much of the author's statement
as is needful in the original:—

"Nouvelle Méthode pour calculer la Valeur d'une Assurance
au Survivant Désigné.

"L'assurance d'une somme s'obtient en retranchant de cette somme sa
rente servie jusqu'à l'année à la fin de laquelle doit être payée l'assurance;
ainsi, qu'il s'agisse de 1 fr. payable au décès de A, r étant le taux d'intérêt,
et A l'annuité viagère sur la tête A, le prix de cette assurance sera—

$$\frac{1-Ar}{(1+r)}.$$

"Les formules d'assurances doivent donc en général se déduire de
celles des annuités viagères; c'est ce que nous nous sommes proposé de
faire pour l'assurance au *survivant désigné*, qui semblait faire exception.

"Supposons qu'il s'agisse de deux têtes A et B.

"Soit x la valeur de l'assurance de 1 fr. payable au décès de A s'il
meurt le premier, et y la valeur de l'assurance de 1 fr. payable au décès de
A s'il meurt le dernier.

"En ajoutant ces deux valeurs, x et y, on obtient évidemment le prix
de l'assurance de 1 fr. payable sans conditions au décès de A; ainsi on a—

$$[1]\quad x+y=\frac{1-Ar}{(1+r)}.$$

"Tel est le principe bien simple sur lequel nous nous appuyons: x est
précisément la valeur de l'assurance de 1 fr. payable à B au décès de A, or
on tire de l'équation [1],

$$x=\frac{1-Ar}{(1+r)}-y.$$

"Reste à déterminer la valeur de y.　Pour y parvenir, supposons que
A et B conviennent d'acheter une *assurance payable au dernier décès*,
y sera la part de A dans cet achat; or, sans nous préoccuper de la question
pratique qui pourrait conduire à chercher la proportion dans laquelle A et
B doivent contribuer au paiement de la dite assurance, remarquons que si
ce paiement était fait par primes annuelles, la quotité de ces primes serait—

$$[2]\quad q=\frac{1-(A+B-AB)r}{(1+r)(1+A+B-AB)}.$$

" Tellc est la rente qui doit être payéc d'avance et jusqu'au dernier décès des têtes A et B pour acquitter l'assurance, et la part de A dans une rente de 1 fr. (servie dans ces mêmes conditions) étant $\frac{1}{2}+A-\frac{1}{2}AB$ (*voir* Baily, tome 1, chap. 4, § 84), sa part y dans une rente de q francs sera $y=q\left(\frac{1}{2}+A-\frac{1}{2}AB\right)$; d'où en remplaçant q par sa valeur [2], et portant y dans l'équation [1], on tire—

$$[3] \quad x=\frac{1-Ar}{(1+r)}-\left(\frac{1}{2}+A-\frac{1}{2}AB\right)\left[\frac{1-(A+B-AB)r}{(1+r)(1+A+B-AB)}\right].$$

" C'est la formule que nous proposons pour le calcul des *assurances au survivant désigné.*

" Si l'on y remplace A par B, et B par A, on aura la valeur de l'assurance de 1 fr. payable à A ou décès B; désignons là par x'. On trouve

$$x'=\frac{1-Br}{(1+r)}-\left(\frac{1}{2}+B-\frac{1}{2}AB\right)\left[\frac{1-(A+B-AB)r}{(1+r)(1+A+B-AB)}\right],$$

x et x' représentent respectivement les parts de B et de A dans l'assurance au survivant quelconque; comme vérification, leur somme $x+x'$ doit donc se réduire à $\frac{1-ABr}{(1+r)}$, valeur de l'assurance de 1 fr. payable au premier décès.

" Avant d'effecteur, remarquons que la quantité [3] étant symétrique par rapport aux lettres A et B, n'a pas changé par leur permutation, et peut être mise en facteur commun, ce qui donnera—

$$x+x'=\frac{1-Ar+1-Br-(1+A+B-AB)\times\dfrac{1-(A+B-AB)r}{(1+A+B-AB)}}{(1+r)},$$

d'où, en supprimant $(1+A+B-AB)$, facteur commun, et développant, on obtient

$$x+x'=\frac{1-Ar+1-Br-1+Ar+Br-ABr}{(1+r)}=\frac{1-ABr}{(1+r)},$$

expression qui en effet se réduit à

$$x+x'=\frac{1-ABr}{(1+r)}.$$

" Les réductions que comporte le développement de la formule [3] ne permettant pas de la rendre plus élégante, nous nous abstiendrons des trans-formations, et nous la laisserons sous cette forme, qui nous a paru la plus commode pour le calcul.

" Il est important de remarquer que cette formule [3] ne renferme que quatre quantités, A, B, AB, et r, et que la manière dont elles sont com-binées entre elles permet d'arriver trés-simplement au résultat, qui même se déduirait de deux opérations seulement, si l'on puisait les valeurs de q et de $\frac{1-Ar}{(1+r)}$ dans les tarifs d'assurances.

" La formule que l'on trouve dans Baily (chap. 8, prob. 27, § 231) est défectueuse et mal commode, car elle ne se prête pas du tout au calcul logarithmique; la méthode employée pour l'obtenir n'est pas rigoureusement exacte, puisqu'elle s'appuie sur cette hypothése évidemment fausse à savoir: que dans le cas où les deux têtes s'éteignent dans le cours de la même année, l'une en particulier a autant de chances que l'autre de mourir la première; de sorte que la probabilité de cet évènement est représentée par $\frac{1}{2}$ dans tout le cours du calcul; de plus, l'introduction de têtes auxiliaires amène dans la formule quatre quantités nouvelles qui la compliquent en aggravant la probabilité d'erreur.

"Enfin, les huit quantités contenues dans la formule de Baily doivent être déterminés avec une approximation beaucoup plus grande que celle qu'on veut conserver au résultat, puisqu'il s'obtient en retranchant l'une de l'autre deux nombres qui ne diffèrent pas en général par leurs trois premières figures.

" Aussi les vérifications ne réussissent-elles pas à beaucoup près aussi bien que par l'emploi de la formule [3], comme on peut en juger par les comparaisons qui suivent (les calculs sont faits pour 100 fr.):—

	Méthode nouvelle.	Méthode de Baily.
Assurance à B 25 au décès de A 20 .... =	18·527	18·691
Assurance à A 20 au décès de B 25 .... =	20·923	21·438
Somme .... =	39·450	40·129
Au lieu de (assurance au survivant quelconque) =	39·452	39·452
Erreur .... =	−0·002	+ 0·677
Assurance à B 60 au décès de A 20 .... =	9·979	9·142
Assurance à A 20 au décès de B 60 .... =	51·737	52·223
Somme .... =	61·716	61·365
Au lieu de (assurance au survivant quelconque) =	61·715	61·715
Erreur .... =	+ 0·001	−0·350

&c. &c. &c.

" *Exemple du calcul de l'Assurance au Survivant désigné de deux têtes, par les deux Méthodes.*

" £100 payables à B 30 au décès de A 25, à 4°/₀, et d'après Deparcieux.

" *Méthode de Réboul :*—

$$\text{Formule}: \quad x = \frac{1-A r}{1+r} - (\tfrac{1}{2} + A - \tfrac{1}{2}AB)\left(\frac{1-(A+B-AB)r}{(1+r)(1+A+B-AB)}\right).$$

" *Méthode indiquée par Baily :*—

$$\text{Formule}: \quad \frac{s}{2}\left\{\frac{1-AB r}{1+r} - \left(\frac{(1+A'B)a'}{1+r} - AB.a\right)\frac{1}{a}\right\}."$$

The writer here works out a particular case by each formula, and proceeds :—

" Ici le calcul ne se compose que d'additions et de soustractions, en tout on a huit logarithmes à chercher, tandis que dans la méthode de Baily, il faut employer de plus la multiplication et la division; car, outre que la formule ne se prête pas au calcul logarithmique, sept figures ne suffisent souvent pas pour les nombres correspondants.

"Ajoutons que, s'il s'agit de construire un tarif entier, notre formule parait bien préférable, puisqu'en l'employant on peut construire en même temps trois tarifs qui ne donnent pas autant de peine qu'un seul construit par la méthode ordinaire.

"Savoir: 1°. Assurance d'un capital payable au premier décès de deux têtes.

"2°. Assurance d'un capital payable au dernier décès de deux têtes.

"3°. Assurance d'un capital payable au survivor désigné de deux personnes."

Such is M. Réboul's statement as to the construction of his new formula, which, if it were true for all values of A and B, would, no doubt, deserve to be looked at with all the partiality with which he evidently regards it; unfortunately, however, this is not the case, the results given by the formula being nothing more than approximate ones, excepting only when the ages of A and B are the same and the values which they represent equal, under which conditions alone the results obtained by it are quite accurate.

Translating M. Réboul's formulæ into a notation with which most of our readers are more familiar, it will be seen that the expressions [1], [2], and [3], are equivalent to the following, viz.:—

$$\underset{1}{\mathfrak{A}\mathfrak{B}} + \underset{2}{\mathfrak{A}\mathfrak{B}} = \mathfrak{A}, \quad q = p, \quad \text{and} \quad \underset{1}{\mathfrak{A}\mathfrak{B}} = \mathfrak{A} - p(\tfrac{1}{2} + A - \tfrac{1}{2}AB),$$

(p being the annual premium for assurance of £1 at decease of the survivor of A and B); and, since $\underset{1}{\mathfrak{A}\mathfrak{B}} = \mathfrak{A} - \underset{2}{\mathfrak{A}\mathfrak{B}}$, it follows, according to M. Réboul's reasoning, that

$$\underset{2}{\mathfrak{A}\mathfrak{B}} = p(\tfrac{1}{2} + A - \tfrac{1}{2}AB).$$

To show that this is not so, let us denote, by $v'd'_{xy}$, $v''d''_{xy}$, $v'''d'''_{xy}$, &c., the value of £1 receivable at the end of the first, second, third, &c., years, in the event of two lives, aged x and y, both dying in them respectively; and by $v''d''_x \delta_y$, $v'''d'''_x \delta_y$, &c., and $v''d''_y \delta_x$, $v'''d'''_y \delta_x$, &c., the value of the like sum so receivable in the event of one dying in the second, third, &c., year, the other having died previously. We have, then, for the value of £1 payable at the end of the year in which the survivor of A and B, aged x and y respectively, shall die, the expression

$$v'd'_{xy} + v''d''_{xy} + v'''d'''_{xy} + \&c. \qquad \cdot \quad \cdot \quad \cdot \quad (a)$$
$$+ v''d''_x \delta_y + v'''d'''_x \delta_y + \&c. \qquad \cdot \quad \cdot \quad \cdot \quad (\beta)$$
$$+ v''d''_y \delta_x + v'''d'''_y \delta_x + \&c. \qquad \cdot \quad \cdot \quad \cdot \quad (\gamma)$$

the sum of which three series, continued to the extremity of life, may be represented by $\overline{\mathfrak{AB}}$, so that $\overline{\mathfrak{AB}} = a + \beta + \gamma$.

Reasoning in like manner, it will be seen that $\underset{2}{\mathfrak{AB}} = \frac{1}{2}a + \beta$.

Now, $\overline{\mathfrak{AB}} = p(1 + A + B - AB)$, and $\frac{1}{2}\overline{\mathfrak{AB}} = p(\frac{1}{2} + \frac{1}{2}A + \frac{1}{2}B - \frac{1}{2}AB) = p(\frac{1}{2} + A - \frac{1}{2}AB) + p(\frac{1}{2}B - \frac{1}{2}A) = \frac{1}{2}a + \frac{1}{2}\beta + \frac{1}{2}\gamma$.

But $\underset{2}{\mathfrak{AB}} = \frac{1}{2}a + \beta$, and is therefore, obviously, equal to

$$p(\tfrac{1}{2} + A - \tfrac{1}{2}AB) + \tfrac{1}{2}\beta - \tfrac{1}{2}\gamma + p(\tfrac{1}{2}B - \tfrac{1}{2}A),$$

and not to $p(\frac{1}{2} + A - \frac{1}{2}AB)$ only, as M. Réboul would have us believe. Hence, too, it appears that $\underset{2}{\mathfrak{AB}} = \frac{1}{2}\overline{\mathfrak{AB}} + \frac{1}{2}\beta - \frac{1}{2}\gamma$, and that $\underset{2}{\mathfrak{AB}} = \frac{1}{2}\overline{\mathfrak{AB}} - \frac{1}{2}\beta + \frac{1}{2}\gamma$.

It will be observed that M. Réboul does not give any demonstration of his formula, but he assumes, in effect, that the quantity $p(\frac{1}{2} + A - \frac{1}{2}AB)$ (which is nothing more than the value of half p payable during the life of A, and of the whole of p during his survivorship) represents truly the single premium for an assurance of £1 at decease of A, provided his life be the second that fails.* He appears, in short, to have argued, that since

$$\overline{\mathfrak{AB}} = p(1 + A + B - AB),$$

or, what is the same thing, that since

$$\underset{2}{\mathfrak{AB}} + \underset{2}{\mathfrak{AB}} = p(\tfrac{1}{2} + A - \tfrac{1}{2}AB) + p(\tfrac{1}{2} + B - \tfrac{1}{2}AB) \quad . \quad . \quad (a)$$

that, therefore,

$$\underset{2}{\mathfrak{AB}} = p(\tfrac{1}{2} + A - \tfrac{1}{2}AB),$$

$$\text{and } \underset{2}{\mathfrak{AB}} = p(\tfrac{1}{2} + B - \tfrac{1}{2}AB),$$

an assumption plainly inadmissible in the absence of a direct demonstration, and, in the present instance, untrue; the fact being, that the quantities on either side of equation (a), are, for the most part, merely complementary to each other; so that, dealing with their sum, the results are truthful, whilst the use of either in the way suggested leads to error. Thus, if we make an addition of the two following equations, viz.:—

---

* The quantity in question represents, in fact, the value of A's share of the premium paid annually for assurance of £1 at decease of the survivor of A and B, supposing it to be arranged that each is to pay half the premium while both are living, and that the survivor is to pay the whole. How far its value differs in particular cases from what M. Réboul supposes it to be, will be seen in the following instances, which are based on the Three-per-cent. Northampton Table—the first column being derived from M. Réboul's expression, and the second from the $\mathfrak{A} - \underset{1}{\mathfrak{AB}}$ of the tables —:

A *aged* 30, *and* B *aged* 60.		A *aged* 45, *and* B *aged* 65.
$\underset{2}{\mathfrak{AB}} = 2994$ 3025		$\underset{2}{\mathfrak{AB}} = 3531$ 3584
$\underset{2}{\mathfrak{AB}} = 1378$ 1347		$\underset{2}{\mathfrak{AB}} = 1766$ 1713
$\overline{\mathfrak{AB}} = 4372$ 4372		$\overline{\mathfrak{AB}} = 5297$ 5297

$$\underset{1}{\mathfrak{AB}} = \mathfrak{A} - p(\tfrac{1}{2} + A - \tfrac{1}{2}AB),$$
$$\underset{1}{\mathfrak{AB}} = \mathfrak{B} - p(\tfrac{1}{2} + B - \tfrac{1}{2}AB),$$

we find that their sum, viz., $\mathfrak{AB} = \mathfrak{A} + \mathfrak{B} - \overline{\mathfrak{AB}}$, expresses what is universally true, although each of the equations composing it is defective, except in the particular case of A and B being of equal ages, when it will be seen that the series β and γ become identical, and that those quantities vanish altogether in the equations given above. The addition, accurately stated, would, of course, be

$$\underset{1}{\mathfrak{AB}} = \mathfrak{A} - \tfrac{1}{2}\overline{\mathfrak{AB}} - \tfrac{1}{2}\beta + \tfrac{1}{2}\gamma$$
$$\underset{1}{\mathfrak{AB}} = \mathfrak{B} - \tfrac{1}{2}\overline{\mathfrak{AB}} + \tfrac{1}{2}\beta - \tfrac{1}{2}\gamma$$

the sum being . . . $\mathfrak{AB} = \mathfrak{A} + \mathfrak{B} - \overline{\mathfrak{AB}}$ as before.

M. Réboul shows (*see* p. 8) that the summation of the figures obtained by his method gives a more consistent result than does that of the figures obtained by Baily's formula. That his should be consistent, though erroneous, will not appear surprising after what has been said. That those arising from Baily's formula should be inconsistent, is, no doubt, attributable to the circumstance that the values of the annuities used in the calculation do not contain a sufficient number of fractional parts to bring out the assurance-values more exactly. The authors of *Assurance and Annuity Tables* have shown how great an influence this insufficiency exercises.* As to the objection raised by M. Réboul, that the method in question is not exact, on the score of the adoption of the fraction $\tfrac{1}{2}$ as the measure of the probability that A will die before B in any particular year, we think those who have read Mr. Peter Gray's unanswerable essay† will attach but little importance to it. It must, perforce, be admitted that Mr. Gray has *proved* that the adoption of such fraction for the purpose is rigorously true, on the hypothesis of a uniform distribution of the deaths of each year,‡ and that he has shown that the errors arising under that hypothesis are, for the most part, very insignificant. Mr. Francis Baily's formula cannot, we submit, be successfully assailed on this score, or, indeed, on any other.

In a note subsequently addressed to the Institute, M. Réboul discusses the several formulæ for the values of an assurance payable earlier than at the end of the year in which the life assured fails. A good deal in reference to these formulæ has already appeared here,

---

* *See* page 6 of the Introduction to that Work.
† "On the true Measure of the Probabilities of Survivorship between Two Lives (*see* vol. i., p. 137).
‡ *See* also Milne, vol. i., article 110.

and the matter would seem to be nearly exhausted; nevertheless, we shall hope to recur to M. Réboul's observations upon it in a future Number.

From all which has preceded, the readers of this *Journal* will, no doubt, observe that there is nothing in M. Réboul's communication which adds materially to our knowledge of the subject to which it refers. They will, however, we think, not be of opinion that an undue prominence has been given to it in these pages, when all the circumstances under which it has originated are taken into consideration.—ED. *A. M.*

On some Considerations suggested by the Annual Reports of the Registrar-General, being an Inquiry into the Question as to how far the Inordinate Mortality in this Country, exhibited by those Reports, is controllable by Human Agency. (Part I.) By H. W. PORTER, ESQ., B.A., *Assistant Actuary to the Alliance Assurance Company, Fellow of the Institute of Actuaries and of the Statistical Society.*

[Read before the Institute the 30th January, 1860.]

UPON perusal of the Annual Reports of the Registrar-General, and of the valuable analyses by Dr. Farr that accompany them, we can scarcely fail to be struck with the alarming fatality which appears to prevail in this country from certain classes of diseases; and the inquiry naturally arises in our minds, whether we have any control over this excessive mortality: is, in fact, any proportion thereof preventable?

It is, I believe, an admitted fact, that the maladies most fatal in England are those which are classed under the general head of pulmonary complaints—of which phthisis, or consumption, contributes the greatest number of victims.

Life Assurance Companies have a great enemy to contend with in this class of diseases, as the returns of those Companies which have published their experience in respect of mortality very clearly show. Unfortunately, but few Societies have thought it desirable to make public the information they possess. It is to be regretted that this should be the case, as, by a combination of the statistics of the large number of Assurance Companies which now exist, some light could not fail to be thrown upon the subject, and the means might possibly be afforded of enabling Assurance Companies to deal with the very numerous cases in which an hereditary taint of consumption is traceable, and which are now necessarily ruth-lessly rejected by them, owing to the absence, at present, of any reliable data on the subject.

It is important, however, not to lose sight of the fact, that it is not among the class of life assurers, but among the labouring community, that the intensity of mortality from consumption is to be found ; and when the inordinate mortality to which we are now referring is made clearly apparent, and the matter is forced upon the attention of the public, as it is likely in the course of time to be through the medium of the Registrar's Reports, a very serious responsibility will, I apprehend, be thrown upon our legislators if they neglect to consider this question with the view to the pre-vention of some portion at least of the unnecessary sacrifice of human life, both from disease and accidental causes, which now takes place every year.

With reference to phthisis.   This disease, Dr. Farr tells us, " is the greatest, the most constant, and the most dreadful of the dis-eases that afflict mankind."   "It is the cause," he says, "of nearly half the deaths that happen between the ages of 15 and 35 years."

It will not, perhaps, be considered that I am asserting too much when I say that the mortality under this head is, in a great mea-sure, produced by causes which are more or less under human control ; and the classification of the occupations followed by those who fall a prey to consumption, which it will come within the pro-vince of the Registrar to make, will serve, in course of time, to point out in what direction public attention must be turned, with the view to ascertain what may be the influence of the unhealthy nature of certain occupations and of peculiar local influences upon the duration of life, and to show our legislators how it is within their power to ameliorate the unhappy condition, in respect to health and longevity, which may, by such means, be made appa-rent, and possibly, moreover, to place in the hands of the medical profession the means of grappling with a disease which is the great terror of this country, and which has hitherto defied all human skill, and entirely baffled medical science.

It is only by bringing these subjects before the public that we can hope to find any remedy for evils which, though they are now beginning to be known to exist around us, it seems to be no one's particular business to help to remove.

The first step towards providing a remedy for an evil, is to promulgate a knowledge of it.

Unfortunately, the Reports of the Registrar, which supply data, the reliability of which, from the elaborate system now in operation for recording facts, is undoubted, are not of a character likely to come into the hands of the general public.

It may, I imagine, be considered to fall peculiarly within the province of the Institute of Actuaries to endeavour to draw attention to matters of this nature, and to place before the public information in a shape in which alone it can be made available for general information.

The Institute, by making itself a medium for the diffusion of facts of the class under consideration, will confer a lasting benefit on the public, who will have good reason to be thankful for the establishment of a Society which, I may be allowed to say, has already, to no inconsiderable extent, assisted in the promulgation, by means of its published transactions, of much useful knowledge; and an acquaintance with facts is clearly essential before any remedy can be suggested with the view to check existing evils.

This Institute may have the proud feeling of having always promulgated its information for the use and advantage of the whole assurance body, an example that might, with advantage, be more generally followed.

Now, however valuable the voluminous returns of the Registrar's office may be, they are, in fact, practically, almost useless, unless the conclusions that may be drawn from them are made clearly apparent.

The public have to thank the indefatigable Dr. Farr for conducting them through the almost Egyptian darkness of those masses of figures, and for acting as an interpreter of the hieroglyphics of the Registrar. But, when we have deduced our inferences from the data presented to us, any deduction we may draw would prove, like the diagnosis of the physician, highly injurious unless it were strictly correct, and, like that diagnosis, however logically it might be made and however correct it might be, would be entirely useless unless we can contrive to combine with our skill in diagnosis the additional resources of the therapeutic art. Supposing, then, that Dr. Farr has most skilfully interpreted the facts recorded in the Reports, and that he has scientifically diagnosed the disease from which the country is suffering, we next have to look to the means of cure. This rests with the Legislature to effect, but, as is the case with most reforms, it is to public opinion that we must really look for taking the initiative in the matter—and how are we to influence public opinion?

It is, probably, by the diffusion of sanitary knowledge only that we are likely to gain our end; and by that means alone that we shall be enabled to introduce remedies—to force them upon our rulers, in fact—for the many preventable evils which are found to exist, and which, if even suspected before the establishment of the registration department, were certainly not positively known to be so bad as they turn out to be.   I allude principally to the mortality caused by pulmonary and other cognate diseases, and by fevers and small-pox—to the excessive mortality among infants—and to deaths from accidental causes.

I will beg to point out, however, that my paper is intended to be more in the nature of an essay than of a statistical analysis of the returns of the Registrar-General, a work which, if I felt inclined to undertake, would be rendered supererogatory by the elaborate inquiry into the subject by Dr. Greenhow, in his capacity of Lecturer on Public Health at St. Thomas's Hospital.

In endeavouring to bring about any sanitary reform, we have a great amount of ignorance to contend with, and it is painful to observe how great a tendency ignorance unfortunately has to reproduce itself.   This fact we see very clearly established by the statistics of marriage.   The results observable from some of the returns give rise to the consideration, the Registrar tells us, " whether the ignorant have a tendency to marry the ignorant in a greater or less proportion than the learned, up to the writing point, marry the ignorant, or than those so far learned marry the learned."   The Registrar, after investigating this question, comes to the conclusion that " the ignorant evidently intermarry, by choice and the force of circumstances, to a much greater extent than would be inferred from their numbers; and this," he goes on to say, " is important, as the result is, that of 24 in every 100 of the families that are now constituted every year by marriage, the children are without the advantage of having either the father or the mother able to write."   Hence the reproduction of ignorance.

In any steps that may be taken by the philanthropist for the amelioration of the sanitary condition of the population, prejudices, the result of ignorance, are a great stumbling-block in the way; and, probably, until we are enabled to improve the education of the people, we shall not make great progress in improving their position in respect to health and longevity.

Our legislators are, perhaps, not free from blame in allowing the present state of things to continue; and I fear it is a slur upon us as a nation, that no general plan of national education has, up to this period, been determined upon; but most sad it is to contem-plate, that, while the evils which result in so many different forms from the existing ignorance of the masses are so patent, the whole question remains at a standstill—not because we are unable to organize a scheme of national education, but because we allow the conflicting strife of different religious sects to interfere with the progress of education among the people. These sects, like the rival physicians of Le Sage, allow the patient to die while they argue about systems.

The urgency of the case, which is continually becoming more apparent, seems almost to warrant, if not to demand, that, to some degree at least, the compulsory education of the young should be made the subject of legislative enactment—notwithstanding the objections to what has been termed "meddling legislation."

There is, probably, no doubt that the failure in the satisfactory settlement of this question has arisen from the difficulty of dealing with the religious element involved; but surely it is time that the contentions of the religionist and the secularist should cease—looking to the urgent necessity for increased education, so very fully made apparent by the annual marriage returns. From the last published Annual Report, the 20th—viz., that for the year 1857—it appears that the number of persons signing the marriage register with marks was no less than 105,778—the proportions per cent. being of the men 72, and of the women 61 only, who were able to write their names, notwithstanding the progressive educational improvement which we are told has been going on since the passing of the Reform Bill in 1832.

The Dean of Carlisle, Mr. Close, no mean authority on the subject of education, attributes the failure of the several attempts to settle this question to the circumstance that too much has been attempted in all Bills that have hitherto been introduced into the House. He objects to one general plan of education, and is opposed to the institution of local rates for the purpose of supply-ing the necessary funds, whether such rates be voluntary or com-pulsory. In a pamphlet on the subject, he enters fully into the question; but it will be sufficient here to state, that a voluntary rate would clearly be most likely to be objected to by a majority of ratepayers in the very localities in which education was most re-quired—in small towns and villages—where the ratepayers were but slightly in advance, in point of education, of those intended to be educated; and that the great argument against a compulsory rate is, that the ratepayers in that case would have a fair ground to insist that the management of the schools which they supported should remain in their own hands, and with this they could not be trusted at present.

Mr. Close considers that the course which would be most easily carried out in practice would be the extension of the present system of Government aid now actually in operation, leaving it to a trust-worthy executive to determine in what manner, and to what extent, the ample public funds with which they might be entrusted should be distributed among the schools, in such manner as the very many different circumstances in which schools in large towns and small villages, in rich or poor neighbourhoods, and affected by manifold local influences, might seem to demand.

In some cases, the total expense of the schools might advantageously be borne by the Government; in others, such a course might be productive of evil, and have the effect of destroying the school; in such cases, either no assistance should be given, or some partial aid only afforded.

With due respect to Mr. Close, however, I must be allowed to express an opinion that his idea of "leaving it to a trustworthy executive to determine in what manner and to what extent the ample public funds with which they might be entrusted should be distributed," is just begging the whole question, and is, in fact, supposing the matter he has to provide for already decided. How is this "trustworthy executive" to be appointed, and of what religious party is this executive to consist, which is to have the disposal of large national funds? It would appear, from the statement of this gentleman—and it must be remembered that the name of the Dean of Carlisle has been intimately associated with the educational question for many years past—that, in practice at least, the religious element is not found to create the difficulty that might be imagined; and his plain common-sense view of the case would seem to show that the difficulties in the way of the satisfactory settlement of the matter are, to say the least, not of an insuperable character.

In all existing schools (of course, those assisted by Government funds are referred to), the Bible, we are told, is used, with one or two exceptions only; the Church Catechism, too, is taught in many schools in which the children of Dissenters are among the pupils; more or less tolerance upon points of religious teaching is displayed by the managers, and it is stated that, except in a few instances, those engaged in the actual instruction of the children of the working classes cannot complain of having met with any serious obstacles in preserving the religious portion of education.

In the hands of this executive—which would be composed, I apprehend, of all religious sects—Mr. Close, a clergyman, re-member, of the Established Church (and one who may therefore be fairly considered to have his own bias in favour of keeping all education which is the result of a Government grant in the hands of the clergy), would be content to leave the power of determining, according to the local circumstances of each case, whether any par-ticular school should be Church of England, Dissenting or mixed, or in the principles of the British and Foreign; and how, by whom, and in what proportion, the Bible should be read and the religious instruction given for the good of the whole community.

But, above all, he considers that the education of the children of the working classes should be made compulsory. "Slowly and reluctantly, and after struggling against this necessity for nearly 40 years," this practically-experienced authority says, "I am an absolute convert to this necessity;" and, looking to the figures just quoted from the marriage records, will anyone present say that this necessity is not apparent? The same influence that rendered legis-lation an absolute requirement of the age with the view to pre-venting the too early employment of children in factories—the covetousness, Mr. Close calls it, but I should prefer to say the necessities, of the parents—will ever be at work to restrain them from voluntarily giving their children the advantage of education, and probably, too, this effect will be increased in proportion to the improved education that the schools may provide; for it seems that it has been proved, and the opinion of one of Her Majesty's Inspectors of Schools is quoted in confirmation of the fact, that, "in many of our best schools, where there is every possible attraction of superior cultivation, where the best attention is devoted to religious train-ing, and where every encouragement to success is held out, there the children are taken away at an earlier age than ever"—and why? because, by reason of the improved standard of education, boys of 9 years of age are more advanced than those of 12 used to be, and they are accordingly removed by their parents from school, to be placed at work—a strong reason, I submit, for the compul-sory system of education.

It is satisfactory to find, by the last Report of the Inspectors of Factories, "that some instances have recently occurred of mill-owners giving a preference to the employment of young persons who are able to read."

Lord Brougham—who has done more for education than, per-haps, any man living—stated, in the House of Lords, towards the end of last Session, after alluding to the increase in the number of Day and Sunday Schools, and of the number of pupils who attended them, since the year 1818, that the difficulties of the question were increased by religious differences, and expressed his own opinion that he should prefer education even under the Pope of Rome to no education at all. As a means of improving the defective education of the next higher class in the social scale, he proposed that the Privy Council should have the power of inspection of the middle-class schools throughout the country, in addition to the inspection they now have the power to make of all schools assisted by a Government grant. This proposition, however, was objected to on the part of the Government, on the ground of the additional labour that it would impose if carried out. The suggestion was a most important one, for the recent middle-class examinations that have been established have shown that the middle-class schools of this country are far from imparting generally as sound an *elementary education* as even some of our national schools which are attended by the children of the working classes—the failures in the first of these examinations which was held having shown that a very much less number of candidates than could possibly have been expected gave evidence of being even decently grounded in elementary knowledge of the humblest character.

The National Education question at present stands thus :—

It having been, up to this time, found to be impossible, for the reasons adverted to in the foregoing remarks, to organize any permanent system, the education of the children of the working classes of this country has been carried on by means of Government grants, subject to certain conditions, which have from time to time been increased, in aid of the funds collected by private efforts in different parts of the kingdom.

The amount of Parliamentary grants voted for education in the year 1857 was £541,233.

A commission is now sitting, with the view to inquire into the whole question of Government grants for educational purposes; and the result of the report of the commissioners will possibly lead, at last, to some legislative measure for settling this long-debated question.

This may appear a lengthened digression from the exact subject of this paper, but I am anxious to show that the extension of education is what we must very much regard if we desire to amend the sanitary condition and so to improve the health of the labouring community, of which the bulk of the population is composed, and to diminish the unnecessary mortality that is shown at present to exist in this country, seeing that our efforts are so much checked by the ignorance and prejudices that we have to contend against.

The efforts of the Industrial Pathology Committee, in connection with the Society of Arts, are stated, in a paper by a member of that Society, to have almost entirely failed in their object, which was, to endeavour to mitigate some of the evils at present inherent in certain unhealthy employments, owing, I apprehend, to the apathy resulting for the most part from ignorance of the power of science to influence the matter.

In immediate connection with the too early employment of children in factories, just referred to, as affecting the question of education, the consideration arises as to the effect upon their health and longevity, which we shall consider hereafter.

Now, not only does the existing want of education tend to promote and engender disease, because parents, owing to their ignorance of the evils that are likely to result from such a course, persist in placing their children far too early at laborious employments; but the higher standard of education now happily beginning to be afforded to the present generation, tends, as has been shown, to increase the evil as affects the parents. It is not, therefore, probable that, until the last generation has passed away altogether, we shall advance as we ought; but, then, the present generation, having had the means of obtaining a higher class of information in their youth, will hardly fall into the errors committed by their fathers—of placing in turn their children so early at work as to render their proper education impossible; and so the tendency will be to increase the standard of education, and to diminish the evil consequences which result, in so many different ways, from a want of knowledge among the working classes. This is so far as the ignorance of the parents affects the question—how far the greed of gain, or the necessities of the heads of families, may still tend to bring about the same results, is another question; and hence the crying necessity that education should be compulsory, at least for many years to come.

Whether such a state of things should be allowed to exist, in which the labour of the young children of a family should be an actual necessity to enable the whole family to live, as now really is the case in some manufacturing districts, is a problem in political economy that cannot be discussed here.

To return to phthisis: the total number of deaths registered in the year 1857 under this head was 50,106.

If to this number we add the deaths which are classed under other heads, but which are likewise of a tubercular character—viz., hydrocephalus, tabes mesenterica, and scrofula—we obtain a total of deaths from tubercular diseases of 65,762; adding to this total the deaths from other diseases of the respiratory organs—viz., bronchitis, pneumonia, asthma, disease of the lungs, laryngitis and pleurisy, all of which are more or less of a character cognate to phthisis—we increase the total to 124,082; and if we further add one-half of the number of deaths ascribed to influenza, and one-half of those set down to hæmorrhage, we obtain a final total of 125,385. My object in including a proportion of the deaths under the two last heads is, as regards influenza, because this complaint is scarcely likely to prove fatal except in cases where there is a tendency to pulmonary disease, which influenza tends to develop; and as regards hæmorrhage, because a great proportion of the deaths recorded under this head no doubt resulted from hæmorrhage of the lungs following disease of those organs. It appears, therefore, that more than half a quarter of a million of the population—that 1 in every $5\frac{1}{3}$, or nearly 19 per cent. of the whole number born in a year, fall a prey annually to that class of diseases which is so peculiarly fatal in this country. In other words, 1 in every $3\frac{1}{3}$, or nearly 30 per cent. of the whole number of deaths occurring in this country in the course of a year, arise from one class of diseases alone.

Looking to such facts as these, it can hardly, I submit, be considered otherwise than an undoubted duty incumbent upon every one to render any assistance that may lie in his power, however slight his opportunities may be, in the investigation of the causes that may tend to produce such fatal results as the Reports of the Registrar-General annually record.

Phthisis, the indigenous disease of this country—a disease that has hitherto baffled our most eminent physicians—we do, I fear, our utmost to foster. Hear Dr. Guy—no mean authority on statistical and medical subjects.

In connection with a paper read before this Institute on the subject of "The analogy existing between the aggregate effects of the operations of the human will and the results commonly attributed to chance," Dr. Guy gave us some interesting verbal information on the subject of the mortality among the printers in the metropolis—a class that helps to swell largely the returns of deaths from phthisis. From some notes I took at the time, I am enabled to state the result of his investigation into the subject.

He had visited, he told us, the principal printing offices in London, which are, most of them, situated in the parish of St Clement's, in the Strand, or in that vicinity.

The offices, he found, were all close and dark, as most of the business premises in that locality are, and a great deal of gas is necessarily consumed, particularly in the winter months, a good light being required for printing operations, more especially in the composing-rooms, and every compositor requires a gas-burner to himself. The rooms in which the men work are, for the most part, small and low, and are heated with hot water. All the fire-places in the composing-rooms are found to be blocked up, for the twofold reason of saving space and of avoiding the risk of fire, so that all ventilation from the chimneys and open fire-places, which form so ready and satisfactory a kind of ventilation, is rendered nugatory. The men employed in the process of printing suffer very much from colds and are very sensitive to draughts of air, and dread even the slight current produced by a ventilator, if the proprietor of the establishment has been induced to fit one up, to such a degree, that they very generally take steps to check its action altogether. Above all, many printers are much addicted to drinking.

Looking to such facts as these, we cannot be surprised that phthisis is very prevalent among this class of operatives—the peculiar position in which they are placed, or in which they place themselves, being exactly one which, as medical men perfectly well know, fosters this disease, and thus printing offices become actual hot-beds for its propagation; and we find, that while 1 in 5 or 6 of the whole population die of phthisis or other cognate diseases —and this is a fearful proportion—among compositors no less than 2 in 7 fall victims to these maladies, the disease being in some cases brought on, and in others accelerated in its progress, by the bad local influences surrounding its victims.

With the view to obtain additional information on this subject, and to test the facts above stated from my own observation, at the present time—Dr. Guy's information having been obtained some years ago—I recently visited a large printing establishment em-ploying about 300 hands, that of Messrs. Petter and Galpin.

The premises of this firm have lately been rebuilt upon an improved plan, and form a favourable contrast to those to which I have referred—and, no doubt, as the old printing offices come in course of time to be rebuilt, the evils now existing will gradually disappear.

, From the information I obtained from the intelligent foreman of these works—Mr. Wilson—I am quite satisfied that printers' compositors generally, in London, are in a very unfortunate position as regards health and longevity.

Phthisis is their great enemy, brought on to a positive certainty if the faintest taint exists in the system, and in an incredibly short space of time; and generated, as it were, spontaneously in numerous cases where no such taint is discernible, but in which, possibly, there might be a latent hereditary disposition to the disease.

The principle causes operating to produce this result appear to be—

1. The closeness of the work rooms, the imperfect ventilation, and the impurity of the atmosphere of the small, low, and overcrowded apartments in which, for the most part, printers work—the impurity of the air being increased (in fact, rendered, I may say, deadly) by the quantity of gas—often very impure in quality—consumed, and the fact that little attention is paid to the necessity of providing an outlet for the products of its combustion.

2. The quantity of drink imbibed by the printers in some departments of the business. Many drink, it seems, not only beer, but gin, rum, and brandy, and no surer means than these could be suggested for increasing the bad results likely to ensue from the evils under the first head.

3. The inhalation of dust from the types, which are composed of antimony and lead. This dust is being constantly disturbed, owing to the necessity of shaking up the masses of distributed type in order that the compositor may take quickly in his fingers the letters he requires to use, and that he may be enabled to place them in the necessary position for performing the operation of composing easily and quickly—a great desideratum in printing.

With reference to No. 1, it is clear that the evils complained of under this head are quite capable of remedy; and the same may be said of No. 2. Some inconvenience under No. 3 must, I suppose, continue to exist.

Thackrah, in his treatise *On the effects of Arts, Trades, and Professions, on health and longevity*—a work published as long ago as 1832, and which might, therefore, be considered not to be applicable to the state of things at the present time—refers to the injurious effect produced by the heated types upon the health of the workmen. Although many of the processes of manufacture to which he alludes are certainly now obsolete, and some of his remarks therefore do not now apply, it is remarkable how applicable some of his observations on many points still continue to be, showing how little has been done towards remedying those evils which affect the health and longevity of the operatives, and which were pointed out nearly 30 years ago.   Thackrah mentions that printers suffer injury from the types, which, being composed of lead and antimony, emit, when heated, a noxious fume injurious to respiration, and which produces partial palsy of the hands; and adds, that all danger may be avoided by abstaining from using the types until they are cold.   This evil, I believe, no longer exists—in London, at all events—though it has taken many years to abolish. It is well known to intelligent workmen in the trade that this consequence would ensue from handling the types when in a heated state, owing to the antimony in their composition; but there is no necessity in the world for their doing so.

The types are cleaned with lye, an alkaline mixture composed of potash, pearlash, and soda—in which composition there is nothing injurious—and the lye is washed away with cold water, so that the types are not necessarily heated at all, and it is only when heated that antimonious fumes are generated.

It used to be the practice of some printers to dry the types when wet (and it is necessary that they should be wetted, for in that state they can be more easily distributed) upon a stove, and there is no reason, I believe, to doubt that if this were done to any extent, and the types used while still heated, paralysis of the hands and arms would be the result; but this fact is now so well practically known to the trade, from cases that have occurred, that, I am credibly informed, the practice has long since been discontinued.

Compositors frequently fall victims to phthisis at the early age of 30, and probably 40 would be too high a mean age to assign to them as a body.*

---

* Since writing the foregoing, I have been favoured by Mr. David Arnot, the Secretary of the Metropolitan Typographical Widows' Fund, with the Annual Reports of that Society extending over 10 years, from which I find that the average age at death of the printers who were members of the Fund was 48 years.

The number of deaths caused by phthisis and other diseases of that class among the members, in the ten years ending 31st December, 1859, was 101 out of a total number of 173, being 58¾ per cent. of the whole.

We find, accordingly, from these returns, that 4 in 7 die of these diseases.

Dr. Guy, it will be remembered, ascertained that in the cases he investigated among printers, 2 in 7 fell victims to phthisis; probably, however, he did not, as I have done, include other diseases of the respiratory organs, which would account for some difference in our results.

It must be remembered, however, that men who become members of provident societies are naturally of a superior class.   They are, besides, in the enjoyment of good health at the time of entry; and, accordingly, more or less select lives.

Fortunately, they have their sick clubs and houses of call, which are not now, as formerly, public-houses, but private establishments supplied with reading rooms and libraries, where the men, when out of work, are to be heard of if wanted at a moment's notice, as is often the case in the printing business.

I say they have such establishments to which they can go in their leisure hours, but, unfortunately, many still prefer to frequent the public-houses.

This sort of evil is, no doubt, much encouraged by the arbitrary regulations enforced by the combinations that exist among the men engaged in printing, in common with many other trades, as the recent strike among the builders has shown, and by which the hands are compelled to obey certain rules and conditions in taking engagements, many of which encourage and necessitate the expenditure of money in drink—the object of the combination being to keep up wages, which free and independent competition, it is thought, might tend to lower. The sight of compositors is frequently very much weakened; ostensibly, by the constant close application to minute type, but, possibly, in a measure, as well by the quantity of snuff that they take, which cannot fail to be injurious in its effect. The practice is commenced with the view of stimulating the faculties, and particularly the sight; this effect, however, which it does seem to have at first, soon ceases, but the habit then has become confirmed.

Besides the injury from the use of tobacco as snuff, the adulteration, particularly of the commoner sorts, with sundry compounds and, amongst others, with ground glass, tends to increase the injurious effect, and renders the use of snuff most pernicious. The ground glass is mixed with the snuff to make up for the want of pungency, which is diminished by other adulterations. Carbonate or sub-carbonate of lead, too, is found in snuffs in considerable proportions, and is formed by a chemical combination of carbonic acid, which is stated to be gradually evolved from the snuff, with the metallic lead in which the article is very generally packed. The carbonate thus produced penetrates into the mass, and has been found in the interior of a packet of snuff to the extent of ·951 per cent., and on the outside layer of a sample which had been in actual contact with the metallic lead, to the extent of 2·743 per cent., according to the experiments of M. Wicke. Taking, then, the mean of these two percentages, we find nearly 2 per cent. of this pernicious salt in any given portion of the mass. This carbonate or sub-carbonate of lead, which is also known by the name of cerusse or white lead, is probably the only deleterious salt of that metal, and is of such a nature that not only is it seldom used internally in medicine, but when used externally, in the form of a powder or ointment, great caution is necessary in its application; and yet the habitual snuff-taker, daily, without fear of the consequences, takes this deleterious poison into his system. A moderate snuff-taker will consume about two ounces of snuff a week, so that, allowing for waste, he will take say 100 ounces in the year; of this, accordingly, nearly two ounces will consist of subcarbonate of lead, a highly-poisonous salt of that metal. This is a moderate estimate; some snuff-takers consume considerably more. The Reports of the Analytical Sanitary Commission, recorded in the *Lancet*, showed that of forty-three samples of snuff, chromate of lead was found to be present in nine, and oxide of lead in three cases. The presence of the lead was attributed by the commission to intentional adulteration. Saturnine disease has been clearly shown to be caused by the use of snuff thus adulterated, no less than by other well-known means by which the system becomes impregnated with lead. Printers begin the habit of snuff-taking by using Grimstone's eye-snuff, which is the mildest form in which snuff is manufactured, and so get on until they come to take Gillespie's, the strongest preparation made.

It has been considered that the digestive functions may suffer injury by reason of the long-continued standing posture which habit has necessitated in the printing trade. I have not, however, been able to ascertain that printers suffer in this respect, or from rupture or varicose veins, more than other classes of operatives that are obliged to stand long at a time—all such being liable to injury in these respects, but to a much less serious extent, I apprehend, than those whose labour necessitates the stooping posture or any pressure against the chest, as in the Sheffield grinders, in shoe-making and wig-making, for example. There is, however, no reason for the maintenance of the standing posture by compositors; contrivances have been invented to support the body and relieve the pressure on the legs entirely, but from the habit of a sort of circular motion round the case, during the process of composition, being acquired by the men, they seem always to prefer to work standing; and at times they work very long hours—thirty-six hours, perhaps, without cessation, except for meals. This practice, however, is only, I believe, owing to habit, and it is extraordinary to observe how the working classes cling to old habits and customs, at the expense often of their health, and all owing to a want of knowledge of the evil results that are certain to be engendered by a perseverance in them; and this affords another instance of how much an extension of education is called for among the working classes.

So with the shoe-makers and wig-makers: these workmen are very likely to be injured by the pressure of the last and block against the chest, but they cannot be made to see it.

An arrangement to prevent the necessity for stooping over their work, and to obviate the pressure of the last, has long since been invented for shoe-makers; and I have ascertained that a similar contrivance, arising perhaps out of the former, has been arranged for wig-makers, consisting of an apparatus for fixing the wig-block, to remedy the injurious effects that had been observed to exist; but neither of these inventions are in general use.

Numerous instances of this class might be adduced, to show the indifference exhibited by artizans in availing themselves of means that may be suggested by the ingenious and philanthrophic for remedying the pernicious results that arise from some species of occupations. This " sure it was always so " and " never heed it " spirit will be found to prevail, with greater or less intensity, in proportion to the indigence and ignorance of the unhappy sufferers.

Some injury to health is likely to be caused by attendance in the drying-room of printing offices, where the newly struck-off sheets are hung up to dry, and which is heated by hot water or steam to a temperature of about 70° Fahrenheit. The men who are much exposed to the high temperature and moist air of this department, if constitutionally disposed to suffer from heat and damp, are not unlikely to be injured by such exposure.

The laborious movement of the heavy levers by which the hand presses in printing offices are worked, would strike an observer as being likely to be injurious, from the tendency to contract the chest, which the action certainly has; but this labour is alternated with that of inking the types by passing the roller over them, a far from heavy work; and since at each press two men are always employed, the heavy mechanical labour is divided between them, so that no injurious result is likely to happen to strong men from this cause.

Of late years, owing to the street improvements in the populous neighbourhood in which printing offices abound, the small houses where printers, as well as other men engaged in works in the vicinity, used to live, have, in many instances, been, and are being, pulled down; and the consequence has been, that the operatives have been driven over the bridges, to the evident improvement of their health and social position ; for the walk to and from the scene of their labours cannot fail to conduce to health, both on account of the exercise they obtain in the open air, and by the change of scene which acts on the mind. The fact of their living in a purer air too, and the tendency to resort less to the public-house, which is brought about by their having a more airy and more pleasant residence a little way over the water, and a garden, perhaps, with which to amuse themselves, has tended to improve, to some extent, both the health and habits of this class of men.

I have mentioned that, in course of time, as the old printing offices come to be rebuilt, more attention will, no doubt, be paid to promote the health and comfort of the workmen. This is evidenced in the establishment of Messrs. Petter and Galpin, who carry on a general printing business in which no such appearance as that which Dr. Guy describes is visible. In the first place, the men are more favourably situated in an ordinary printing office than in the office of a newspaper, where the principal work is at night. The men engaged on newspapers are always, so to speak, working at high pressure, since the papers must be published within a very few hours after the events they record have occurred; and night work is always more trying than work by day. In the establishment just referred to, the composing-room is 8 feet in height, and its dimensions 86 feet by 50; and in it every possible attention seems to have been paid to ventilation, by means of siphon ventilators in the roof, the effect of which is to ensure a proper amount of ventilation without at the same time creating draughts of cold air, which printers seem so much to dislike, owing, of course, to their having been so much accustomed to be shut up all day, and perhaps half the night as well, in low-pitched and heated apartments. The ventilation is also increased by a large opening in each floor for the purpose of working lifts from the basement to the top of the building. A bad effect of this, however, is, that the foul and heated air is carried up to the composing-room, which is in the upper story ; and it is, of course, the most vitiated air that finds its way up there. These openings, however, cannot be done without, as it is absolutely necessary, in a large establishment, to have certain powerful steam or hydraulic lifts, for the purpose of raising the heavy forms of type which require to be conveyed from the composing-room to the machine-room. Under these circumstances, then, it would appear to be advantageous to increase the altitude of the composing room, making it at least as high as the machine-room, which is 14 feet in height—6 feet higher than the former. The position in the building of these two rooms it would scarcely be possible to reverse, as the great weight of the machinery employed renders the desirableness of placing it on the ground floor, if possible, clearly apparent. A slight alteration in the method of constructing the roofs of printing offices and factories generally, and one that could be carried out very inexpensively, would tend materially to obviate an inconvenience of another kind, viz., that arising from the heat during the summer months, caused by the sun's rays shining directly on the heads of the workmen through the skylights with which the upper stories of such buildings are usually lighted. It is only when the sun's rays are so strong as to prove injurious in this way, that the gas, which has so evil an effect, and which it is necessary to use at all other times, is capable of being dispensed with, and another source is thus opened up whereby suffering is unnecessarily entailed upon the operatives. This inconvenience may be remedied by means of dormar windows in the roof, in place of the obnoxious skylights. I have seen this plan carried out at the printing office of Messrs. Wertheimer, to the great comfort of the men employed. Of course, when the roof is sloping, as is often the case, these dormars are more than ever desirable, as increased height is gained by their adoption. Care should be taken not to glaze those sides of the dormars which, from their position, are most exposed to the sun's rays; or, at least, they should be fitted with shutters. I have been induced to dwell somewhat on this subject, because the same remarks apply to factories of all descriptions. The suggestions I have thrown out, it will be borne in mind, are not those of an architect, but of an unprofessional man, who, from personal observation, has seen and regretted the unnecessary injury to health and longevity to which large classes of our labouring community are often most cruelly subjected.

If the metropolitan inspectors of buildings were vested with some discretionary powers to see to the appropriateness of the buildings under their *surveillance,* as respects their answering satisfactorily, in a sanitary point of view, the objects for which they were intended, it is possible that some diminution of the evils I have pointed out might be the result.

It is very difficult to ascertain satisfactorily the causes that tend to produce phthisis, and to reconcile the apparently anomalous results of the observations of different writers on the subject.

Even with respect to the effect produced by the inhalation of dust, which has generally been considered to be the most fertile cause of this complaint, great difference of opinion seems to prevail, and the statistics of the most celebrated investigators of the question are far from agreeing in their results.

Dr. Knight, of Sheffield, has shown that the grinding and polishing of steel causes phthisis in an unusually short period; and that out of 250 workmen engaged in the occupation of polishing steel, 154 suffered from affections of the chest; and that there was no case of a person engaged in polishing forks reaching his 36th year—magnets, wire masks, currents of air, and moisture, having been successively tried for the purpose of arresting the passage of the metallic particles to the lungs, but without diminishing the mortality: while Dr. Cowan, in his notes to the translation of the celebrated work on phthisis, by M. Louis—the greatest authority, probably, in Europe, on the subject of phthisis—considers that the evidence brought forward on this point tends to an opposite conclusion. Dr. Cowan seems to consider that the grinders, who, he says, were not an unhealthy set of men previous to the commencement of the last century, have become so from the fact of their now pursuing their trade in close rooms, whereas they used to carry on their business in the country in large open rooms; from their working 16 hours a day at grinding only, instead of working 4 or 5 hours a day, and combining with the operation of grinding other less injurious departments of the trade, such as hafting and forging. The division of labour is now carried to great excess in this trade. He attributes much of the injury caused by grinding to the posture in which the grinders labour. They work with the shoulders elevated, the elbows resting on the knees, and the body inclined forwards. Thackrah, again, considers that the angular or pointed form of the particles detached is the chief cause of the injury to workers in iron; for the dust from old iron, which is thrown off, he tells us, so copiously as to deposit a thick brown layer on the dress of the dealers in the article, produces no inconvenience. M. Lombard, from inquiries made at Geneva, found that one of the principal causes tending to produce phthisis was the inhalation of air in which mineral and other dusts were floating; and showed that, among polishers, sculptors, stone-cutters, plasterers, watch-hand makers, and others of that class, the proportion of consumptive complaints was 177 in 1,000—the general average number of such cases, in all the professions in Geneva combined, being 114. Again, with reference to mineral, vegetable, and animal particles, the results of the inquiries of

MM. Benoiston and Lombard appear to be exactly the reverse of each other—the former making animal particles the most injurious and mineral the least; and the latter entirely reversing the order. The inquiries of Drs. Young, Leblanc, and Alison, are generally considered to determine the question of the liability of stonemasons to consumption. Dr. Cowan, however, does not entirely assent to their dicta. Certainly, from the observations of M. Benoiston, it would appear that the proportion of cases of phthisis among quarrymen, stonecutters, and marble-workers, was less than the general average; and it is stated, that in a department of France (Meusnes) celebrated for the fabrication of millstones, in a period of 17 years no increase of mortality from phthisis, as compared with other departments where stone-cutting was not general, was to be detected.

M. Lombard, however, found sculptors, stonecutters; and plasterers, afflicted with the disease in a high degree; and Thackrah found this class of persons to be phthisical and short-lived, dying generally before attaining the age of 40.

Upon the whole, Dr. Cowan considers, that, "were the existence of tubercular disease ascertained, the influence of dust in its production is still undecided; and that the hard nature of the material, the partial exercise of the body, exposure to varieties of temperature, their habits of intemperance, &c.; probably exert a far more powerful influence."

M. Lombard considers, on the whole; "that the influences which modify the *system in general,* such as a sedentary life, &c., &c., are more active in the production of phthisis than those acting *locally* upon the lungs, as dust, vapours, &c.; and if we reflect that the *latter* agents are never present independently of the former, their *real* activity, as causes of pulmonary tubercles, is, at best, but problematical."

Dr. Clarke, who has written on this subject in the *Cyclopædia of Practical Medicine,* observes, that "in almost every instance the sufferers are exposed to causes, independently of dust, fully adequate to the production of the tuberculous cachexia."

It will be seen, from these considerations, how difficult it is to disunite and sever the manifold predisposing causes that may lead to the production of phthisis; the effect of this is, that we find among the conditions tending to create this disease, many which, to say the least, we are accustomed to consider as most favourable to ensure immunity from the complaint.

We are likely more especially to observe this in the returns of

M. Lombard, since his investigations comprehend so large a number of cases, and, in proportion as the number of facts recorded increases, so do we more nearly approximate to an investigation of the mortality of the whole population from this disease; for, of course, if we were to inquire into the number of deaths from phthisis in every calling, we should in fact be investigating the mortality from phthisis of the entire community, and it is absurd to suppose that there is necessarily a predisposing cause to the disease in question in *every* possible condition of life; hence it follows that, in any very extensive inquiries into the relative effect of different occupations in the production of phthisis, a certain proportion of deaths from that disease is attributed to local influences, which may possibly have no effect whatever on their production; and this the anomalies apparent between the statistics of different inquirers into the subject seem clearly to show. In many conditions of life, therefore, we ought perhaps to endeavour to discover other causes for the disease than those set down by the statist, who merely records the fact that, out of a certain number of deaths from phthisis, so many occurred to persons who were exposed to peculiar influences, such as, for example, "muscular exercise, and active life." We should hardly say this could be the cause of the large percentage of deaths recorded under that head in the returns of M. Lombard. In point of fact, all writers on the question appear to unite in recommending muscular exercise and active life as amongst the best safeguards against the ravages of phthisis. We may therefore look for some other cause, and may probably fairly consider, from what we know as to the hereditary nature of this complaint—and this fact is not made an element in the classification of deaths, under the head of occupations —that of 100 persons dying of phthisis, in a certain percentage of the number no particular predisposing cause could be traced; but it was observed, that this percentage of persons was placed in a position to conduce to the use of muscular exercise, and to the leading of an active life, which, so far from being the *cause* of their death from phthisis, had, in all human probability, the effect of retarding the fatal issue of a disease which had, perhaps, been hereditarily transmitted to them, and in which the taint was too strong to be overcome by the favourable circumstances in which the victims were placed—such an hereditary taint being, unhappily, too well known to exist in families, and to be capable of transmission from generation to generation, as we shall presently see.

With reference to the effect on the constitution of the most laborious muscular exercise, I may mention, that from an inquiry into the subject, instituted by Mr. John Whitterom, into the health of the forgemen in the low moor district, it was shown, in a paper read at the recent meeting at Bradford of the Society for promoting Social Science, that the mortality of that class of men whose muscular labours were of the most laborious nature, was but slightly in excess of the general average of the country; and further, that notwithstanding the slight excess apparent in respect of mortality, they were not exposed to the same amount of disease as the ordinary out-door labourers employed in the same district.

Similarly of the "exercise of the voice": when not carried to a most unusual extent, provided the lungs are not seriously advanced in disease, the due exercise of the voice has always been considered as likely to improve their strength, and to increase, not only the capacity of the chest, but the muscular development of the whole of the upper part of the body. This we see frequently evidenced in singers; but even in persons with weak lungs, where actual disease does not exist, judicious exercise of those organs is one of the best methods of increasing their strength.

It is related of Cuvier, the naturalist, that he ascribed his own exemption from phthisis, with which disease he was threatened at the period of his appointment as Professor at the College of Surgeons, to the practice of public lecturing, which this position necessitated. This result was due, no doubt, to his lungs being in such a state as to be improved by the exercise of speaking aloud. Had the disease been further advanced, its progress would doubtless have been accelerated.

We may, perhaps, be justified in considering that the exercise of the voice, which appears as a cause of disease in the records of M. Lombard, was excessive, or that the hereditary taint was implanted in the lives under observation, or, perhaps, that both reasons aided in bringing about the result recorded, if, after what has been said, we are to take any note of this condition as a cause of the complaint at all. "Living in the open air" has always been considered to be most conducive to health, and persons whose avocations necessitate this mode of life are usually healthy and long-lived; and this is the case, I apprehend, even with persons exposed to such vicissitudes of the weather as might be deemed to be excessive —exposed, in fact, as much as they can be. We can only, therefore, imagine the existence of the hidden hereditary taint in this class too, supposing, in this as well as in the previous cases, that no other more patent reason for the creation of the disease were to be discovered. In confirmation of the opinion as to the advantage to be derived from open-air employments, I may refer to the Reports of the Registrar-General, in one of which it appears that of twelve classes in the tables—consisting of tailors, shoemakers, farmers, carpenters, butchers, persons engaged in wool, cotton, and silk manufactures, bakers, inn and beershop keepers, grocers, miners of all descriptions, labourers, and blacksmiths — among which a comparison was instituted, the farmers were found to be the longest lived.

The Registrar, in alluding to this class, and in speaking of the advantage to Assurance Offices and to Friendly Societies of a knowledge of the facts contained in these tables, states that "it is evident that the lives of farmers, for example, may be safely insured at much lower rates than the lives of licensed victuallers."

Now, with all due deference to the Registrar, this is not quite the way to put the case; and probably the actuaries present will agree with me, that the experience of Life Offices shows, that if the lives of farmers may be safely assured at the ordinary rates, the lives of licensed victuallers, if they are taken at all, should be assured only at an extra premium—the rate of mortality among the class of innkeepers being in excess of that among all the other eleven classes just referred to, at all ages, except the class of butchers in one particular decade of ages, viz., those between 55 and 65, the butchers dying at the rate of 41 in 1,000, while the innkeepers died at the rate of 39 in 1,000; so that even in this decade the innkeepers were not much longer lived than the butchers, and at all other ages the mortality among the innkeepers was the highest of the twelve classes under observation.

The rate of mortality among the whole population at the ages just quoted, 55 to 65, was only 30 in 1,000.*

Of late years, Assurance Companies have been on their guard against this class of persons, and some Companies decline to entertain proposals for assurance on the lives of innkeepers altogether.

As a class, they are exposed, the Registrar-General tells us, "by their business, to unusual temptation, live intemperately, and enjoy less quiet than the rest of the community. They are exposed also to zymotic diseases, by intercourse with large numbers of people."

The result of this investigation points very clearly to the fact, that intemperance may have no inconsiderable effect in unduly swelling the records of the Registrar-General.

With reference to the deaths set down to the cause of "animal emanations," much difference of opinion, I find, prevails as to whether

* *Vide* the 14th and 15th Reports of the Registrar-General, p. xv.

such a cause as this may be considered generally to affect health injuriously.

The deaths from phthisis, among the class of persons, according to the statement of M. Lombard, exposed to animal exhalations, though not so numerous as those that are stated to have occurred from the other causes quoted, are nevertheless considerable in number. The Registrar seems to consider that the great mortality among butchers may be owing to what he states to be the most probable cause, viz., the element of decaying matter by which they are surrounded in the slaughter-house and its vicinity.

This is, however, nothing but matter of opinion, though no one perhaps is more competent to form a correct one on such a subject than the Registrar-General. It certainly is not found, however, that those persons who are exposed to animal exhalations of this nature are generally *unhealthy.* We may, therefore, fairly look for some other cause as tending to induce the great mortality of this class.

The high death-rate among butchers could perhaps be traced to other causes than those just referred to; indeed, the Registrar only throws out for consideration the suggestion, whether what he mentions may not be the most powerful reason for the high mortality among them. The other probable causes that he mentions are, the excess of animal food in their diet, the fact that they often drink to excess, and that they are much exposed to heat and cold.

It is not impossible that the actual inoculation of the systems of butchers with decaying animal matter may be the cause of not a few deaths; and the constant exposure in their open shops to draughts of air—the extent of which, in windy weather, is very apparent from the effect on the gas-burners—may not improbably have an injurious effect; while the quantity of foul gas consumed in the unprotected burners will hardly tend to purify the atmosphere. The Registrar observes, that the red injected face of the butcher is an indication of disease—to the ordinary observer this might be an indication of robust health. Similarly with respect to brewers' draymen: their appearance would indicate that they were blessed with strong constitutions; this is not, however, the case. It is found, in the hospitals, that these men are a very difficult class to treat when attacked with inflammatory symptoms; and they are very prone to such attacks, and succumb readily to them. Their horses, which are good matches often for their masters in appearance, have very much the same attributes as the men; and as the draymen probably acquire their peculiarity of appearance and of constitution from the quantity of beer they imbibe, and owing to the insufficient amount of active walking exercise they take in proportion to their potations, so, perhaps, are the horses affected by much the same causes, by being fed to some extent on the brewers' grains, which act, no doubt, on their livers as the beer does on those of their masters. These horses are far from strong, though they look so fat and comfortable as to have given rise to the expression, "I would I were a brewer's horse;" and they can endure about the same amount of bodily labour, proportionally, as the draymen, and that is not much.

In connection with the subject of the inhalation of foreign particles, I may state that the workmen employed in flour mills are grievous sufferers from a spasmodic affection of the lungs, caused by the inhalation of minute particles of dust with which the atmosphere of most flour mills, as usually constructed, is impregnated; and I have been informed by the proprietor of some extensive flour mills in Birmingham, that if the workmen who suffer in this way continue too long in the business, a fatal result would be certain to ensue. I have great pleasure, however, in adding, that this gentleman—Mr. Evans—has introduced into his mills an apparatus for remedying the evil consequence just mentioned. He was kind enough to take me over his works, and personally to explain the action of the improvements. In the portion of the mill into which the ground grain is delivered into bins from the upper part of the mill, where the grinding process goes on, the atmosphere in which the men have to work had, prior to the introduction of his improvements, been completely impregnated with fine particles of flour dust; and these, from being the portion that was affected by the air, you will understand, consisted of the very finest particles of the flour—being, in fact, an almost impalpable powder—and, of course, by consequence, the more penetrating and the more likely to be inhaled and to be carried into the minute air cells of the lungs in exact proportion to its impalpability.

The effect of such inhalation is the production of a spasmodic action, which is excited by the physical contact of these minute particles with the delicate surfaces with which they are thus brought in contact. The lower parts of Mr. Evans' mills are now free from dust, and the effect is produced as follows :—

The ground grain descends from the upper portion of the mill through tubes of wood, and is delivered into bins placed on the ground floor. As the flour runs into these bins, in mills of ordinary construction, a cloud of fine dust is emitted, of the before-described injurious nature.   In these mills, however, by means of a fan, worked by the steam-power used for driving the ordinary mill machinery, a current of warm moist air is made to take such a direction as to cause the light obnoxious particles of flour-dust to ascend up a wooden tube to an upper floor, where the dust accumulates, and is swept up twice a day.

There is, accordingly, no reason whatever why flour mills should continue to remain the death-dealing receptacles many still are. Surely, looking to the disastrous results entailed under the ordinary system, it should be made compulsory on every proprietor of such an establishment to introduce improvements of this nature, when it is shown by actual trial how fully they answer in practice, and by how inconsiderable an amount of expense the decimation of a portion of our labouring community may be prevented!

It is required, by legislative enactment, that the machinery used in driving mills should be so fenced off as to prevent accidents to the lives and limbs of the operatives—why should not the preservation of their lungs, and thus of their health and lives, be equally attended to?   We can exist without our legs and arms, but the free action of our respiratory organs is an essential requirement of our being.

Mr. Evans has set a noble example, in spontaneously taking the initiative in a sanitary mill-reform of so much importance.

Injurious, however, as the effects of the inhalation of soft vegetable dusts may be, they are probably less serious than those resulting from that of metallic and stony particles.

With reference to stone-cutters, to which class previous reference has been made, the late Hugh Miller, in the history of his life, thus speaks as to the effect of the inhalation of the dust of stone.

It will be remembered that this eminent geologist was originally a mason.   He found, he tells us, that during his apprenticeship the dust of the stone had an injurious effect upon his lungs; and that after he had been engaged in the occupation for two years, he began to suffer so severely that he was obliged, though a poor man, to discontinue his occupation for a time.

This observant writer states, as the result of his own personal knowledge of the subject, that, at the stage of the malady at which he had arrived, poor workmen who may be unable to give up their employment sink, in six or eight months, into the grave. So general, indeed, is the affection, that few of the Edinburgh stone-cutters pass their 40th year unscathed, and not one out of every 50 ever reaches the 45th year.

In his case, after being engaged in hewing stone only two seasons (I should mention that the masons in the north only work during the six summer months), the dust of the stone inhaled at every breath had exerted the usual weakening effects on the lungs; and, after working day after day with wet feet in a water-logged ditch, he began to be sensibly informed, by a dull depressing pain in the chest and by a blood-stained mucoidal substance expectorated with difficulty, that he had already taken harm from his employment.

Hugh Miller, however, seems to have been blessed with a good constitution, and being enabled to give up his occupation for some months, he threw off the disease with which he was threatened.

The disease which causes death in these cases seems to be created by the mechanical irritation of the fine and hard particles of stone, which fret the delicate air passages of the respiratory organs, which accordingly become charged with matter and stone dust, and the eventual result appears to be death by phthisis or some allied complaint.

I may mention a very curious circumstance bearing on this question, and one which it is difficult to understand, but yet which is found to be a fact, viz., that the most sober workmen engaged in this occupation are the first to suffer from the results just detailed.

Hugh Miller was a most temperate man, and it was confidently predicted by his comrades that he would suffer for his sobriety, by being the sooner affected by the stone-cutters' malady. "A good bouse," they said, gave a wholesome fillip to the constitution, and "cleared the sulphur off the lungs;" and that his would suffer for want of the medicine which kept theirs clean. Whether there was any virtue in their remedy or not, it is just possible— Hugh Miller considered—that the shock given to the constitution by an over-dose of strong drink might, in certain cases, be medicinal in its effects.

Judging, however, from the information possessed by Life Assurance Companies, and from what the Registrar-General tells us about innkeepers, very few in this room would be willing, I imagine, to try the prescription of the stone-cutters of the north.

In this case, however, they were right in their predictions as to their fellow-workman being the first of his party to be affected by the malady; but his recovery, after a few months' cessation from the work, and his ultimate power of contending with and conquering the disease, arose, no doubt, as well from his extreme temperance as from the original strength of his constitution.

Having read in an article by Mr. William Jones, treating of the effect of different occupations on health, published some time ago in the *New Monthly Magazine,* that a preparation of the lungs of a mason who had died from phthisis was deposited in the anatomical museum of the Edinburgh College, showing the air cells literally blocked up with fine particles of dust, I have taken some trouble to ascertain whether this is really the fact, as a strong opinion appears to prevail that, in cases of this nature, the stony character of the deposit in the lungs is due to disease, and not to inhalation. I have positively ascertained that there is no foundation whatever for this statement.

In the Museum of the Royal College of Surgeons of Edinburgh, there are two preparations, consisting of sections of the lungs of stonemasons, put up in spirit.

In both there is a deposit of tubercle, and cavities have been formed, lined by false membrane.

These have just been examined by Dr. Sanders, the Conservator of the Museum, at the request of Dr. Douglas Maclagan, the President of the College, with the special object of determining the point now under discussion.

I have to record my obligations to both these gentlemen, for their courtesy, and to the latter, in particular, who has been so good as to enter very fully into the subject with me.

Portions of the lung tissue, of the tubercular matter, and of the false membrane of the cavities, were examined by Dr. Sanders, under the microscope, and with the addition of nitric acid, &c., but in none of these was there any appearance of particles of stone-dust. The sediment, moreover, that had fallen to the bottom of the jars, was likewise examined, and was found to consist of organic matter only, chiefly of cholesterine, but no particles of silica or carbonate of lime were found to be present.

It did not, therefore, appear to Dr. Sanders that there was any stone-dust in the portions of the lungs he examined.

There is, also, a preparation of the lung of a stonemason in the University Museum, which is, probably, the one referred to by Mr. Jones.

Dr. Maclagan informs me, that this specimen does not present any appearance different from that of an ordinary tubercular lung. This gentleman has taken the trouble to investigate the origin of the story, which he considers to be a pathological romance.

It seems that the deceased was a mason who worked in granite; he died phthisical; and his lungs were found to contain some grey tubercles, mixed with some of the usual carbonaceous deposit seen in most adult lungs. The Museum porter was an acquaintance of the deceased, and remarked that his friend's lung *looked like* a bit of the granite with which he worked.

It will be seen, from this investigation, how careful it is necessary for inquirers into such subjects as these to be, to ascertain the truth of the statements that are met with in unscientific writings.

Dr. Maclagan informs me, that he has had an opportunity of examining the lung of a Sheffield fork-grinder, in the museum of Dr. W. Gardner, who exhibited to him a sample of the dust separated by the grindstones. The latter was composed entirely of black particles, some of an irregular shape, and others in round globules, which were entirely attracted, Dr. Maclagan says, by the magnet—being either metallic iron or the metallic oxide. There were also a few particles of silicious matter, from the grindstone, mixed with the metallic particles.

The section of this lung was examined microscopically by Dr. Cleland, Demonstrator of Anatomy in the University, who could see, it appears, none of the iron particles.

Dr. Maclagan further examined this lung, by drying, pulverising, and then bringing the magnet over it, but not a trace of iron was separated from it.

Any attempt, therefore, on the part of these gentlemen to find inhaled particles was fruitless.

Dr. Maclagan, however, is of opinion, that this is no proof that such particles may not have been inhaled, and thus have been an exciting cause of disease; for he does not see, he informs me, how we could expect particles inhaled into the pulmonary cells to remain there in a disease in which there is so much expectoration, and in which the victims must be off work for some time before death.

This opinion, it will be seen, agrees exactly with what I have stated elsewhere, as the result of other inquiries into the subject.

It may also be open to conjecture whether a deposit of such a nature may not be the result of some chemical action taking place in the system, it being well known that carbonic acid will hold a large quantity of lime in solution, forming carbonate of lime, which would be deposited upon the acid being set free.

Patissier seems to consider that the stone-dust is deposited bodily in the lungs. He says, in his *Traité des Maladies des Artisans,* " La poussière qui se détache des pierres pénètre par la bouche dans les poumons, s'arrête dans les voies respiratoires, se mêle aux mucosités dont ces organes sont lubrifiés, *et forme par-fois de vrais calculs,* qui donnent naissance à la toux aux hémopty-sies, et peuvent même occasionner des pleurésies des péripneumonies dangereuses.''

Deslandes, on the other hand, seems to hold an opinion exactly the reverse of that of Patissier. He says, in his *Dictionnaire de Médecine et de Chirurgie Pratiques,* "Longtemps les matières crétacées que l'on rencontre dans le voisinage des bronches, et même les tubercules des poumons, ont été attribués à des agglomé-rations, et même à l'absorption de molécules pulvérulentes. Cette opinion ne mérite plus aujourd'hui qu'on s'arrête pour la réfuter. Quant à séjour des poussières dans les voies aériennes, il y a lieu de penser, avec Laënnec, que le plus souvent, il doit être fort court.''

Dr. Hodgkin, in his *Lectures on the Morbid Anatomy of the Serous and Mucous Membranes,* mentions that "An idea has been advanced, that earthy collections in the lungs are sometimes pro-duced by the inhalation of impalpably fine dust, to which some persons are, by their occupations, exposed; and the workmen of the quarries of Montmartre have been cited as instances of this fact. Although there can be no doubt that the inhalation of the atmo-sphere highly charged with mineral particles, however minute, must be a highly deleterious source of irritation, and, consequently, be very liable to become a fruitful cause of those affections which I have mentioned as producing calculi in the lungs; yet it appears extremely unlikely that, under ordinary circumstances, the earthy particles should either penetrate profoundly into the substance of the lung, or be so partially collected as to form concretions. It is more probable that they would be arrested by the mucus of the air-passages, and voided by the efforts at expectoration excited by the irritation which they induce.''

Probably, on the whole, the opinion of the medical profession generally is in favour of the assumption that any calcareous *deposits* that may be found in the lungs after death are not the result of inhalation, but are either due to change of structure, produced by disease, or to chemical deposit, as previously suggested.

It is not, however, only particles of a hard nature, or of a high specific gravity, that cause injury to the delicate tissues of which the respiratory organs are composed; soft and light woolly matter, which floats in the air, is found to have a highly-injurious effect, when inhaled into the bronchial tubes and lungs.

(*To be continued.*)

The Life Assurance Companies of Germany; their Business and Position in the Year 1858. *By* G. HOPF, *Manager of the Gotha Life Insurance Bank, Corresponding Member of the Institute of Actuaries in London, of the Commission Centrale de Statistique at Brussels, and of the Useful Sciences at Erfurt.*

[Read before the Institute, 26th March, 1860.]

WIDOWS' Funds and Burial Societies have existed in Germany already for more than 200 years, and in very great numbers. There was no town of any importance that did not possess one or more establishments of that kind. The guilds (corporations) of artisans usually maintained such institutions for their members; seeking in them, at the same time, a tie of stronger combination, and uniting their corporate interests. The love of these institutions, and the endeavour to provide, by means of them, for those left behind, spread from the guilds of artisans to the other classes of the people, in consequence of which, many hundreds, even thousands, of those little Burial Societies and Funeral Funds are to be found in Germany, from which, at the death of a member, a certain sum of money is paid to the survivors, to defray the expenses of interment, and to provide for the family immediately after the death of the subscriber. An extensive use was also made at all times of these establishments by such classes of people as were not so wealthy, because even the poorest are ambitious to take measures that they may have after their death a decent burial furnished at a fixed expense. This motive induces many, and particularly women, to subscribe to a Funeral Fund in Germany.

The number of Widows' Funds, from which, after the death of the husband, a yearly pension is paid to the widow, is much smaller in Germany than the number of Burial Societies. Such institutions were, principally, founded by Governments and Princes for their public functionaries and officers, and there is scarcely one country in Germany that has not one or several Widows' Funds for the officers of state, and all the persons in the service of the sovereign, generally also for clergymen, schoolmasters, and military officers. Duke Ernest the Pious, of Gotha, the ancestor of all the princes of the Ducal Saxon houses, who has deserved very well of the churches and schools of his country, laid the foundation of a Widows' Fund for the clergy of the churches of his country, as far back as the year 1645, and a similar one for the schoolmasters, in 1662 —establishments which subsisted until recent times, with many changes in their previous regulations, until they were united, about

40 years ago, with the Widows' Fund, founded in 1775, for all public functionaries and officers of the Duchy.

All those establishments, which had their origin in ancient times, were very imperfectly regulated, and many of them, which did not require the forced accession of the members of certain corporations, have been ruined for this reason. The scientific elements were totally wanting at the time of their foundation. Partly the laws of mortality were not yet sufficiently investigated, partly the mathematicians had not yet paid the necessary attention to the parts of their science relating to life contingencies. In both respects, little was done till the eighteenth century. There were, indeed, facts already collected relating to mortality in Germany; and it is known that Halley calculated his table of mortality from observations collected in the town of Breslau during the years 1687–91. But even though the defective method followed by him in this calculation had not then been perceived, yet it was evident that the law of mortality could not be found out from such a limited experience as could be gathered in one town only, during the above-named short period, and that, therefore, no confidence was to be bestowed upon the table of Halley. It has, therefore, never been made use of in Germany for calculations relating to Burial or Widows' Funds, or similar establishments.

It was not till the middle of the last century that the inquiry into the law of human mortality made great progress. The man who distinguished himself, not only in this branch of statistics, but who examined into all the theory of population, and raised it to a science, was Süssmilch. The work published by him, *Die göttliche Ordnung in den Veraenderungen des menschlichen Geschlechts*, appeared first in 1741, and was followed by several much enlarged editions. It forms the basis of all the later inquiries in this wide subject. Not long after a table of mortality had been constructed by Süssmilch, the mathematical rules for the calculations as to life insurance and annuities, founded on such observations, were also made public by several learned men of Germany. The most prominent, at that time, were Euler,[1] Florencourt,[2] Karstens,[3] and Tetens.[4] The latter treated the whole method of these compu-

[1] *Histoire de l'Academie Prusse, Année* 1760; Berlin, 1761; p. 163. *Neues Hamburger Magazin;* Leipzig, 1770; No. 43.

[2] Florencourt: *Abhandlungen aus der Juristischen und Politischen Rechenkanst;* Altenburg, 1781.

[3] Karstens: *Theorie der Wittwencassen;* Halle, 1784.

[4] Tetens: *Einleitung zur Berechnung der Leibrenten und Anwartsohaften, die von Leben und Tode einer oder mehrerer Personen abhaengen;* 2 vols.; Leipzig, 1785-6. Mr. Hendriks has drawn attention to this work in a clever paper "On the early History of Auxiliary Tables for the Computation of Life Contingencies," *Ass. Mag.*, vol. i., p. 1.

tations with such sagacity and judgment, and gave, at an early period (1785), such practical rules for performing the calculations, that his directions are not, even to this day, to be considered as antiquated, and they form the foundation of all the later labours in this field. But it was long before the principles laid down by those learned men were employed by the institutions depending on human life, and a great many Widows' and Burial Funds were consequently established with the old imperfect regulations, even after the appearance of their works.

The Burial Funds are Life Insurance Societies upon a small scale. Every member pays a premium, either once only or periodically, to the fund, and after his death a certain sum is paid out of it to his heirs. Though the idea of establishing Insurance Societies for higher sums, and for the admission of members from more extended localities, was made very intelligible by the great number of Burial Funds in Germany; yet it required the observations made in similar Societies in England to stimulate the Germans to found such institutions. This was done at first at Gotha, where, after the establishment of the Mutual Fire Insurance Bank for Germany, which is still flourishing there, the idea was seized and the resolution taken to create a Life Insurance Company, based on similar principles. After various preparations and calculations, in which the work of Charles Babbage (*A Comparative View of the various Institutions for the Assurance of Lives;* London, 1826; translated into German: Weimar, 1827) did useful service, the first plan was published in 1827, and from that date members were enlisted for it. A considerable number of members was soon found, notwithstanding that their policies were not delivered till the 1st January, 1829; consequently, a period of 30 years, closing with the year 1858, has passed since the opening of this branch of Assurance on German soil. Since the establishment of the Gotha Company, scarcely any year has passed without the formation of one or more similar Companies.

Table I. shows what results all these Companies have had, up to the end of 1858; and Table II., what premiums are payable to them for an assurance for life. Both the tables are arranged like those which we published in the former volumes of this *Journal.* If we compare Table I. with the same table for the year 1857 (p. 163, vol viii.), we see that since then two new Life Insurance Companies were established ι in Germany — the Life Assurance Branch of Nuova Società at Trieste, and the Anker at Vienna. They are both Share-Companies. The latter is founded with a share capital of 2,000,000 Austrian florins, and has been called into existence principally by Belgian resources and influence ; the former is founded with a capital stock of 4,000,000 florins.

By comparing the two tables, it appears also that one of the older Companies, the Hammonia of Hamburgh, has ceased to exist. It was reduced to such a sadly ruinous condition by the unsoundness of its operations, especially by too high expenses of administration, that it could no longer meet its obligations, and was obliged to declare itself insolvent. The state of this Society was very unsatisfactory. The share-capital was very small, and did not exceed the sum of £35,714, of which not more than £4,587 were paid up. Many shareholders now refuse to pay up the balance of their shares, for which reason many actions must be brought against them, the result of which is very doubtful. It is doubted whether the creditors of the Hammonia will receive more than 30 per cent. of their claims, at the settlement of the bankruptcy.

The number of German Life Insurance Companies, which were in active operation during the year 1858, amounted to 26. The number of the English, French, and Belgian Companies, which have extended their business to Germany, is as great, if not greater. Of these, the English Companies especially do a considerable business in the large commercial and seaport towns. In the greater part of the States of Germany, the permission of Government is necessary for establishing an agency, and this licence is often refused, or connected with troublesome conditions. Legislation is, in this respect, very different in the different German States. Whilst some Governments, as those in the Hanseatic towns, in the kingdom of Saxony, and in the Thuringia countries, proceed liberally, and leave free the establishment of agencies for Life Insurance, or require only the observation of formalities easily to be performed, it is very difficult in others to obtain the requisite permission. And these prohibitive measures extend as much to German Companies as to foreign ones. The severest restrictions are in Austria, which has not allowed admission into its States to any foreign Assurance Company, whereas Austrian Companies have been allowed to carry on operations in many German countries. But the protection afforded, by this method of proceeding, to the few Life Insurance Companies in Austria against foreign competition, has by no means brought them to a flourishing condition. On the contrary, we maintain that it has turned rather to their prejudice than to their advantage; and to prove this, we refer to Table I. Of the five older Life Insurance Companies in Austria, only two have hitherto published reports of their results; and of these, only one, the Assicurazioni Generali of Trieste, has made remarkable, but by no means brilliant, progress—the operations of the other having remained till now of very limited extent; and we think we are not mistaken in affirming, that the affairs of the three Companies which have not yet made a report are still less satisfactory. Likewise, it cannot be denied, that the legislation which excludes entirely all foreign Assurance Companies from Austria, turns also to the prejudice of the Austrian people. From want of competition, they have not been duly made acquainted with the nature, use, and the varied applicability of Life Insurance; and those who are acquainted with it, have often felt the want of an institution which suited their necessities and wishes; and their people have been obliged to disregard an important means of increasing their prosperity.

Casting a glance on Table I., we see that, in 1858, 14,645 persons have newly entered the 20 German Life Assurance Companies which, out of the existing 26, have made public reports; and that, the subsequent assurances (Nachversicherungen) included, they have insured, upon the whole, a sum of £2,340,298 in case of death—for our table comprehends only the insurances effected for the case of *death,* the life assurances in the proper sense of the word. The state of assurance of those 20 Companies was—

At the beginning of the year, £12,875,128 on the lives of 80,523 persons.
At the end of the year   .   14,383,010   „   „   90,128   „

Showing an increase of £1,507,882   „   „   9,605   „

According to this statement, the total amount of assurances increased, in the course of 1858, by 11·71 per cent. of its amount at the beginning of the year, and the number of assured persons by 11·93 per cent. The average sum assured on one life is about £160.

The receipts of premiums and interest increased in a greater degree than the assurances. According to Table I. on p. 163, vol. viii., these receipts were—

In the year 1857   .   .   .   £509,007
„   1858   .   .   .   618,382

Increase equal to   .   £109,375, or 21 per cent.

Whilst £251,545 were paid for claims on 1,510 persons who died in the year 1857, a payment of £303,438 was made for 1,900 cases of death in 1858.

The total of assurance funds was—

At the end of 1857   .   .   £2,258,088

      „   1858   .   .   2,437,844

           Increase   .   .   £179,756, or 8 per cent.

As every new enterprise in Germany always requires time for its development, and people learn but by slow degrees to accustom themselves to new institutions and to make use of them, such also was the case with life assurance. It requires considerable time to give the German a clear idea of the nature and manifold utility of life assurance, and to inspire him with confidence in it; but when this is once done, the interest in this means of advancing the welfare of families increases from year to year. During the last few years, life assurance has developed itself in Germany in the following rapid manner :—

Year.	Number of the reporting Offices.	New Assurances in the course of the Year.		Assurances at the end of the Year.	
		Persons.	£	Persons.	£
1852	12	5,236	841,844	46,980	8,224,130
1853	13	5,558	939,854	50,019	8,750,239
1854	14	5,224	841,459	52,876	9,150,885
1855	18	9,366	1,361,711	61,832	10,411,549
1856	18	12,778	1,633,272	71,169	11,487,487
1857	19	13,601	1,930,649	81,348	12,893,086
1858	20	14,645	2,340,298	90,128	14,383,010

It results from these figures that the use people make of life assurance has increased, especially in the latter years, since the termination of the oriental war, during which a stagnation took place on account of the doubtful security of our position, and the predilection for insuring has made considerable progress. It is to be expected that, with security, this rate of increase will continue a long time before it reaches its culminating point.

The Gotha Company has reached the greatest development of all the Companies. Table I. shows that their insurances comprehend more than one-third of the sums assured in all German Life Insurance Companies. As it is founded on mutuality, all surplus, annually ascertained, is returned as dividend to the assured, and deducted from the premiums, or paid in cash if the assurance is extinguished—a proceeding adopted also by the other German Companies which give dividends to the assured. On each premium paid a dividend is granted—generally five years after payment of the premium—in proportion to the surplus of the corresponding year for which the premium is paid, no matter whether the assurance be still in existence or not at the time of distribution. The practice of adding the bonus to the sum assured is not found in any German Life Office. The dividends in the Gotha Office amounted, upon an average, to 28 per cent. of the premium, and, conse-quently, the cost of assuring in this Company is the lowest of all German Offices, as Table II. shows. It is also the Company which is managed at the most reasonable rate. All the expenses of ad-ministration, the commission for the agents included, do not now amount to quite 4 per cent. of the annual receipts of premiums and interest.

The Gotha Company passed through the third decennium at the end of 1858. Table III. shows the working of the business in each one of these periods. The new assurances amounted in the

1st decennium (1829–38), in-cluding the preliminary period of 1827 and 1828, to	10,648	persons, with	18,387,200	thalers.
2nd decennium (1839–48), to	10,404	„	16,692,000	„
3rd decennium (1849–58), to	12,380	„	20,959,000	„
Total . .	33,432	„	56,038,200	„

The new members were, therefore, considerably more numerous in the last period than in either of the two former; and as they exceed the withdrawals by more than double, we may surely hope that the assurances in the Gotha Office will reach a still greater extent than they have already done.

The calculations of the Gotha Office are founded on the table computed by Charles Babbage according to the experience of the Equitable Society, with the modification that the mortality of the higher ages above 55 years, which that table seemed to represent as too small, was augmented. It will not be without interest to see how true and safe that table has proved in the 30 years now past. This is shown by the Table IV.; according to it—

Should die	6787·68	persons, with	11,215,898	thalers assured sum.
Died .	6628·00	„	10,988,974	„ „
Therefore, *less* died	159·68	„	226,924	„ „

There have been, then, disbursed in the 30 years past, 226,924 thalers, or about 2 per cent. *less* for deaths than might have been spent according to the basis of calculation adopted; and the number of the claims was about $2\frac{1}{2}$ per cent. below the expectation. The adopted table of mortality has therefore proved, upon the whole, con-venient and safe. If, however, we follow mortality into the single classes of age, there appear, indeed, greater deviations; for mortality at the ages under 55 years was regularly less, and at the ages above 55 years greater, than the table in use led us to expect, as this was formerly (vol. iv., p. 54; vol. v., p. 58; vol. vi., p. 295; vol. vii., p. 177) proved by the results of single years. A nearer examination has shown that the greater mortality at the ages above 55 years is not caused by such of the assured as are admitted at these ages, but by those who, having been assured at the younger periods of life, attain to the higher ages. The health of the latter, examined at earlier periods, is naturally, on an average, not so good at their entrance into the higher ages, as the health of those who did not undergo the examination till later in life. The examination of health, however, just at the period from the 40th to the 55th year is of great consequence, because at that period of life essential change takes place in the constitution of man—not only of females, but also of males—which have a decided influence on their health. Of those who pass an examination as healthy lives at the age of 40, a greater number is doubtful and bad after 10 years than of those who have been selected as healthy at ages 30 and 50, and are reconsidered after an equal period.

During the 30 years past (1829–58), 6,779 cases of death occurred amongst the persons assured in the Gotha Office, which are distributed in the different classes of age, as shown in Table V. This table has been constructed in the same manner as the table published vol. v., p. 329, for the first 25 years (1829–53), during which period 4,521 cases of death had happened. The experience has, therefore, been enriched by 2,258 cases in the last five years. However, the results formerly obtained have not been essentially modified thereby, as is seen by a comparison of cols. 9 and 10 in Table V. This table is founded on the experience gathered from one calendar year to another. Beside the registers destined for them, on the basis of which the annual balances are made and the reserves are computed, there are arranged registers of the persons assured and the cases of death according to the years of assurance (vol. vi., p. 1), which do not require a correction for admission and withdrawal, as was necessary in Table V., cols. 6 and 7, and they furnish, therefore, more exact data. A table of mortality deduced from these data will be computed as soon as the experience in the younger and higher ages has become somewhat fuller, and, in consequence, more certain in its results.

Table VI. exhibits a comparative view of mortality according to different tables. It results from this table that the mortality in the Gotha Company in the ages under 55 years agrees almost entirely with the mortality in the English Life Insurance Companies, as it is expressed in the Actuaries' Table, founded on the experience of 17 Companies. Beyond the 55th year of age, however, the mortality is essentially greater among the assured in the Gotha Office than in the English Companies. This difference has a general reason, because, in our opinion, mortality is in general less at the higher ages in England than on the continent. It may be that people enter there the higher ages with less weakened vitality, or that they are preserved from adverse influences at this period of life more in England than on the continent, or that both causes co-operate. The fact itself seems no longer doubtful, and the same result will be noticed on a comparison of the tables of mortality founded on the experience of the mixed population, though these tables are less accurate and certain than those which are based on the experience of the Assurance Companies. We refer to the data furnished by Nos. 8–12 of Table VI. According to them, the proportions for the higher ages seem to be more unfavourable, particularly in Germany (Hanover and Saxony), because mortality after the 50th year is greater there, even considerably greater, than in England. The mortality at the higher ages does not appear so favourable either in Belgium or in France, as in England, which country is, therefore, a particularly fortunate locality for aged persons.

It is rather remarkable that mortality amongst the members of the Friendly Societies in England is essentially less than amongst the members of Life Insurance Companies, though they are not obliged to undergo, at their entrance, so severe an examination respecting their health as the latter. The correctness of this fact is not liable to any doubt whatever, in consequence of the carefulness of the investigations made by the two highly-respectable authorities, Finlaison and Neison, the results of which are shown by Nos. 4 and 5 in Table VI. This observation, however, is of weight only for England—it is just the reverse in Germany.

In conclusion, we must not omit to draw attention to the mortality amongst the members of the Mutual Life Assurance Society at New York, which is *less* at most ages than amongst the members of the English and German Life Insurance Companies, though part of them live in the southern regions of North America, which are injurious to health. We owe the statement of this interesting fact to the elaborate researches of the able actuary of that Society, Mr. Homans, in his valuable report on the experience of the Society during the first 15 years of its existence.

TABLE I.—New Business and Position of the Life Assurance Companies of Germany in the Year 1858.

Established	Name of Company	Assurances existing at the beginning of the Year — Persons	Sums	New Assurances during the Year — Persons	Sums	Assurances existing at the end of the Year — Persons	Sums	Income (Premiums and Interest)	Claims paid — Persons	Sums	Expenses of Management — Absolute	Per Cent. of Income	Per £1000 of Assured Sums at end of Year	Assurance Fund In General — Amount	Per Cent. of Assured Sums at end of Year	Reserve and advanced Premiums	Clear Surplus	Average Dividend during the last ten years in per Cent. of Premium	Share Capital — Nominal	Paid up
			£		£		£	£		£	£			£		£	£		£	£
1827	Gotha	20,841	4,792,614	1,316	331,329	21,470	4,944,023	231,597	517	125,871	9,208	3.98	1.87	1,324,212	26.78	1,080,968	216,177	28· (0.251 On Sum Assured)	Mutual	Mutual
1828	Lubeck	6,936	984,571	2,396	272,512	8,777	1,171,793	48,062	133	24,872	6,217	12.94	5.31	122,656	10.47	?	?		72,857	7,286
1830	Leipzig	4,881	786,786	357	51,886	5,041	807,171	33,924	133	21,086	2,404	6.18	2.98	233,135	28.88	198,838	27,126	16½	Mutual	Mutual
1830	Hanover	*2,889	251,685	*105	10,100	*2,878	251,671	9,186	59	4,979	?	?	?	44,943	17.86	?	?	None	Mutual	Mutual
1834	Trieste (Generali)	†4,400	971,429	†2,000	471,428	*5,600	1,285,714	57,143	†110	†25,714	?	?	?	?	?	?	?	None	420,000	?
1836	Berlin	8,390	1,469,486	480	105,843	8,510	1,508,043	72,532	256	42,671	5,286	7.29	3.51	336,486	22.31	284,872	51,614	14⅖	142,857	28,571
1836	Munich	2,424	262,278	318	40,424	2,570	286,016	6,286	47	4,718	?	?	?	46,441	16.24	?	?	None	Capital of the Loan Bank.	
1839	Vienna	10,419	426,963	1,322	56,819	11,142	455,916	18,013	184	8,337	2,143	11.90	4.70	48,181	10.57	36,815	11,366	13⅜	Mutual	Mutual
1842	Brunswick	1,020	79,314	70	6,086	1,063	83,243	3,286	19	1,329	?	?	?	14,285	17.16	?	?	c. 6	Mutual	Mutual
1844	Frankfort	1,336	256,637	322	53,009	*1,600	285,714	13,428	15	2,343	1,857	13.83	6.50	17,149	6.00	?	?	6⅚	244,898	24,439
1847	Hamburg (Janus)	*6,235	716,880	*1,270	131,196	*7,032	792,878	28,550	84	10,284	4,955	17.36	6.25	61,185	7.72	?	?	7⅔	71,428	7,143
1852	Leipzig (Teutonia)	*1,259	95,447	*327	32,463	*1,440	117,181	6,375	53	2,661	1,875	29.42	16.00	12,286	..	{21,222 Debit	None	None	85,714	12,857
1853	Cologne (Concordia)	†3,400	973,262	†1,000	221,812	†4,200	1,112,365	32,857	†50	16,548	?	?	?	109,433	9.84	79,351	23,341	None	869,714	173,945
1853	Schwerin	*292	51,900	*73	14,600	*349	63,886	1,888	5	1,114	?	?	?	3,586	5.61	2,059	1,527	..	14,286	1,428
1854	Halle (Iduna)	*1,868	182,956	575	61,736	*2,232	217,349	10,455	62	4,932	3,046	29.14	14.02	7,392	3.40	{12,870 Debit	None	None	28,571	?
1854	Stuttgart	1,539	247,460	516	87,640	2,002	323,795	12,102	19	1,796	734	6.06	2.27	27,728	8.56	16,594	10,354	..	Mutual	Mutual
1855	Darmstadt	861	49,720	283	14,825	1,073	62,744	2,149	42	820	?	?	?	4,435	..	?	None	None	Capital of the Annuity Society.	
1856	Magdeburg	*933	154,311	*573	111,233	†1,353	252,608	13,015	94	1,526	4,689	36.03	18.56	8,864	..	8,837	3,056	None	282,000	56,400
1857	Stettin	†600	121,429	†1,150	228,571	1,606	323,832	11,251	16	1,614	?	?	?	9,798	..	14,362	6,184	None	428,571	85,714
1857	Zurich	..	..	192	36,786	190	36,563	1,283	2	223	513	?	?	5,649	..	?	202	None	Capital of the Credit Bk.	
	Total	80,523	12,875,128	14,645	2,340,298	90,128	14,383,010	618,382	1,900	303,438	..	..	..	2,437,844	..	..	..	..	..	..

* Policies.   † Approximated.

TABLE II.—*Showing the Gross Premium per Cent., the Average D*

Established.	Name of Company.	Average Dividend per Cent.	Age 25.			Age 30.			Gross Premium.	Dividends.
			Gross Premium.	Dividends.	Net Premium.	Gross Premium.	Dividends.	Net Premium.		
1827	Gotha	28	2·356	0·660	1·696	2·633	0·737	1·896	2·969	0·831
1828	Lubeck	*Old Tariff:* 0·251 on sum assured *New Tariff:* non profit, 1857	2·403	0·251	2·152	2·669	0·251	2·418	2·992	0·251
			..	..	1·828	..	..	2·106	..	..
1830	Leipzig..........	16½	2·356	0·389	1·967	2·633	0·434	2·199	2·969	0·490
1834	Trieste (Assicurazioni Generali)	Non profit	..	..	2·14	..	..	2·42	..	..
1836	Berlin	*Old Tariff:* 14⅖ *New Tariff,*1854: Not yet divided	2·406 / 2·133	0·346 / ?	2·060 / ?	2·672 / 2·433	0·385 / ?	2·287 / ?	2·992 / 2·817	0·431 / ?
		Non profit	..	..	2·000	..	..	2·283	..	..
1836	Munich..........	Non profit	..	..	2·167	..	..	2·433	..	..
1839	Vienna (Mutual)..	c. 13	2·150	0·279	1·871	2·483	0·323	2·160	2·883	0·375
1842	Brunswick	c. 6	2·125	0·127	1·998	2·382	0·143	2·239	2·722	0·163
1844	Frankfort........	Non profit	..	..	1·992	..	..	2·242	..	..
1847	Hamburg	Non profit	..	..	1·950	..	..	2·225	..	..
1851	Trieste (Azienda Assicuratrice)..	Non profit	..	..	2·12	..	..	2·40	..	..
1852	Vienna (First Austrian Assurance Company)	Non profit	..	..	2·100	..	..	2·383	..	..
1852	Leipzg (Teutonia)	Not yet divided	2·181	?	?	2·500	?	?	2·861	?
1853	Trieste (Riunione Adriatica di Sicurtà)	Non profit	..	..	1·96	..	..	2·25	..	..
1853	Cologne	Non profit	..	..	1·783	..	..	2·042	..	..
1853	Schwerin	Only two dividends	2·203	?	?	2·521	?	?	2·906	?
1854	Halle (Iduna)	Not yet divided	2·192	?	?	2·500	?	?	2·861	?
1854	Stuttgart	Only one dividend	2·230	?	?	2·457	?	?	2·760	?
1855	Darmstadt	Non profit	..	..	1·750	..	..	2·017	..	..
1856	Magdeburg	Non profit	..	..	1·867	..	..	2·083	..	..
1856	Erfurt (Thuringia)	Non profit	..	..	1·758	..	..	2·008	..	..
1857	Stettin (Germania)	Non profit	..	..	1·817	..	..	2·075	..	..
1857	Frankfort (Providentia)	Non profit	..	..	1·883	..	..	2·125	..	..
1857	Zurich	Non profit	..	..	2·174	..	..	2·439	..	..
1857	Trieste (Nuova Società Commerciale)	Non profit	..	..	2·050	..	..	2·330	..	..
1858	Vienna (Anker) ..	Non profit	..	..	1·920	..	..	2·240	..	..
	Average Premium	..	2·249	..	1·958	2·535	..	2·221	2·885	..

An average dividend is only computed

e Net Premium after Reduction, in the German Life Assurance Companies.

Premium	Age 40.		Age 45.			Age 50.			Age 55.			Age 60.		
Premium	Dividends.	Net Premium.	Gross Premium.	Dividends.	Net Premium.	Gross Premium.	Dividends.	Net Premium.	Gross Premium.	Dividends.	Net Premium.	Gross Premium.	Dividends.	Net Premium.
386	0·948	2·438	3 961	1·109	2·852	4·733	1 325	3·408	5 742	1·608	4·134	7·161	2 005	5·156
400	0 251	3·149	3·894	0·251	3·643	4·531	0·251	4·280	5·317	0·251	5 066	6·367	0·251	6·116
.	..	2 928	..	..	3·564	..	..	4·417	..	..	5 561	..	..	7·133
886	0 641	3·245	3 961	0·654	3·307	4·733	0·781	3·952	5·742	0·947	4·795	7·161	1·182	5·979
.	..	3·21	..	..	3·81	..	..	4·66	..	..	5·78	..	..	7·25
400	0·490	2·910	3·898	0·561	3·337	4·533	0·653	3·880	5·317	0·766	4·551	6·367	0·917	5·450
333	?	?	3·950	?	?	4·750	?	?	5·858	?	?	7·417	?	?
.	..	3·150	..	..	3·758	..	..	4·558	..	..	5·650	..	..	7·142
.	..	3·183	..	..	3·750	..	..	4·517	..	..	5·533	..	..	6·933
333	0·433	2·900	3·950	0·513	3·437	4·733	0·615	4·118	5·667	0·737	4·930	6·967	0 906	6 061
187	0·191	2·996	3·805	0·228	3·577	4·660	0·280	4·380	5·854	0·351	5·503	7·479	0 449	7·030
.	..	2·950	..	..	3·483	..	..	4·192	..	..	5·142	..	..	6·422
.	..	2·950	..	..	3·425	..	..	4·100	..	..	5·125	..	..	6·492
.	..	3·24	..	..	3·84	..	..	4·64	..	..	5·70	..	..	7·23
.	..	3·167	..	..	3·767	..	..	4·550	..	..	5·633	..	..	7·183
300	?	?	3·900	?	?	4·667	?	?	5·624	?	?	6·833	?	?
..	..	2·99	..	..	3·62	..	..	4·35	..	..	5·55	..	..	7·07
..	..	2·825	..	..	3·417	..	..	4·208	..	..	5·267	..	..	6·692
391	?	?	4·005	?	?	4·802	?	?	5·906	?	?	7·661	?	?
317	?	?	3·850	?	?	4·600	?	?	5·750	?	?	7·200	?	?
182	?	?	3·757	?	?	4·572	?	?	5·688	?	?	7·077	?	?
..	..	2·800	..	..	3·433	..	..	4·300	..	..	5·550	..	..	7·317
..	..	2·758	..	..	3·333	..	..	4 108	..	..	5 033	..	..	6·300
..	..	2·750	..	..	3·317	..	..	4·083	..	..	5·267	..	..	6·717
..	..	2·842	..	..	3·417	..	..	4·192	..	..	5·208	..	..	6·592
..	..	2·821	..	..	3·396	..	..	4·167	..	..	5·212	..	..	6·671
..	..	3·226	..	..	3·846	..	..	4·545	..	..	5·556	..	..	7·143
..	..	3·100	..	..	3·710	..	..	4·500	..	..	5·660	..	..	7·160
..	..	2·990	..	..	3 600	..	..	4·390	..	..	5·430	..	..	6·830
..	..	2·980	3·903	..	3·526	4·665	..	4·271	5·679	..	5·285	7 063	..	6·670

panies who have divided at least five dividends.

TABLE III.—*Assurances in the Gotha Life*

Year.	Assurances proposed.		New Assurances effected.		Assured in the course of the Year.		DECREASED. By discontinued, expired, and lapsed Policies.	
	Pers.	Thlrs.	Pers.	Thlrs.	Pers.	Thlrs.	Pers.	Thlrs.
1829	1,581	2,746,600	1,285	2,379,200	1,285	2,379,200	..	..
1830	607	1,121,700	504	1,044,700	1,777	3,408,800	14	26,200
1831	1,491	2,501,400	1,244	2,348,400	2,991	5,696,700	46	126,000
1832	1,574	2,513,500	1,165	2,168,100	4,083	7,687,500	212	490,100
1833	1,351	1,975,300	1,041	1,738,100	4,857	8,832,800	120	358,700
1834	1,218	1,749,800	902	1,424,500	5,593	9,825,700	154	335,500
1835	1,318	1,877,900	989	1,577,600	6,361	10,958,000	155	298,300
1836	1,635	2,390,400	1,213	1,932,500	7,333	12,422,800	170	383,800
1837	1,581	2,435,600	1,151	1,886,200	8,204	13,743,300	170	290,200
1838	1,555	2,224,900	1,154	1,837,900	9,061	15,114,200	156	338,200
Total 1829–38	13,911	21,537,100	10,648	18,337,200	51,545	90,069,000	1,197	2,647,000
1839	1,419	2,143,000	1,024	1,743,200	9,803	16,304,700	195	378,900
1840	1,485	1,942,800	1,089	1,568,400	10,570	17,283,500	193	354,600
1841	1,414	2,002,100	983	1,604,800	11,217	18,255,700	178	292,400
1842	1,352	1,939,800	1,013	1,594,700	11,888	19,264,100	167	348,600
1843	1,503	2,082,600	1,049	1,658,400	12,572	20,259,200	193	398,600
1844	1,493	2,101,700	1,073	1,747,500	13,249	21,281,800	183	291,800
1845	1,412	2,046,000	1,018	1,646,800	13,884	22,281,000	178	327,100
1846	1,484	1,961,300	1,096	1,648,500	14,564	23,218,700	208	388,800
1847	1,706	2,425,200	1,235	1,948,200	15,361	24,412,400	240	396,500
1848	1,151	1,790,200	824	1,531,500	15,652	25,070,700	277	543,200
Total 1839–48	14,419	20,434,700	10,404	16,692,000	128,760	207,631,800	2,012	3,720,500
1849	1,377	1,946,500	1,011	1,629,800	16,047	25,641,000	239	449,500
1850	1,537	2,138,000	1,138	1,860,500	16,609	26,479,600	196	402,300
1851	1,667	2,369,200	1,262	2,018,800	17,342	27,522,000	190	359,000
1852	1,868	2,671,600	1,423	2,326,600	18,278	29,028,900	217	421,200
1853	1,804	2,576,900	1,324	2,153,900	19,039	30,182,300	213	418,800
1854	1,534	2,240,700	1,073	1,855,900	19,500	30,971,100	252	443,400
1855	1,699	2,466,300	1,156	2,032,600	20,014	31,927,200	180	304,300
1856	1,910	2,768,800	1,314	2,183,100	20,677	33,071,800	169	309,800
1857	1,913	3,090,400	1,363	2,578,500	21,461	34,637,900	167	332,300
1858	1,901	2,938,200	1,316	2,319,300	22,157	35,867,600	151	341,700
Total 1849–58	17,210	25,206,600	12,380	20,959,000	191,124	305,329,400	1,974	3,782,300
Total ...	45,540	67,178,400	33,432	56,038,200	371,429	603,030,200	5,183	10,149,800
Average for one year..	1,518	2,239,280	1,114	1,867,940	12,381	20,101,007	173	338,327

Assurance Bank during the Years 1829–58.

By Death.		Total Decrease.		Net Increase.		Existing Pers.	Existing Thlrs.	Average Sum on one Life.	Average Age of Assured.	
Pers.	Thlrs.	Pers.	Thlrs.	Pers.	Thlrs.	Pers.	Thlrs.	Thlrs.	Yrs.	Ms.
12	15,100	12	15,100	1,273	2,364,100	1,273	2,364,100	1,857	42	7
16	34,300	30	60,500	474	984,200	1,747	3,348,300	1,917	42	9
27	51,300	73	177,300	1,171	2,171,100	2,918	5,519,400	1,892	41	10
55	102,700	267	592,800	898	1,575,300	3,816	7,094,700	1,859	42	4½
46	72,900	166	431,600	875	1,306,500	4,691	8,401,200	1,791	42	9
67	109,800	221	445,300	681	979,200	5,372	9,380,400	1,746	43	1
86	169,400	241	467,700	748	1,109,900	6,120	10,490,300	1,714	43	3½
110	181,900	280	565,700	933	1,366,800	7,053	11,857,100	1,681	43	5
127	226,800	297	517,000	854	1,369,200	7,907	13,226,300	1,673	43	8
126	214,500	282	552,700	872	1,335,200	8,779	14,561,500	1,659	44	0
672	1,178,700	1,869	3,825,700	8,779	14,561,500	..	..	17,789	..	
127	210,700	322	589,600	702	1,153,600	9,481	15,715,100	1,658	44	4½
143	278,000	336	632,600	753	935,800	10,234	16,650,900	1,627	44	8¾
164	293,900	342	586,300	641	1,018,500	10,875	17,669,400	1,625	45	1
198	314,700	365	663,300	648	931,400	11,523	18,600,800	1,614	45	5
203	326,300	396	724,900	653	933,500	12,176	19,534,300	1,604	45	10
200	355,800	333	647,600	690	1,099,900	12,866	20,634,200	1,604	46	0
238	383,700	416	710,800	602	936,000	13,468	21,570,200	1,602	46	4
230	365,700	438	754,500	658	894,000	14,126	22,464,200	1,590	46	7
293	476,700	533	873,200	702	1,075,000	14,828	23,539,200	1,588	46	10
339	516,300	616	1,059,500	208	472,000	15,036	24,011,200	1,597	47	4
2,135	3,521,800	4,147	7,242,300	6,257	9,449,700	..	..	16,109	..	
337	572,400	576	1,021,900	435	607,900	15,471	24,619,100	1,591	47	7
333	574,100	529	976,400	609	884,100	16,080	25,503,200	1,586	47	10
297	460,700	487	819,700	775	1,199,100	16,555	26,702,300	1,584	48	0
346	579,300	563	1,000,500	860	1,326,100	17,715	28,028,400	1,582	48	2
399	648,300	612	1,067,100	712	1,036,800	18,427	29,115,200	1,580	48	3
390	633,100	642	1,076,500	431	779,400	18,858	29,894,600	1,585	48	6
471	734,200	651	1,038,500	505	994,100	19,363	30,888,700	1,595	48	9
410	702,600	579	1,012,400	735	1,170,700	20,098	32,059,400	1,595	48	11
453	757,300	620	1,089,600	743	1,488,900	20,841	33,548,300	1,610	49	0
536	917,700	687	1,259,400	629	1,059,900	21,470	34,608,200	1,612	49	2
3,972	6,579,700	5,946	10,362,000	6,434	10,597,000	..	..	15,920	..	
6,779	11,280,200	11,962	21,430,000	21,470	34,608,200	..	..	49,818	..	
226	376,007	399	714,333	716	1,153,607	..	..	1,661	..	

TABLE IV.—*Claims payable during* 1829–58.

Year.	Amount Pers.	Amount Thlrs.	Average for one Life Thlrs.	Per cent. of the Assurances Pers.	Per cent. of the Assurances Thlrs.	Expected Pers.	Expected Thlrs.	Average for one Life Thlrs.	Plus Pers.	Plus Thlrs.	Minus Pers.	Minus Thlrs.
1829	12	15,100	1,258	0·93	0·63	15·19	27,979	1,842	..	..	3·19	12,879
1830	16	34,300	2,144	0·90	1·01	22·84	43,688	1,913	..	..	6·84	9,388
1831	27	50,300	1,863	0·90	0·88	32·38	63,254	1,954	..	..	5·38	12,954
1832	53	102,300	1,930	1·30	1·33	50·25	94,681	1,884	2·75	7,619	..	..
1833	46	71,032	1,544	0·95	0·80	64·81	117,700	1,816	..	..	18·81	46,668
1834	65	108,300	1,666	1·16	1·10	77·19	135,909	1,761	..	..	12·19	27,609
1835	85	165,400	1,946	1·34	1·51	89·34	156,092	1,747	..	9,308	4·34	..
1836	108	179,200	1,659	1·47	1·44	103·42	178,234	1,723	4·58	966	..	..
1837	123	217,417	1,768	1·50	1·58	117·99	202,162	1,713	5·01	15,255	..	..
1838	124	211,900	1,709	1·37	1·40	133·88	228,054	1,703	..	..	9·88	16,154
Total 1829–38	659	1,155,249	17,487	11·82	11·68	707·29	1,247,753	18,056	12·34	33,148	60·63	125,652
1839	123	206,400	1,678	1·25	1·26	149·74	254,226	1,699	..	..	26·74	47,826
1840	136	256,200	1,884	1·29	1·48	164·25	276,936	1,686	..	..	28·25	20,736
1841	158	278,525	1,763	1·41	1·53	179·02	299,402	1,672	..	..	21·02	20,877
1842	191	300,800	1,575	1·61	1·56	193·93	322,746	1,664	..	..	2·93	21,946
1843	201	323,800	1,611	1·60	1·60	208·55	345,219	1,655	..	..	7·55	21,419
1844	197	340,400	1,728	1·49	1·60	224·81	370,748	1,649	..	..	27·81	30,348
1845	235	381,200	1,622	1·69	1·71	241·13	397,277	1,648	..	..	6·13	16,077
1846	224	353,800	1,579	1·54	1·52	257·03	421,986	1,642	..	..	33·03	68,186
1847	289	470,000	1,629	1·88	1·93	275·53	450,174	1,634	13·47	19,826	..	..
1848	326	496,500	1,523	2·08	1·98	290·27	474,505	1,636	35·73	21,995	..	..
Total 1839–48	2,080	3,407,625	16,592	15·84	16·17	2184·26	3,613,219	16,585	49·20	41,821	153·46	247,415
1849	335	566,500	1,691	2·09	2·21	302·28	494,798	1,638	32·72	71,702	..	..
1850	326	556,600	1,707	1·96	2·10	318·57	520,490	1,634	7·43	36,110	..	..
1851	293	459,000	1,567	1·69	1·67	337·52	550,166	1,630	..	..	44·52	91,166
1852	335	554,900	1,656	1·83	1·91	359·54	584,449	1,626	..	..	24·54	29,549
1853	392	635,800	1,625	2·06	2·11	382·07	618,850	1,620	9·93	16,950	..	..
1854	375	608,600	1,623	1·99	2·04	400·52	649,828	1,622	..	..	25·52	41,228
1855	466	721,450	1,548	2·41	2·34	417·83	681,135	1,630	48·17	40,315	..	..
1856	406	692,450	1,706	2·02	2·16	437·73	715,080	1,633	..	..	31·73	22·630
1857	444	749,700	1,688	2·06	2·16	459·28	752,017	1,637	..	..	15·28	2·317
1858	517	881,100	1,704	2·33	2·46	480·79	788,113	1,639	36·21	92,987	.	..
Total 1849–58	3,889	6,426,100	16,515	20·44	21·16	3896·13	6,354,926	16,309	134·46	258,064	141·59	186,890
Total....	6,628	10,988,974	50,594	48·10	49·01	6787·68	11,215,898	50,950	196·00	333,033	355·68	559,957
Average for one Year	221	366,299	1,687	1·60	1·63	226·26	373,863	1,698	6·53	11,101	11·86	18,665

TABLE V.—*Persons Assured in the Gotha Life Bank, and Deaths during 1829-58.*

Ages.	Number of Persons assured.	Number of Persons entered.	Number of Persons gone in lifetime.	Total of 3 and 4.	Half of the Numbers in Col. 5.	Number exposed to the Risk of Mortality for a whole Year.	Deaths.	Rate of Mortality per cent. 1829-58.	Rate of Mortality per cent. 1829-53.
1.	2.	3.	4.	5.	6.	7.	8.	9.	10.
15–25	2,833	1,241	165	1,406	703	2,130	12	0·56	0·46
26–30	14,643	4,625	557	5,182	2,591	12,052	100	0·83	0·87
31–35	37,584	7,304	896	8,200	4,100	33,484	290	0·87	0·92
36–40	56,680	7,180	873	8,053	4,027	52,653	508	0·97	1·00
41–45	64,213	5,201	850	6,051	3,025	61,188	628	1·03	1·04
46–50	61,802	3,715	799	4,514	2,257	59,545	868	1·46	1·45
51–55	52,195	2,464	523	2,987	1,493	50,702	963	1·90	1·82
56–60	38,599	1,446	323	1,769	884	37,715	1,014	2·69	2·77
61–65	23,926	241	133	374	187	23,739	917	3·86	3·83
66–70	12,145	15	43	58	29	12,116	734	6·06	6·08
71–75	4,944	..	15	15	7	4,937	482	9·76	9·04
76–80	1,562	..	6	6	3	1,559	190	12·19	11·35
81–85	281	..	..	..	..	281	65	23·13	23·94
86–90	24	..	..	..	..	24	8	33·33	33·33
Total	371,431	33,432	5,183	38,615	19,306	352,125	6,779		

TABLE VI.—*Comparative View of the Annual Rate of Mortality according to different Tables.*

	26–30.	31–35.	36–40.	41–45.	46–50.	51–55.	56–60.	61–65.	66–70.	71–75.	76–80.	81–85.
1. Gotha Life Experience (1829-58)	0·83	0·87	0·97	1·03	1·46	1·90	2·69	3·86	6·06	9·76	12·19	23·13
2. Actuaries' Table (17 English Life Offices, 1762-1840)	0·81	0·89	0·99	1·13	1·43	1·91	2·65	3·79	5·55	8·13	11·88	17·22
3. Equitable Experience (1762-1840, Morgan)	0·78	0·88	1·03	1·18	1·38	1·85	2·68	3·72	5·48	7·89	11·18	17·97
4. Friendly Societies' Experience (1836-40, Nelson)	0·73	0·80	0·89	1·04	1·29	1·70	2·24	3·05	4·62	6·85	8·84	11·97
5. Ditto ditto (1846-50, Finlaison)	0·75	0·80	0·95	1·13	1·37	1·77	2·45	3·12	4·75	6·70	10·32	15·31
6. New York Mutual Society Exp. (1844-58, Homans)	0·85	0·89	0·93	0·96	1·20	1·52	2·03	3·07	4·85	7·66	13·01	18·96
7. Carlisle Table (1779-87, Heysham—Milne)	0·87	1·02	1·15	1·45	1·41	1·61	2·50	3·84	4·66	7·67	11·08	14·88
8. Hanover (1825-43) Tellkamps	1·11	1·25	1·35	1·48	1·77	2·50	3·61	5·70	7·91	9·34	12·74	18·50
9. Saxony (1840-49, Heym)	0·84	0·99	1·11	1·32	1·63	2·24	3·22	4·81	7·35	11·31	15·94	21·58
10. England, Males (1841, Farr)	0·97	1·10	1·25	1·42	1·62	1·87	2·71	3·95	5·75	8·32	11·94	16·90
11. France, Males (1817-31, Demonferrand)	0·88	0·95	0·94	1·17	1·43	1·93	2·50	4·10	5·60	9·25	12·79	18·62
12. Belgium, Males (1841-45, Quetelet)	1·50	1·58	1·72	2·16	2·43	2·47	3·15	4·24	5·83	8·65	12·86	18·01

NOTICES OF NEW WORKS.

An Essay on Life Assurance; being a popular Exposition of the Subject, and a Plea for its more general Adoption. By H. W. PORTER, B.A. Charles and Edwin Layton, Fleet Street.

IN this essay Mr. Porter points out the great advantages to be derived from the practice of life assurance, touches briefly upon the principles on which that practice is founded, describes the nature of the several institutions established for the assurance of lives and some of the various modes in which contracts of assurance are framed, combats successfully certain false and very prevalent notions with regard to the system itself, and concludes with such forcible observations as to the desirableness of persons in ordinary circumstances resorting to it, as must, we should think, have considerable influence upon any reader of them. We have good reason to believe that there are several classes of our vast population by which this essay would be read with advantage. There are many educated persons still who are quite ignorant, not only of the theory, but of the practice of assurance. Some doubt the soundness of the principles, and some the advantages of the system, whilst many are led to neglect the opportunities it affords by sheer carelessness and inattention.

Writing on this point, Mr. Porter aptly says:—" I sincerely believe that the reason why life assurance is so little resorted to, is not owing to any reluctance on the part of those who have wives and children depending upon them for support—and whose widows and orphans would be reduced to poverty by their premature decease—to make a small present pecuniary sacrifice to prevent so cruel an alternative, but because their attention has not been called to the subject, or, if it has, because they have not fully understood it or reflected sufficiently on the matter. It is hardly to be imagined that the father of a family, who works hard day by day for the support of his wife and children, and cheerfully exerts every effort for their advantage, would fail to make provision for them by the assurance of his life, supposing that he were fully cognizant of the advantages to be derived therefrom by his family, and that he clearly understood in what manner he might secure an independence for them in the event of his early death, or—if his means would not by possibility allow of his making such a provision for them as would constitute an independence—at least, a sufficient sum of money to enable them to provide for all demands at the time of his decease, to satisfy all expenses connected with his death and probable antecedent illness, and to enable them to live in comfort till such time as arrangements could be made for providing for his family, who might otherwise be left destitute, and to enable them to start fair again in the race of life. It is hardly to be imagined, I repeat, that his moral sense would fail to induce him to perform the duty so highly incumbent upon him, of making the best provision in his power for those dependent upon him, and who have clearly the same claim upon him *morally* to make provision for their support after his death, as they have *legally* upon him for their maintenance during his life."

We quite agree with Mr. Porter. It is, indeed, hardly to be imagined!

INSTITUTE OF ACTUARIES.

PROCEEDINGS OF THE INSTITUTE.

Second Ordinary Meeting, Session 1859-60.—Monday, 2nd January, 1860.

W. B. HODGE, Vice-President, in the Chair.

The minutes of the first ordinary meeting were read and confirmed.
The Secretary announced several donations to the library.
The following candidates, duly nominated at the last ordinary meeting,
were elected members of the Institute, viz. :—

Official Associate—A. G. Ramsay.

Associates.

Charles Graham Carttar. | Sydney King.
John James Holliday Chapman. | John Thomas Minett.
William Pateman.

The Chairman announced the following as the results of the annual
examinations :—
Out of five candidates for the matriculation examination, three passed, in
the order of merit indicated below :

A. H. Green.
W. M. Makeham.
Sydney King.

Four candidates presented themselves for the second year's examination,
and all passed, in the following order :

A. G. Finlaison.
W. C. Mullins.
C. G. Laing.
C. Bischoff, Jun.

For the third year's examination, two candidates appeared, and both passed,
as follows :

W. P. Pattison.
James Terry.

These gentlemen are now entitled to certificates of competency.
Papers were read—" On a formula for calculating the value of a survivorship assurance," by M. Réboul; " On the purchase of life assurance policies as
an investment," by Archibald Day, Esq.
Thanks having been voted to M. Réboul and Mr. Day, the meeting adjourned to Monday, the 30th inst.

Third Ordinary Meeting, Session 1859-60.—Monday, 30th January, 1860.

CHARLES JELLICOE, Vice-President, in the Chair.

The minutes of the last ordinary meeting were read and confirmed.
The Secretary announced several donations to the library.

The undermentioned gentlemen, duly nominated at the last ordinary meeting, were elected members of the Institute, viz. :—

Francis Archibald Corrie.
Alexander A. Green.

A paper was read " On some considerations suggested by the Reports of the Registrar-General; being an inquiry into the question as to how far the inordinate mortality in this country, exhibited by these Reports, is controllable by human agency." Part I.   By H. W. Porter, Esq., B.A.

Thanks having been voted to Mr. Porter, the meeting adjourned to Monday, the 27th February.

Fourth Ordinary Meeting, Session 1859-60.—*Monday,* 27th *February,* 1860.

W. B. HODGE, Vice-President, in the Chair.

The minutes of the last ordinary meeting were read and confirmed.
The Secretary announced various donations to the library.
The undermentioned gentlemen, duly nominated at the last ordinary meeting, were elected members of the Institute, viz. :—

Official Associates.
Edward Butler. | F. Ferguson Camroux.

Associates.
Edward Plumbridge Clark. | John Edward Primrose Gage.
John Elphick Leyland.

Letters were read from Charles Babbage, Esq., M.A., and Professor Sylvester, acknowledging, with thanks, their election as honorary members of the Institute.

A paper was read, " On some considerations suggested by the Reports of the Registrar-General; being an inquiry into the question as to how far the inordinate mortality in this country, exhibited by these Reports, is controllable by human agency."   Part II.   By H. W. Porter, Esq., B.A.

Thanks having been voted to Mr. Porter, the meeting adjourned to 26th March, 1860.

THE

ASSURANCE MAGAZINE,

AND

JOURNAL

OF THE

INSTITUTE OF ACTUARIES.

On the Rates of Interest for the use of Money in Ancient and Modern Times. (Part IV.) By WILLIAM BARWICK HODGE, *Vice-President of the Institute of Actuaries, and Fellow of the Statistical Society.*

[Read before the Institute, 30th April, 1860.]

IN the former papers upon this subject[1] which I had the honour to lay before the Institute, I described the vehement controversy carried on in earlier ages as to the propriety of allowing any interest to be taken for the use of money, and I alluded to the discussion which, after interest to a limited extent had been allowed by law, arose as to the advantages to be derived from a reduction of the legal rate.

I now propose to give some account of this latter discussion, and of the writers who took part in it. Among the foremost of them was Sir Josiah Child, whose works upon commerce were long held in high esteem, and even now retain a great reputation.

To advocate a reduction to 4 or even 3 per cent. of the legal rate of interest (then at 6 per cent.), he wrote, in 1665, a *Discourse concerning Trade*. This tract was not published until 1668, and was then without the author's name—bearing the initials "J. C." only. It opens thus:—"The prodigious increase of the Netherlanders in their domestick and foreign trade, riches, and multitude of shipping, is the envy of the present, and may be the wonder of

---

[1] *Assurance Mag.*, vols. vi., vii., and viii.

all future, generations. And yet the means whereby they have advanced themselves are sufficiently obvious, and, in a great measure, imitable by other nations."[1]

The writer then proceeds to enumerate the causes of their success, attributing it, principally, to their reputation for integrity in trade; their thrifty mode of living; their care to bring up their children, both sons and daughters, to have a full knowledge of "arithmetick and merchants' accompts;" the superiority of their commercial code, with their use of a public register for recording transfers and mortgages of lands; and, lastly, "the lowness of interest of money with them." "This," he says, "in my poor opinion, is the *causa causans* of all the other causes of the riches of that people; and that if the interest of money were with us reduced to the same rate as it is with them, it would, in a short time, render us as rich and considerable in trade as they now are."[2]

In support of this opinion, he urges the extraordinary increase in the prosperity of the country which followed the reduction of interest in the reign of James I. and in that of Charles II.

The proofs adduced of this increased prosperity are abundant and striking enough, and form, indeed, the most valuable portion of the book. There is, however, little attempt to show the connection between the result and the causes assigned to it, beyond the expression of the writer's conviction—"that the bringing down of interest, from 6 to 4 or 3 per cent., will necessarily, in less than 20 years' time, double the capital stock of the nation."[3]

The theory thus laid down was by no means new. Indeed, its resemblance to one propounded, with much greater ability, half a century before, by Sir Thomas Culpepper the Elder, in the *Small Tract against Usury*, already mentioned,[4] might lead us to suppose it to have been derived from that source, did not Child assure us his own *Discourse* was written before he saw Sir Thomas's tract. He spoke of it, however, with great admiration, and reprinted it with the second edition of his work in 1670, as well as with the collected edition of his writings in 1693.[5]

That the matter had already been much discussed is obvious, from the objections cited by Child and his attempts to refute them. He quotes also " Sir Henry Blount (an honourable member of His Majesty's Council of Trade), as having well said before the Lords, *at the debate,* that abatement of interest was the *unum magnum* towards the prosperity of this kingdom." [6]

---

[1] Page 1.   [2] Page 9.   [3] Page 16.   [4] *Assurance Mag.* viii. 75.
[5] London; 12mo.   [6] *See* Preface.

Sir Thomas Culpepper appears to have published, in 1641, a second *Tract against Usury*, to enforce the doctrines contained in the first. It is referred to in Child's *Interest of Money Mistaken*,[1] and quoted in Macpherson's *History of Commerce*;[2] but I have not seen it elsewhere mentioned.

The same writer's son, Sir Thomas Culpepper the Younger, published, in 1668, *A Discourse, shewing the many Advantages which will accrue to this Kingdom by the Abatement of Usury*. The list of promised advantages proves the extraordinary delusions then exist- ing as to the power of legislation to increase national wealth. The reduction of interest was to " supply His Majesty's present wants ; in a short time double, if not treble, the yearly fruit and pro- duct of our lands ; revive our dying manufactures ; plentifully relieve the poor, by setting all our heads and hands to work ; pre- vent the fatal destruction of our timber ; pay the debts of the whole gentry by timely sales ; make monie so easy to be borrowed that the lender must shortly pay the broker's conveyance ; rebuild London, profitably to the builder and speedily for the public ; and, lastly, inviolably establish the Crown of England, by the advance- ment of His Majesty's revenues, by the welfare of his subjects, and by making land the overbalancing scale of wealth and power." With amusing inconsistency, the author asserts, in another part of his book, that the reduction of interest to 6 per cent. during the Commonwealth was passed " by the grandees of that junta with a view to the utter oppression of the nobility and gentry."[3] For- getting, too, " the fatal destruction of our timber," he elsewhere asks, " Have we not in many places in this kingdom iron oare without end, with woods adjacent even to a nuisance ?"[4]

Although the great majority of his contemporaries appear to have shared Child's opinions upon the subject, his *Discourse* soon called forth replies. One was published anonymously, in the same year (1668), under the title of *Interest of Money Mistaken*;[5] and, in 1669, appeared *Interest at Six per Cent. Examined, by Thomas Manley, Gent.* The former of these I have not seen ; but the author seems to have pointed out, clearly enough, that the low rate of interest in Holland was not the cause, but the consequence, of the large capitals accumulated there by successful trade. " Low interest," says Manley, too, in his preface, " is both in nature and time subsequent to riches, and he who says low usury begets riches takes the effect for the cause—the child for the mother— and puts the cart before the horse."

[1] Page 10. [2] ii. 242. [3] *See* Preface. [4] Page 13.
[5] *Ency. Brit.;* 7th edit.; art., " Pol. Eco."

Child wrote a special reply to *Interest of Money Mistaken,* under the title of *Trade and Interest of Money Considered,* in the preface to which he noticed Manley's book. In this preface, which was printed with the collected edition of his works (Lond., 1693), as if it had been the general preface to the whole, occurs his celebrated aphorism—" That land and trade are twins, and ever will wax and wane together. It cannot be ill with trade but land will fall, nor ill with land but trade will feel it."

The frequent repetition of this aphorism during the discussions upon the Corn Laws, has led to a popular impression that Child was one of the early advocates of freedom in trade; but this opinion has no real foundation. Indeed, the great authority possessed for so long a period by his name in commercial matters, arose from the fact that he was deeply imbued with nearly all the narrow prejudices upon which the so-called protective system was founded.[1]

In addition to the egregious error of believing that money could be made plentiful by Act of Parliament, he looked upon the Navigation Laws as England's *Charta Maritima,*[2] proposed differential duties of 50 per cent.,[3] and strenuously recommended that Ireland and the colonies should be restricted to trading with England only.[4] The ferocity with which he spoke of the "total ruin and extirpation" of our commercial rivals, the Dutch,[5] could hardly have been exceeded in the worst ages of commercial barbarism. It is true, he endeavoured to shake the popular opinions upon the subject of the balance of trade,[6] but he was unable to perceive the error of the principle upon which those opinions were founded, and his inducement to the attempt, which, after all, was but a feeble one, was merely to defend the trade to the East Indies, from which a great part of his fortune was derived, and which could only be carried on by the exportation of bullion.

That Child was an acute observer is obvious from his works, and it is certain that he had a high character for sagacity with his contemporaries; but this would hardly be inferred from his reasoning in support of his theories on the subject of interest, which is generally weak, oftentimes contradictory, and sometimes contemptible. His notion, that the power, wealth, and populousness of the Jewish nation, in ancient times, were due to the scriptural prohibition of usury, was a strange crotchet to enter the mind of a trader.[7]

In one place, he denies that there was any ground for the com-

[1] *Hist. of Com.* ii. 843.   [2] *See* Preface.   [3] Page 97.   [4] Page 95.
[5] Page 10.   [6] Page 135.   [7] Page 28.

plaints of the scarcity of money, alleging that people complained, partly because it was their nature to complain, and partly because they were " uneasie in matters of their religion."[1]   In another, he asserts that the scarcity was entirely caused by the new trade of hankering, which enabled every man who could scrape £50 or £100 together to deposit it with a goldsmith[2]—as if goldsmiths would pay 6 per cent. upon money, merely to lock it up in their tills !   It is difficult to understand how such an idea could have been adopted by anyone acquainted with the commonest operations of trade.   Manley, who was full of the prejudices of the day, attempted to account for the scarcity of money by the vast consumption of foreign commodities, and instanced the great increase in the importation of wines ; whereupon Child accused him of asserting "that abatement of use money brought in our drinking."[3]

The author of *Interest of Money Mistaken* having objected that a reduction of the legal rate would lead to the withdrawal of large sums of foreign capital, Child's only answer is, " So much the better, for the borrower is always the slave of the lender."[4]

Samuel Fortrey, also spoken of with great approbation by Child, published, in 1673, *England's Interest and Improvement*, recently reprinted by the Political Economy Club.[5]   This writer, who was gentleman of the bedchamber to Charles II., appears to have believed, that, by merely lowering the legal rate of interest, the nation might be converted from borrowers into lenders— asserting, that, by reducing the rate below that of other nations, they would make no profit on us by lending, but rather we on them.[6]

Among the arguments against a high rate of interest urged by Culpepper the Elder, was one, that a hundred pounds, " managed at ten in the hundred, in seventy years multiplies itself to a hundred thousand pounds ;" and he thence inferred, that, if a hundred thousand pounds of foreigners' money be managed here at 10 per cent., it would in threescore and ten years carry out ten millions.[7]   There is rather an important mistake in this estimate, as the correct sum, according to the proportion stated, should have been one hundred millions, instead of ten millions.   Child took up the same theme. He did not repeat Culpepper's error, which, however, he did not

---

[1] *See* Preface.   [2] *Ibid.*   [3] *Ibid.*   [4] Page 17.
[5] *A Select Collection of Early English Tracts on Commerce*, reprinted by the Political Economy Club; London, 1856; p. 211.   I take this opportunity of expressing my high sense of the usefulness of this reprint, as well as of that, even more valuable, of the *Select Tracts on Money*; London, 1856.   Both volumes are ably edited by Mr. McCulloch.
[6] *Ibid.*, p. 246.   [7] Child's Reprint, p. 227.

correct, speaking himself merely of the increase of £100 to £100,000 at the same rate.[1] When the accuracy of his result was called in question, he went into a very elaborate calculation, and, by taking quarterly periods of conversion, showed that money at 10 per cent. doubled itself, within a fraction, in seven years. His mode of doing this was the cumbrous one of separately calcu- culating and adding the interest of each succeeding quarter.[2] Neither he nor his opponents seem to have perceived that there was a very startling answer to this theory.

It is clear that merchants only borrow money to carry on trade which they are unable to carry on without borrowing, and that unless they make a larger profit upon the loan than suffices to pay the interest of it, they must either leave off borrowing or become bankrupts. Merchants' profits are of course fluctuating and un- certain; and, if they were certain, it would be very difficult to ascertain their average amount. According to Adam Smith, double interest was in his time, 1776, considered in Great Britain " a good, moderate, reasonable profit."[3] Manley asserts his belief that, in 1678, all trades gained 15 per cent.,[4] which was considerably more than double the then legal rate of interest. Taking Adam Smith's estimate as correct, the borrowers at 10 per cent., referred to by Child and Culpepper, would make on the average 20 per cent. upon the money borrowed. Now, one hundred thousand pounds accumulated at that rate for seventy years, assuming quarterly periods of conversion, as in Child's calculation, would amount to nearly eighty-six thousand millions of pounds, a sum that leaves the contemptible hundred millions of accumulated debt very far behind. The answer is framed upon exactly the same principle as the assertion, and the extravagance of the one is a sufficient exposure of the absurdity of the other. The truth is, that the elaborate calculations about accumulation have nothing to do with the question; but it is surprising how slight an acquaintance with the mysteries of compound interest—and Child's knowledge upon the subject does not appear to have been deep—will enable a man to bewilder the public.

Among the controversialists of the seventeenth century respect- ing interest, the latest and the most celebrated was John Locke, whose opinions were published in 1691 in a letter to a member of Parliament (most probably Charles Montague, afterwards First Lord of the Treasury and Chancellor of the Exchequer), entitled,

---

[1] *Trade, &c., Considered*, p. 7.   [2] *Ibid.*, p. 9.   [3] *Wealth of Nations*, book i., ch. 9.
[4] *Interest at Six per Cent. Examined.*

Some Considerations of the Consequences of lowering the Interest of, and raising the Value of, Money.[1]

The principal object of this letter was to refute the notions of Lowndes and others, who proposed to enrich the nation by raising the nominal value of the coin. The portion of it relating to interest is said by Locke to have been written twelve years before, when his attention had been called to the subject by the discussion then going on as to the advantage of reducing interest to 4 per cent. He goes fully into the question—" Whether the price of the hire of money can be regulated by law ?"—and, after stating his reasons for believing it cannot, delivers his opinion that the attempt would —1, make the difficulty of lending and borrowing greater; 2, pre- judice none but widows and orphans, whose estates lie in money ; 3, tend only to increase the profits of brokers and scriveners ; and, 4, lead to perjury and evasion. " People," he says, " who have their estates in money, have as much right to make as much of the money as it is worth (for more they cannot) as the landlord has to let his land for as much as it will yield."[2]

The works of Locke and Bacon stand out as gigantic landmarks, showing the progress of the human intellect during the interval between them; and upon no subject had the growth of rational views been more remarkable than upon the one under consideration. The popularity of Locke's other writings, and the authority of his name, gave, no doubt, great weight to his opinions respecting interest; but the effect produced by them would probably have been much more beneficial if it had not been weakened by an incon- sistency that cannot fail to create surprise.

A learned and lamented historian has made it a grave charge against Adam Smith, that he should have maintained, " in direct opposition to his own principles," that the rate of interest ought to be regulated by the State, particularly after Locke had taught " that it was as absurd to make laws fixing the price of money as to make laws fixing the price of cutlery or of broad-cloth."[3]

The opinions respecting the limitation of interest, laid down in the *Wealth of Nations*, are, undoubtedly, among the very few blemishes to be found in that wonderful work; but, strange as it may seem, it is by no means improbable those very opinions were adopted in deference to the authority of Locke. At any rate, there is considerable injustice in this attempt to elevate him at the expense of Adam Smith, who will, perhaps, be held in some measure

---

[1] Locke's *Works*; Lond., 1812; vol. v.   [2] Page 11.
[3] Macaulay's *Hist. Eng.* iv. 631.

excused for remaining unconvinced by Locke's reasoning, when we find that reasoning failed to convince Locke himself.   This eminent writer did, indeed, show, by what appears to us overpowering logic, that laws regulating interest can only be ineffective and mischievous; but he, nevertheless, concluded thus—"What then ! should there (you will say) be no law at all to regulate interest ?   I say not so ; for it is necessary there should be a stated rate : " and the reasons he gave were—1, "that where contracts have not settled it between parties, courts of justice may know what damages to allow;"[1] and, 2, "that dexterous and combining money-jobbers may not have too· great a power to prey upon the necessity of borrowers."[1]   The first of these reasons is quite inadequate, and the second gives up the whole question ; for, with the exception of the disciples of Child and Culpepper, who had made the brilliant discovery that the way to increase the supply of a commodity was to diminish the profits of those who brought it to market, few persons ever contended for limiting the rate of interest, except from the absolute necessity of shielding needy borrowers from the grasping avarice of lenders.   Dr. Johnson, indeed, once urged that the usury laws were required for the protection of creditors as well as of debtors, "for that, if there were no such check, people would be apt, from the temptation of great interest, to lend to desperate persons, by whom they would lose their money;"[2] but this opinion is more likely to be admired for its ingenuity than for its soundness.

. While Locke hesitated thus in adopting the conclusions to which his arguments appeared inevitably to tend, there existed two contemporary writers who perceived the truth of those conclusions more clearly, and adhered to them more boldly.   Sir Dudley North published, in the year Locke's work appeared (1691), *Discourses upon Trade, principally directed to the cases of the Interest, Coynage, Clipping, and Increase of Money.*   This work is among the valuable reprints of the Political Economy Club,[3] without whose aid it would probably have been lost to the public, as the author seems, soon after its appearance, to have consented, for some reason or other, to its suppression ; so that, according to his brother, who wrote his life, it was "utterly sunk, and a copy not to be had for money."[4]

That this suppression was a great evil, there can be no doubt ; for no work has come down to us so well calculated to enlighten the public mind or to produce correct opinions upon the subject.

[1] Page 63.   [2] Boswell; Lond., 1847; p. 502.
[3] *Early Tracts on Commerce;* Lond., 1856.   [4] *Ibid.,* Preface xii.

It is alike admirable for clearness, conciseness, and comprehensive-ness; and is, indeed, a perfect model in every respect.

North pointed out the true cause of the lowness of interest in Holland, and that "as plenty makes cheapness in other things, so, if there be more lenders than borrowers, interest will also fall; wherefore," he continues, "it is not low interest makes trade, but trade, increasing the stock of the nation, makes interest low." [1]

He, in some degree, disputes the facts stated by Child; for, while admitting that the rate of interest upon mortgages in Holland was as low as 3 or 4 per cent., he asserts, that "the current rate between merchant and merchant was 6 per cent." [2]

"To be a landlord," says North, "or a stocklord, is the same thing; and the state may, with as much justice, make a law that lands which heretofore have been lett for ten shillings per acre shall not now lett for above eight shillings per acre, as that money or stock from 5 per cent. shall be lett for 4 per cent." [3]

Alluding to the difference in the values of securities, he asks, "Shall any man be bound to lend a single person upon the same terms as others lend upon mortgages or joynt obligations?" add-ing, that the poor trader with a narrow stock, or none at all, supplies himself by buying goods of rich men at time, and thereby pays interest, not at the rate of 5, 6 or 8, but 10, 12 and more per cent., which no law can prevent; [4] concluding with the sound maxim that, all things considered, it would be found best for the nation to leave the borrowers and the lenders to make their own bargains, according to the circumstances they lie under, and, in so doing, "follow the course of the wise Hollanders so often quoted." [5]

In a postscript to the *Discourses*, the various projects for securing the advancement of trade by legislative enactments are charac-terised as follows:—" Thus we may labour to hedge in the cuckoo, but in vain; for no people ever yet grew rich by policies: but it is peace, industry, and freedom that bring trade and wealth, and nothing else. [6]

There is a great similarity between the reasoning of Locke and that of North, but it is in no degree probable that either of them had derived his opinions from the other. Their works were both pub-lished in the same year, although Locke's had been written so long before, and the difference in their politics would effectually prevent any communication between them. It has been truly remarked by Lord Macaulay, "that if Locke had not taken shelter from tyranny

[1] Page 518.   [2] Page 518.   [3] Page 522.   [4] Page 520.
[5] Page 521.   [6] Page 540.

in Holland, it is by no means impossible that he might have been sent to Tyburn, by a jury which Dudley North had packed.[1]

Sir William Petty, the second of the two writers alluded to, in his *Quantulumcunque* concerning money, addressed to the Marquis of Halifax, contended, that laws limiting the rate of interest were as impolitic as it would be to enact laws limiting exchange, which he defined as interest of place.[2]   This work was published in 1682, after the remarks upon interest had been written by Locke, who could not, therefore, have adopted any hints from the *Quantulum-cunque;* but Petty had advocated the same views at much greater length in his *Treatise on Taxes and Contributions,* published as early as 1662,[3] the fifth chapter of which is dedicated entirely to the consideration of usury.   In this chapter, after giving a similar definition of exchange as local usury, Petty says, in reply to the question, What are the natural standards of usury and exchange?— " As for usury, the least that can be is the rent of so much land as the money lent will buy where the security is undoubted; but where the security is casual, then a kind of insurance must be interwoven with the simple interest, which may advance the usury very conscionably to any height below the principal itself; "[4] and he concludes, " I see no reason for endeavouring to limit usury upon time any more than usury upon place (which the practice of the world doth not), unless it be that those who make such laws were rather borrowers than lenders; but of the vanity and fruitlessness of making civil positive laws against the laws of nature I have spoken of elsewhere, and instanced in several particulars."[5]

There occurs in this *Treatise* a singular and fanciful notion as to the relation between the rent and the selling value of land, which appears very extraordinary when proceeding from a person so well versed in such matters as Petty.   The number of years' purchase of the rent the fee simple is naturally worth, he estimated at as many years as one man of fifty years old, another of twenty-eight, and another of seven years old, all being alive together, might be expected to live.   " Now, in England," he says, " we esteem three lives equal to one-and-twenty years ; and, consequently, the value of land to be about the same number of years' purchase ;" and, further on, he speaks of land selling in other countries for thirty years' purchase, " by reason of the better titles, more people, and, *perhaps, truer opinion of the value of three lives.*"[6]

---

[1] *Hist. Eng.* iv. 630.   [2] *Select Tracts on Money,* pp. 156, 157.
[3] *See* Petty's *Tracts;* Dublin, 1769; pp. 1 to 88.   [4] Chap. v. 3.
[5] Chap. v. 37.   [6] Chap. iv. 19, 20, 22.

Although Petty may not have written upon the subject of interest with the force and clearness of North or Locke, he is undoubtedly entitled to the credit of having been the first English writer—and, so far as I know—the first writer of any nation who advocated the removal of all legal restrictions upon the rate.

He suggested too, in his *Quantulumcunque*, the advantage of "keeping all accompts in a way of decimal arithmetick." [1]   The discovery of decimal fractions towards the close of the sixteenth century, with that of logarithms at the commencement of the seventeenth, seem speedily to have suggested to mathematicians the advantages of a decimal subdivision of all units of quantity. Edmund Wingate, described by Dr. Hutton as the clearest writer upon arithmetic in the English language, pointed out in his *Treatise of Arithmetic*," [2] that under such an arrangement the science would be taught with much more ease and expedition; but added, in a prophetic spirit, "it is improbable such a reformation will ever be brought to pass."

North and Petty were considerably in advance of their contemporaries upon most other subjects, and they were alike remarkable for the soundness of their opinions respecting money.

They not only showed, in the clearest manner, the absurdity and mischievous tendency of the proposals for raising the nominal values of the current coins, made by persons who thought that the wealth of the nation could be increased by merely calling a sixpence a shilling, but they both attacked the laws for preventing the exportation of coin and bullion.

Petty asserted that these laws were contrary to the laws of nature, and impracticable; and boldly laid down the principle, then entirely new, that a nation was not always less wealthy for having less money.[3]   "No man," said North, "is the richer for having his estate all in plate, money, &c., lying by him; but, on the contrary, he is for that reason the poorer." [4]

The numerous statutes enacted, and the various penalties imposed, by the English Legislature, to prevent the exportation of bullion and the melting of the current coin, are sufficient evidence of the importance attached to these objects and the difficulty of attaining them.[6]

These attempts arose out of the delusion that wealth consists of money alone; and that a nation, therefore, can only become rich

---

[1] *Select Tracts*, p. 166.   [2] Lond., 1658; p. 233.
[2] *Select Tracts on Money*, p. 165.   [4] *Early Tracts on Commerce*, p. 525.
[5] *Hist. Com.* i. 463, 464, 474, 511, 529, 530; ii. 43.

by accumulating the precious metals—a delusion lying at the root of nearly all the errors that have prevailed respecting commerce, and one still existing in the minds of many, as is evident from the opinions frequently published.

A long period elapsed before any considerable portion of mankind could be brought to perceive that the value attached to gold and silver arises principally from the facilities they afford for the exchange of commodities, and is founded upon the same principle as the value attached to canals, roads, ships, and carriages, on account of the facilities they afford for the transport of merchandise; and that to accumulate more money than is required for the purposes of exchange, would be nearly as unwise as to construct more canals, roads, and ships or carriages than can be employed; for, as these are worthless where there is nothing to transport, so money is nearly valueless where there is nothing to be exchanged.

Money, in fact, is to trade what oil is to the steam-engine. Without complete lubrication, the engine can never be worked to its full power; but all the oil in the world would be insufficient to keep it in motion if the real prime mover, the steam, were wanting.

A due regard to these considerations will show, that a nation not producing the precious metals, having once procured a sufficient supply for the purposes of exchange, can have no further occasion to import those metals except for repairing the diminution of its coinage from wear or from accidental losses, or for supplying the gradual increase of its circulating medium, required by the growth of its internal trade. The quantities necessary for these purposes must, however, always be insignificant, as compared with the total amount of its commerce.[1]

---

[1] The annual supply necessary to meet the demand for an increasing currency of gold and silver can hardly ever be large. If the quantity required of any article of consumption, such as sugar, were to be gradually doubled in seventy years, the importation of sugar in each year would go on increasing until it became twice as great as at first; but a constant annual importation, or balance of imports over exports, of less than 1 per cent. upon the amount of precious metals in circulation, would double that amount in seventy years, supposing no losses to accrue from waste or accidents. A celebrated writer, Mr. McCulloch, has estimated (*Ency. Brit.*; 7th edit.; art., " Money") the annual cost to Great Britain of a purely metallic currency of fifty millions sterling at three millions and a half, or 7 per cent —allowing 6 per cent. per annum for loss of profit upon the capital sunk, and 1 per cent. for losses " from wear and tear, from fire, shipwreck, and other accidents." This estimate for a currency purely metallic is, I think, much beyond the truth; but, at any rate, it is entirely inapplicable to a mixed currency of gold and paper, such as that of Great Britain, of the cost of which it may be possible to give some idea. We know, from experience, that the loss from the wear of our gold coinage is very small. The sovereigns struck in 1819 were, in the course of twenty-five years, depreciated in value, by diminution of weight, to the extent of about two-pence each. This was at the rate of less than 4 per cent. in a century, or of 0·04 per cent. annually. The risk of the destruction of money by fire must evidently be much less than that of the destruction of immovable property, which, if not specially hazardous, may be insured against for an annual premium of 1*s*. 6*d*. (=0·075) per cent., a rate that, besides the

Importations for objects of luxury, or for use in the arts, are here left out of consideration, as not affecting the circulation.

There are, indeed, some countries, a large portion of whose foreign commerce consists in the importation of bullion in exchange for native productions; but these are among the countries which, in proportion to their population and their natural resources, are the poorest, and have the smallest external trade—countries where the remuneration of labour is at the lowest, and interest of capital at the highest, rates.

The constant flow of bullion to the East is one of the commercial phenomena that has occupied the attention from the earliest period, when the principles of trade became a subject for discussion, and can only be accounted for by the prevalence of the habit of secret hoarding, so common among oriental nations.

This habit may be in some degree attributable to the combina-

fund necessary to meet losses, covers a considerable expenditure, with, sometimes, large profits to Assurance Companies. Destruction of coin by fire, therefore, can hardly exceed 6d. (0·025) per cent. per annum upon the circulation; and its destruction by other modes, exclusive of shipwreck of foreign importations, is, probably, not greater. Thefts of money, although productive of losses to individuals, do not cause any diminution of the coinage, the object of those who steal money being to put it into circulation as soon as possible. Losses from shipwreck are, undoubtedly, the most important, but a small fraction only of these is chargeable to the home circulation. The enormous foreign commerce of this country gives birth to a very large transit trade in bullion, which is certainly liable to considerable losses from shipwreck; but this trade would still be carried on whatever might be the nature of the national currency, which is only fairly chargeable with the losses upon the importations of the precious metals necessary to keep it up. There is, indeed, the loss from shipwreck of coin carried coastways, but this must evidently be trivial, and may be included in the destruction from various accidents, estimated at about equal to the destruction from fire. It would be a high estimate to take the losses from shipwreck upon foreign importations at 2 per cent. for each voyage.

There appears to me to be an important fallacy in charging a metallic currency at so high a rate of interest as 6 per cent. Investments in the precious metals, so far as the safety of the principal is involved, are, undoubtedly, the most secure, and for that reason ought only to be charged at the most favourable rate which the best investments command. This rate in England has been, on the average of the last 130 years, lower than 4 per cent., and for securities immediately convertible, like money, very much less. Four per cent., then, seems to me to be the highest rate of interest that can be charged against the capital sunk in a metallic currency, the total annual cost of which as at present used in Great Britain would, I think, be estimated, with tolerable correctness, as follows:—

Annual loss from wear upon 50 millions, at 0·04 per cent.	£20,000
Annual destruction by fire (=0·025) and other accidents (=0·025), including shipwrecks of coin carried coastways, 0·050 per cent.	25,000
Annual loss from shipwreck upon foreign importations necessary to keep up the circulation, £47,368, at 2 per cent.	947
Annual expenses of importations, 3 per cent.	1,421
Total cost of importations	47,368
Interest on capital employed, £50,047,368, at 4 per cent. per annum	2,001,895
Making the total annual cost	£2,049,263

Which is, upon a circulation of 50 millions (4·099 =) £4. 2s. per cent.; leaving the cost, exclusive of interest upon the capital employed, at 2s. per cent. per annum only.

tion of timidity and faithlessness characteristic of so large a portion of the Asiatic races, such qualities being eminently calculated to engender in the mind distrust and suspicion of others; but it is, no doubt, principally due to the state of insecurity that rendered it for so many ages in all, and still renders it in some, of those countries impossible to embark capital in such extensive under-takings as can only be profitably carried on under regular and orderly governments.

Whatever the motive for the habit, it must be productive of enormous losses to the nations by whom it is practised.

Out of the erroneous ideas alluded to respecting money, grew the notion of the importance of a favourable balance of trade.

The "balance of trade" is a term for the difference in value between the exports and imports of a nation; and it was long held almost universally that prosperity in commerce consisted in having a favourable balance, or a preponderance in the value of exports, as it was supposed that an excess of imports would have to be paid for by the exportation of bullion.

Upon this theory was founded the whole system of protection, as it once existed in our own country, and as it now exists in most other countries, including bounties upon exportation, and differ-ential and prohibitive duties upon importation, with all the long train of devices invented by men in the vain endeavour to sell to other nations without buying of them in return. Yet, as is most remarkable, the fallacy of the theory is one of the few propositions in political economy admitting of complete demonstration.

If an English merchant send a cargo of commodities to a foreign country for sale, and for the investment of the proceeds in goods to be imported into England, it is clear, if the transaction be profitable, that the value of the goods so imported must cover—1, the value of the goods exported; 2, all the expenses attendant upon the transaction; and, 3, the merchant's profit, including the interest of his capital while so employed. Now, if this be true, as it most undoubtedly is, of the transactions of each individual, it must be equally true of the aggregate of all the transactions in which the nation is engaged.[1] Thus what is called an "adverse balance of trade" may exist to a considerable extent without a single

---

[1] Let E_v represent the value of the exports in the transaction referred to in the text, I_v the value of the imports, x the expenses of the speculation, i the interest of the capital employed, and p the profit obtained beyond the interest. Then if, as has been assumed, the transaction be a profitable one, $I_v = E_v + x + i + p$, or $I_v - E_v = x + i + p$; that is to say, the surplus value of the imports is the fund out of which expenses, interest, and profits are to be paid. Now, $\overline{x + i}$ being a fixed quantity, it is evident that p, the profit,

farthing being required to liquidate it; and, so far is it from being an evil, that it is the essential condition of an extensive commerce, for without it such a commerce could not be carried on at a profit, and therefore would not be carried on at all.

It seems astonishing that conclusions so simple and obvious should have been so long neglected; but still more astonishing are the wild and extravagant assertions we constantly meet with, as to the effect of the balance of trade in causing the transference of bullion from one country to another.

Samuel Fortrey, in his *England's Interest and Improvement*, asserted that, in 1673, the trade between England and France was a clear loss to the former of £1,600,000 annually[1]—a sum that, in about 10 years, would have cleared the country of the whole of the precious metals then in circulation, if Dr. Davenant's estimate of their total amount formerly quoted be correct.[2]

The annual balance of trade in favour of this country, for the four years preceding January, 1796, was estimated by Mr. Irving, then Inspector of Imports, at £10,500,000, exclusive of the profits of the fisheries. Macpherson, who copied the statement with perfect faith, expressed, with amusing simplicity, his surprise that so large an influx should not have caused any apparent increase of the money in the country.[3] Sir Archibald Alison, in his *History of Europe* (2nd series), asserts, that in this country the balance of imports over exports " has of late years risen to thirty or forty millions a year," and that " this immense balance must, of course, be paid in cash or bills convertible." [4] In support of this assertion, the writer quotes figures from *Porter's Progress of the Nation*, which, if correct, would prove that the balance referred to amounted

increases as the excess of I_v over E_v increases. If $I_v = E_v + x + i$ only, there will be no profit; and if I_v be less than $\overline{E_v + x + i}$, there will be a loss, which will be greatly increased if I_v be less than E_v, the state of things considered so desirable by the believers in the old doctrine about the balance of trade. In the foregoing equations, $\overline{x + i}$ is supposed to be due to the English exporter; but if the expenses were paid, and the capital furnished by the foreign merchant, $\overline{x + i}$ would be deducted from the return cargo, the value of which (z being taken as the total cost of it in the foreign market) would then be $I_v - \overline{x + i} \cdot \dfrac{I_v}{z} = E_v + p_1$, and the excess in value of the imports, as well as the amount of profits (p_1), would be thereby diminished. It generally happens, however, from the large capitals possessed in England and the facilities they afford, that the greater portion of the expenses and interest upon our foreign trade is payable in this country. The freights, both outwards and homewards, are, in the majority of cases, paid in England; the assurances are principally effected here, and the capital employed being for the most part English, the bulk of the interest thereon has to be included with the profit of the English merchants. All these elements of commercial superiority tend, as has been shown, to swell the excess in value of imports over exports, or the balance of trade against England, as it is often erroneously called.

[1] *Early Tracts*, p. 234. [2] *Assurance Mag.* viii, 68, Note.
[3] *Hist. Com.* iv. 414. [4] Vol. vi., chap. xxxv. 6.

in the course of the five years ending with 1849, to upwards of 158 millions sterling. The errors in this quotation are such as will appear incredible to those unacquainted with Sir Archibald's habitual inaccuracy, and even in those with whom it has become proverbial will excite surprise. To explain fully all his mistakes would be too tedious; I therefore print below, with totals added by myself, Porter's original table, and that published by the historian,[1] from which it will be seen that the latter has jumbled together official and declared values, in a manner that could never have been done by anyone with the slightest pretension to a knowledge of British statistics, or who had even taken the trouble to read Porter's explanation of the table printed on its opposite page. The most extraordinary thing in the case, however, is, the entire omission of all notice of the value of foreign and colonial goods re-exported, which amounted, during the five years quoted, to upwards of 96 millions, and ought, as a matter of course, to have been deducted from Sir Archibald's balance of 158 millions—an amount that, after this correction, would still be entirely untrustworthy.

[1] *Extract from Porter's Table.*—*Progress of the Nation*, 3rd edition, p. 356.

YEARS.	OFFICIAL VALUE.			Real or declared Value of British and Irish Produce and Manufactures exported.
	Imports of Foreign and Colonial Merchandise.	Exports of Foreign and Colonial Merchandise.	Exports of British and Irish Produce and Manufactures.	
	£	£	£	£
1845	85,281,958	16,280,870	134,599,116	60,111,081
1846	75,953,875	16,296,162	132,288,345	57,786,875
1847	90,921,866	20,036,160	126,130,986	58,842,377
1848	93,547,134	18,368,113	132,617,681	52,849,445
1849	105,874,607	25,561,890	164,539,504	63,596,025
	451,579,440	96,543,195	690,175,632	293,185,803

Table published by Sir A. Alison.—*Hist. Europe*, vol. vi., chap. xxxv. 6.

Years.	Exports.	Imports.	Balance.
	£	£	£
1845	60,111,081	85,281,958	25,170,877
1846	57,786,875	75,953,875	18,266,700
1847	58,849,377	90,921,586	32,072,505
1848	52,849,455	93,547,134	40,657,859
1849	63,596,025	105,874,607	42,278,682
	293,192,813	451,579,160	158,446,623

Porter's Progress of the Nation, 3rd edition, p. 356.

These enormous errors may have resulted from mere negligence, but such negligence is hardly less culpable than dishonesty; and it is impossible to speak of it with too much severity, when we reflect upon the solemn duties assumed by historians, from whose statements of facts and of opinions the great bulk of mankind, less happily situated for inquiry, deduce the principles that govern their conduct under circumstances materially affecting their happiness and security, their property, their freedom, and their lives.

I have dwelt at some length upon this instance, because it is one of a class of cases giving rise to a very common, but very foolish assertion, that " anything may be proved by figures." The public frequently find conclusions revolting to their common sense, supported apparently by an array of numbers, the accuracy of which they are unable or unwilling to test, and they impute to a most useful and valuable science defects arising solely from their own want of industry or of intelligence.

The science of statistics is as essential to the statesman as the compass to the navigator; but the science of statistics does not consist, as seems to be supposed by many British writers, and even by some British senators, in merely searching numerical statements for the purpose of picking out such figures only as may appear to favour the inquirer's opinions.

It may be supposed that a question so frequently and so earnestly discussed as the reduction of interest, could not fail to attract the attention of the Legislature.

It remained, however, unnoticed in Parliament from 1668 to 1690. On the 10th October in the latter year (2nd William and Mary), the House of Commons, without a division, resolved that leave should be given to bring in " A Bill for the reducing of Interest of Money from Six pounds to Four pounds per cent.; Sir Edward Hussey, Sir Matthew Andrews, and Mr. Papillon to prepare the same." [1] The Bill was read a first time on the 13th October,[2] but the second reading was negatived on the 26th November following, by a very narrow majority, the Yeas being 155 and the Noes 158.[3] Another attempt was made in the following year, 1691, when the House granted, by a majority of 131 to 105, leave to bring in a Bill to lessen the interest of money. The Bill was introduced on the 14th November, and read a first time. On the 8th January, 1691-2, it was read a second time, without a division, and the question—" That the Bill be committed," was carried by 169 to 153. The third reading took place on the

[1] *Commons' Journals* x. 433.   [2] *Ibid.* x. 440.   [3] *Ibid.* x. 484.

23rd January, after a division, the numbers being—Yeas 150, Noes 101 ;[1] and the Bill was sent up on the 25th to the House of Lords, where it was read a first and second time, and referred to a committee composed of sixteen earls, seven bishops, and thirteen barons. Among the last was Lord Culpepper, a near relative of the two writers of that name, whose works have been described.

No progress having been made with the measure, the House of Commons, on the 17th February, sent a message to their Lordships, reminding them of the Bill. The receipt of the message is entered in the *Lords' Journals*, but no further notice was taken of the subject, which was terminated by the prorogation of the two Houses on the 24th of the same month.[2]

Another Bill, for the same purpose, was introduced into the Lower House on the 25th February, 1697–8,[3] but it did not reach a second reading; and this was the only attempt to reduce the legal rate of interest made for a period of more than twenty years after the publication of Locke's treatise.

It may have been partly from deference to his authority that so little was done upon the subject, but the principal cause was probably the great demand for money produced by the war with France which followed the Revolution. The high rates of interest charged for loans to the Government about this period are evidence ·rather of the low state of its credit than of the scarcity of money. In 1662, according to Samuel Pepys, the only funds available for the purposes of the navy were obtained at 15 and even 20 per cent. from the goldsmiths,[4] who, in 1665, refused, as he informs us, to lend to the King at 10 per cent.[5] Nevertheless, the East India Company, in the latter year, were able to borrow at 4 per cent. ;[6] and, in 1680, obtained money at the low rate of 3 per cent.,[7] although, by an Act of Parliament passed in 1679 (31 Charles II., c. 1), 8 per cent. was offered for advances made to the Exchequer.[8] Towards the close of the 17th century, a striking proof of the accumulation of capital that had taken place was afforded in the first of those examples of monetary excitement, amounting almost to insanity, which since that period have been phenomena of periodical recurrence in the history of English commercial finance. In the year 1680, several projects for the remunerative employment of capital were started; but the mania did not reach its height until about the year 1694, when the public were overwhelmed by projects of every possible description. To go into details upon

---

[1] *Commons' Journals* x. 550, 552, 616, 629.   [2] *Lords' Journals* xv. 49, 53, 56, 80.
[3] *Commons Journals* xv. 23, 129, 145. [4] *Diary;* Lond., 1858; i. 374. [5] *Ibid.* ii. 250.
[6] Child's *Works*, p. 35. [7] *Hist. Commerce* ii. 597.   [8] *Statutes of the Realm* v. 750.

the subject would carry me far beyond the limits of the present
sketch, were I even disposed to venture upon the description of
a scene already depicted by the master-hand of Lord Macaulay.[1]

The peculiarities of the time were brought on the stage
(A.D. 1693) in Shadwell's posthumous comedy of *The Volunteers ;
or, the Stock-Jobbers*,[2] from which we may learn that the tricks and
manœuvres that have sometimes characterised our money transactions in recent times were thoroughly understood and practised
at a comparatively early period. The illustrious historian whose
descriptions I have referred to, points out, in reference to this
play, that, contrary to the general opinion, the arts of· stock-
jobbing arose in this country before one farthing of the national
debt had been contracted.[3] The title of Shadwell's comedy is an
incongruity, and the portion relating to share speculations does not
seem to have formed part of the original design, but to have been
an after-thought, arising out of the circumstances of the time. He
attempted to throw the odium of stock-jobbing practices entirely
upon the Nonconformists ; but we have had sufficient experience in
our own day that the contagion of credulous avarice which leads
to such outbreaks is not confined to any rank, age, sex, or party.

Great as have been the losses, the suffering, and the discredit
incurred by the nation upon the occasions referred to, they have
not been entirely unmixed with good—some most important and
valuable undertakings having arisen out of them.[4]

During the excitement of 1694 was planned and established
the Bank of England, which has, since that date, been inseparably
connected with the financial and commercial history of the country ;
and, although occasionally imperilled by injudicious management,
has, upon the whole, been conducted with wonderful skill and
remarkable integrity.

---

[1] *Hist. of Eng.* iv. 320. Those who wish for further information may also consult
Macpherson (*Hist. Commerce* ii. 670).   [2] London, 1720.

[3] *Hist. Eng.* iv. 323. The etymology of the word " stock-jobber" is sufficient to show
that it was not derived from the national debt. Stock originally meant a capital employed in trade or in productive industry, and to purchase shares in the stock of the
East India Company, of the Bank of England, and so forth, was therefore a perfectly
correct expression; but it was not usual to call taking the assignment of a mortgage or of
a debt " purchasing stock." The debts of the Government, however, becoming transferable
in the same manner as shares, and being dealt in by the same persons, were included in
the general denomination of "stocks," and that term is now, in the money market, confined
entirely to public debts, while shares in the stocks of public Companies are called merely
" shares."

[4] Many years ago, I met, in a public library, with a bulky volume, consisting of the
prospectuses of various projects bound up together, and labelled " Some of the bubbles of
1825." Among the projects thus described, was one that has since been productive of the
greatest and most rapid advance in the social condition of mankind effected since the first
dawn of civilisation; it was the plan of a Company for constructing a railway between
Liverpool and Manchester.

The question of establishing a national bank had been discussed for some time in England, as was natural from the great success attained by some similar institutions upon the continent, particularly by the Bank of Amsterdam,[1] which, although founded so late as 1609, had already, by the magnitude of its transactions and the extent of its resources, excited the wonder of the age.

Sir William Petty, in his *Quantulumcunque* concerning money (1662), recommended a public bank, asserting we should thereby double our coined money, and that we had in England materials for a bank that might furnish stock enough to drive the trade of the whole commercial world.[2]

Doctor Hugh Chamberlayne, who subsequently projected the abortive Land Bank, attempted, but unsuccessfully, in 1683, to establish a bank for the circulation of bills of credit on merchandise to be pawned therein.[3]

The prejudices of a large portion of the public, however, set strongly in a contrary direction. Lord Bacon alludes to the feeling upon the subject in his time in his *Essay of Usury*, where he says, "Not that I altogether mislike banks, but they will hardly be brooked because of certain suspicions."

The public feeling, indeed, was against bankers generally, who were objects of popular dislike as money-lenders, and of popular envy for the great wealth many of them had rapidly acquired.

Strong evidence of this is to be found in an Act passed in 1671 "for raising a supply for His Majesty." The second section of this Act (22 & 23 Charles II., cap. 3), after reciting "that several persons, being goldsmiths and others, by taking or borrowing great summes of money and lending out the same again for extraordinary lucre and profit, have gained and acquired unto themselves the reputation and names of bankers," enacted, that all bankers, for every sum of £100 they possessed of personal estate, should pay fifteen shillings to the king; the same rate was ordered, by the third section, to be paid upon all money lent to the king at a higher rate of interest than 6 per cent. per annum; while the seventh section laid a tax of six shillings only upon every sum of £100 of personal estate of other kinds. Those who might advance money on security of the taxes imposed by the Act were to be allowed interest at 7 per cent., which was specially exempted from the extra charge.[4] Under this Act, then, bankers and persons who had lent the King money at a higher rate than 6 per cent., such

---

[1] *Hist. Com.* ii. 153.   [2] *Select Tracts*, p. 165.   [3] *Hist. Com.* ii. 612.
[4] *Statutes of the Realm* v. 693.

persons being principally bankers also, were taxed upon their personal estates at a rate 150 per cent. higher than the rest of the
community—a most unjustifiable act of spoliation, which appears
to have escaped hitherto the censure of history.

The bankers were accused of causing the scarcity of ready
money by their dealings in exchanges, a charge very skilfully met
and refuted, in his *England's Treasure by Forraign Trade,*[1] by
Thomas Mun, one of the ablest among the early English writers
upon commerce. Sir William Petty, although remarkably free
from the prejudices of his day, appears to have been disposed to
speak slightingly of the integrity of bankers. In answer to the
question, "What is the trade of a banker?" he says, "Buying
and selling of interest and exchange; who is *honest only* upon the
penalty of losing a beneficial trade founded on the good opinion of
the world, which is called credit."[2]

Very little opposition was offered in the Commons to the Bill
for establishing the Bank of England, but it was fiercely contested
in the House of Lords, where, however, it was carried by a
majority of 43 to 31,[3] eight peers signing a protest against the
portion relating to the incorporation of the Bank.[4] It seems very
probable that this portion would have been rejected had it not
formed part of a Bill of Supply, which the peers were unable to
amend, and afraid, from the state of the public finances, to throw
out. The Bill (6 Wm. and Mary, cap. 20) received the royal
assent on the 25th April, 1694,[5] and was entitled "An Act for
granting to their Majesties severall rates and duties upon tunnage
of shipps and vessells, and upon beere and ale and other liquors,
for securing certain recompenses and advantages in the Act mentioned to such persons as shall voluntarily advance the sum of
fifteen hundred thousand pounds towards the carrying on the war
against France."[6] In consequence of the title of the Act, the new
establishment was called, in derision, the "Tunnage Bank."[7] Its
capital was fixed at one million two hundred thousand pounds,
the whole of which was to be lent to the Government at what Lord
Macaulay says was "then considered the moderate interest of 8 per
cent.,"[8] a further allowance being made for management of £4,000
per annum, which increased the interest to 8⅓ per cent. in all.

This accommodation was the real inducement for passing the
measure, both with the executive and the two Houses of Parlia-

[1] *Select Tracts*, pp. 164 to 167.   [2] *Ibid.* p. 167.   [3] Macaulay iv. 501.
[4] *Lords' Journals* xv. 423.   [5] *Ibid.* xv. 426.   [6] *Statutes of the Realm* vi. 483.
[7] Macaulay iv. 499.   [8] *Ibid.* iv. 498.

ment. That the scheme of the Bank was popular among the monied classes is evident from the fact that the whole capital was subscribed in ten days,[1] although the Government found it difficult to raise money at 8 per cent. by any other means.

This loan from the Bank of England was the first instance of the grant of perpetual redeemable annuities, in which form nearly the whole of our national debt has been contracted. It was also the foundation of our permanent funded debt; all previous loans, with the exception of some raised upon terminable annuities long since expired, having been obtained upon obligations of the nature of our unfunded debt, the principal as well as the interest being payable at fixed periods. A small portion of these obligations were Navy Bills, but they consisted principally of tallies, a species of voucher that would probably have long been forgotten but for the singular custom by which the use of them was retained in the Exchequer, down to a very recent period. They are thus described in Madox's *History of the Exchequer* :—

" These tallies were pieces of wood cut in a peculiar manner of correspondency: for example, a stick or rod of hazel (or, perhaps, of some other wood), well dried and seasoned, was cut square and uniform at each end and in the shaft. The sum of money which it bore was cut in notches in the wood by the cutter of tallies, and likewise written upon two sides of it by the writer of tallies. The tally was cleft in the middle by the Deputy Chamberlains, with a knife and mallet, through the shaft and the notches, whereby it made two halves, each half having a superscription and a half part of the notch or notches—a notch of such a largeness signified M*l*, a notch of another largeness C*l*, &c. It being thus divided or cleft, one part of it was called a ' tally,' and the other a ' counter-tally,' . . . and when these two parts came to be joined, if they were genuine, they fitted so exactly that they appeared evidently to be parts one of the other."[2]

This peculiar mode of recording money transactions was very general when the art of writing was confined to few persons; and, before the invention of that art, similar means may probably have been employed to preserve the memory of historical events.

The quipos, or knotted cords of various colours, in use by the Peruvians at the period of the discovery of America, have been supposed to contain regular annals of their empire, although that opinion was discountenanced by Dr. Robertson.[3] Lord Erskine, in one of his most celebrated speeches, the defence of Stockdale, mentions having seen in his youth a naked savage addressing the

---

[1] Macaulay iv. 502.
[2] *Parliamentary Paper*, No. 443 (1858), p. 85. I had recently in my possession one of these tallies, dated in 1811, which resembled more the rude contrivance of the denizens of a wilderness, than a voucher issued from the exchequer of one of the most powerful monarchs reigning over one of the most polished nations of the globe.
[3] *Hist. Amer.*, book vii.

governor of a British colony, "holding a bundle of sticks in his hand as the notes of his unlettered eloquence."[1]

The word "tally," from the French *tailler*, to cut, is evidently employed in the same manner as the word "score," derived from the Saxons, who appear to have been in the habit of making a score, or notch, at every twentieth enumeration, and hence the prevalence in the agricultural districts of the practice of counting by scores or twenties. From the word "tally" is derived the title of teller or tallier of the Exchequer, an officer whose dignity is alluded to by Pope in the couplet—

> "From him whose quills stand quiver'd at his ear,
> To him who notches sticks at Westminster."[2]

We find frequent reference to tallies in our literature. Jack Cade is made to bring forward the following, among other charges, against Lord Say:—"And whereas, before, our forefathers had no books but the score and tally, thou hast caused printing to be used."[3]

The discontinuance of the use of tallies at the Exchequer did not take place until 1826, and was rendered memorable by a very unfortunate event. Orders having been given, in 1834, for burning all the old tallies, the persons employed in doing so used the stoves of the House of Lords for the purpose, and so over-heated the flues as to cause the fire that burned down both Houses of Parliament on the 16th October in that year."[4]

Originally, tallies were merely instruments of acquittance for money due to the Crown and paid into the Exchequer; but they were also, from a very early period, issued as assignments of anticipated revenue. In such cases they were called "tallies of anticipation," or "tallies of *pro*." The latter expression was probably derived from the dealings of the Lombards with the Exchequer, *pro* being an Italian word signifying gain or benefit."[5] It is true that Mr. Chisholme, in his learned and able *Notices, &c., of the Public Debt*,[6] asserts that there is no instance of tallies of *pro* bearing interest before the reign of Charles II.; but a similar apparent contradiction regarding the loans of the Lombards to the

---

[1] *Erskine's Speeches* ii. 263.   [2] *Epistle to Lord Bolingbroke.*

[3] *Henry VI.*, act. iv., scene 7.

[4] The curiosity about tallies excited by this circumstance led to the publication in the *Times* of the 1st November, 1834, of a very full and accurate account of them, written by that sturdy politician and studious antiquary, William Hone.

[5] The Venetian Government having, in 1171, raised a forced loan, to which all the citizens were compelled to contribute, it was ordained that the subscribers, who were subsequently incorporated as a company and became the celebrated Bank of Venice, should receive interest *a raggione del quattro per cento di* "*pro.*" This is said to be the earliest known instance of the use, now so general, of the expression "per cent.," to denote the ratio between two numbers.

[6] "Notices of the various Forms of the Public Debt; its Origin and Progress:" *Parliamentary Papers*, No. 443 (1858), Appendix.

Crown has been already noticed. There can be no doubt, however, that, in both cases, interest was paid in some form; it may have been included in the principal sum acknowledged, like the discount on a bill of exchange, which instrument the tally in its nature very much resembled.

The first Parliamentary sanction to the issue of negociable public securities bearing interest, was the 17 Charles II., cap. 1,[1] which provided that every person advancing money to His Majesty, or furnishing any " wares, victual, necessaries, or goods," on security of the taxes granted by the Act, should have a tally of loan struck, and be furnished with a warrant for the repayment of the amount, with interest thereon, at the rate of 6 per cent. per annum, payable half-yearly, until the discharge of the principal; and that any person entitled to money by virtue of the Act might transfer his claim by endorsement of the warrant duly registered.[2] The warrants thus became the real securities, rather than the actual tallies; but these, nevertheless, continued to be struck, from which it will be readily inferred that the tally-cutters were paid by fees."[3]

The rate of interest offered by the 17 Charles II. was increased to 7 per cent. in the 29 & 30 Charles II., cap. 1 (1677 and 1678); and both the 30 Charles II., cap. 1 (1678), and 31 Charles II., cap. 1 (1679), authorized the payment of 8 per cent. per annum for similar advances.[4] It must be understood that these Acts were only passed to facilitate the king's obtaining money in anticipation of the taxes voted, and did not in any way pledge the faith of the Legislature for the repayment of the loans to be advanced. The House of Commons was so far from considering these as made to the Government of which it formed a part, that, in the height of the contest on the subject of the Exclusion Bill, when a dissolution was impending, it passed, *nemine contradicente*, on the 7th January, 1680, resolutions declaring that whosoever should thereafter make any advances on any branches of the king's revenue, or purchase any tally of anticipation, or whosoever should pay any such tally thereafter to be struck, should be adjudged to hinder the sitting of Parliaments, and should be responsible to Parliament therefor.[5]

---

[1] Chisholme's *Notices*, &c., p. 90.   [2] *Statutes of the Realm* v. 165.
[3] The fees upon payments by tallies appear to have been higher than 10s. per cent. Pepys, in his *Diary* (12th May, 1665), mentions tallies for £17,500 struck for the use of the navy, and adds, " But to see how every little fellow looks after his fees, to get what he can for everything, is a strange consideration—the king's fees, that he must pay himself for this £17,500, coming to more than £100." This virtuous indignation is in amusing contrast with the entry on the following day:—" Received my watch from the watchmaker—and a very fine one it is—given me by Briggs, the scrivener." Briggs having made the present to secure the Secretary's official influence (ii. 229).
[4] *Statutes of the Realm* v. 858, 870, 928.   [5] *Commons' Journals* ix. 702.

Notwithstanding the high rates of interest paid upon tallies, we find constant mention of their being at discounts that must have afforded considerable profits to those who purchased them upon such terms. A 6 per cent. perpetual stock, bought at 10 per cent. below par, would only secure the holder interest at £6. 13s. 4d. per cent. per annum; but a tally payable in two years, and bearing interest at 6 per cent., would produce, if bought at 10 per cent. discount, more than 11 per cent. per annum. Pepys mentions that he held, in 1666, tallies yielding him 10 per cent.;[1] and that, in 1667, he was offered some as a security at 12 per cent.[2] The heavy interest paid upon tallies was not the only burden they inflicted upon the public; they were largely issued for wares, victuals, and so forth, supplied for the use of the Government; and it was justly remarked by Michael Godfrey, the first Governor of the Bank of England, in a pamphlet published in 1693, and reprinted in Francis's *History of the Bank*,[3] "that those who were to allow 15 or 20 per cent. upon their tallies, as was the case before the establishment of the Bank, would make provision accordingly in the prices they were to have for their commodities."[4]

Manley, in the preface to his *Interest at Six per Cent. Examined*, made a similar observation.

From the regulations inserted in the various Acts referred to, for securing the regular payment of tallies in their order of priority, it seems probable that there was much jobbing and corruption in that respect. The sum due upon tallies at the Revolution was only £84,888. 6s. 9d., which constituted the whole of the public debt at the time, with the exception of about £300,000 arrears due to the army and navy.[5]

In 1693, the year before the establishment of the Bank of England, the national debt had increased to the following sums[6]:—

	Unredeemed Capital.	Annual Charge.	Annual Rate per Cent.
Loans upon tallies	£4,472,400	£302,000	£6 15 0
Navy bills	1,430,439	85,826	6 10 0
	5,902,839	387,826	6 11 5
Terminable annuities		119,275	
Total annual charge		£507,101	

(END OF PART IV.)

[1] *Diary* ii. 412.   [2] *Ibid.* iii. 240.   [3] 3rd edit.; Longman, London; no date.
[4] Page 243.   [5] *Parliamentary Paper*, No. 443 (1858), p. 93.   [6] *Ibid.* p. 2.

*On an unfair suppression of due acknowledgment to the writings of
Mr. Benjamin Gompertz.* By Professor De Morgan.

In 1839, in the *Penny Cyclopædia* (article "Mortality"), I wrote
as follows, referring to Mr. Gompertz's well-known hypothesis on
the law of mortality :—

"We enter into some detail of it the more readily, that it is necessary
as an act of justice to Mr. Gompertz, whose ideas have been adopted by a
recent writer on the subject, without anything approaching to a sufficient
acknowledgment."

Mr. Gompertz gave his theory in the *Philosophical Transactions*
for 1825, p. 513, in a paper *On the Nature of the Function expressive of the Law of Human Mortality*.

The writer of whom I assert that he gave a suppressive account
of what Mr. Gompertz had done is Mr. T. R. Edmonds, in a work
entitled *Life Tables founded upon the Discovery of a Numerical
Law, &c.;* London, 1832; 8vo.

My attention has been again drawn to this matter by two
things—first, the account given of Mr. Edmonds's alleged discovery by his proposer, in the list of candidates for admission
into the Royal Society; secondly, the following mention of Mr.
Edmonds by Dr. Farr, in a paper *On the Construction of Life
Tables (Phil. Trans.,* 1859, part ii., p. 844) :—

"I shall now notice briefly the application of this hypothesis, first suggested by Mr. Gompertz, and applied by him to the interpolation of the
Northampton and other tables. Mr. Edmonds, in 1832, extended the
'Theory,' and applied it to the construction of three Life-Tables. He gave
an elegant formula, similar in principle to that of Mr. Gompertz, from
which the curve of a Life-Table can be deduced, upon the above hypothesis."

The first of these circumstances occasioned my pointing out to
a fellow of the Royal Society what I had written in 1839, with
explanation. Some correspondence followed between him and a
friend of Mr. Edmonds, I having first stated my view of the case
in writing. On May 31st, I was shown a letter from Mr. Edmonds's
friend, recommending* a reference; and also a letter from Mr.
Edmonds himself, in reply to my statement just mentioned. As
the reasoning in the first paragraph of this letter showed me clearly
that publicity must be the result, I declined to read any further,
and expressed my intention of taking the course which I now
do take.

I shall not trouble the reader with any account of what I think

* This recommendation I adopt : my referees are all who are competent to judge and
who choose to read.

about the character or motive of the suppression, nor of my own motive in bringing it forward. I will put the case before him, desiring him to consider it no case at all unless, true quotation being assumed, he find it one to which he cannot imagine a sufficient answer. I will add that I have had no communication whatever with Mr. Gompertz on the subject, direct or indirect, past or present.

The account which Mr. Edmonds gave of Mr. Gompertz's theory is as follows (pp. xvii., xviii.) :—

"The honour of first discovering that some connexion existed between Tables of Mortality and the algebraic expression (a^{b^x}) belongs to Mr. Gompertz: but to arrive at this single common point, his course of investigation differs so widely from mine, that appearances will be found corresponding to the reality,—that my discovery is independent of the imperfect one of Mr. Gompertz."

It is here asserted :—

1. That Mr. Gompertz discovered that *some connexion* existed, and that this—namely, that *some connexion* existed—is the *single common point*. It shall be shown that *all* the points of Mr. Edmonds's alleged discovery had been published by Mr. Gompertz ; the only difference being that Mr. Edmonds takes in the period of infancy, to which Mr. Gompertz pays no attention.

2. That the two courses of investigation differ widely. If the *published* courses of investigation be intended—and no others can present "appearances" to the reader—it shall be shown that they closely agree.

3. That the discovery of Mr. Gompertz is imperfect—meaning, of course, as compared with that claimed by Mr. Edmonds. It shall be shown that there is no difference between the two, except in the introduction of the period of infancy by Mr. Edmonds, according to the method of Mr. Gompertz.

I now give the quotations which substantiate these points :—

Mr. Gompertz (p. 8).	*Mr. Edmonds* (pp. xvi., xvii.).
"Art. 5.—If the average exhaustions of a man's power to avoid death were such that at the end of equal infinitely small intervals of time, he lost equal portions of his remaining power to oppose destruction which he had at the commencement of those intervals, then at the age x his power to avoid death, or [used in the sense of	"The force of mortality is a simple function of the age, or time from birth, and is always of the form (ap^x) during each of the three periods of Infancy, Manhood, and Old Age. . . . Let, now, (y) represent the number Living or Surviving at any time (x). The force of mortality at that time $= ap^x =$ decrement in unit of time on unit of life; the

and also] the intensity of his mortality might be denoted by aq^x, a and q being constant quantities; and if L_x be the number of living at the age x, we shall have $aL_x \times q^x . \dot{x}$ for the fluxion of the number of

deaths $= -(L_x)^{\boldsymbol{\cdot}}$; $\therefore abq^x = -\dfrac{L_x}{L_x}$,

$\therefore abq^x = $ —hyp. log.* of b [misprint for q] $\times$ hyp. log. of L_x, and putting the common logarithm of $\dfrac{1}{b}$ [here b should be q again] $\times$ square of the hyperbolic logarithm of $10 = c$, we have $c . q^x = $ common logarithm of $\dfrac{L_x}{d}$; d being a constant quantity, and therefore L_x, or the number of persons living at the age of $x = d\bar{g}^{|q^x}$; g being put for the number whose common logarithm is c."

finite decrement of (y) at that time $= y \times ap^x$; and the true decrement, or the decrement in an infinitely small given time, $= yap^x dx$; that is, $-dy = yap^x dx$. Using (l) to signify hyperbolic logarithm, and (e) to denote the base of that system, we obtain by integration, $l\dfrac{g}{y} = \dfrac{a}{lp}p^x$ and $\dfrac{g}{y} = e^{\frac{a}{lp}p^x}$.

If it be assumed that $y = 1$ when $x = 0$, then $g = e^{\frac{a}{lp}}$, and the equation becomes $y = e^{\frac{a}{lp}} \times e^{-\frac{a}{lp}p^x}$, or $y = e^{\frac{a}{lp}(1 - p^x)}$.

And calling the modulus of the common system (k), and using (λ) to signify common logarithm, the equation will finally become—

$$y = 10^{\frac{k^2 a}{\lambda p}(1 - p^x)} ."$$

Thus it appears that having a single point in common—namely, that a^{b^x} has some connexion with tables of mortality—arrived at by widely different courses of investigation, means as follows:—

Mr. Gompertz assumes that the intensity of mortality is aq^x.

Mr. Gompertz arrives at the fluxional equation

$$-\dot{L}_x = aL_x \times q^x \dot{x}.$$

Mr. Gompertz integrates this into

$$L_x = d\bar{g}^{|q^x}.$$

Mr. Edmonds assumes that the force of mortality is ap^x.

Mr. Edmonds arrives at the differential equation

$$-dy = yap^x dx.$$

Mr. Edmonds integrates this into

$$y = 10^{\frac{k^2 a}{\lambda p}(1 - p^x)}.$$

The difference is the difference between intensity and force; between a and a; between q and p; between L_x and y; between fluxions and differentials; between d and $10^{\frac{k^2 a}{\lambda p}}$; between g and $10^{-\frac{k^2 a}{\lambda p}}$; and between a formula which gives dg when $x = 0$, and another which gives 1.

In applying his formula, Mr. Gompertz finds it necessary to change the constants at a period depending upon the table to be verified. Thus, taking life from 10 years of age, and not consider-

* The preceding b is a superfluous constant, which is useless. It seems to have originated in the idea that the diminution of the number living is *proportional* to the intensity of mortality, and therefore represented by that intensity multiplied by a constant. But the constant a, already introduced, is sufficient.

ing infancy, he finds one set of constants to represent the Carlisle Table from 10 to 60, another from 60 to 100. Mr. Edmonds finds the same necessity, but he also takes in the period of infancy, and applies the law to the periods from birth to 8 years of age, from 12 to 55, and from 55 to the end of life, acknowledging that the periods vary with circumstances.

Mr. Gompertz considers, as appears by one of his examples, that the constant q is very slowly varying during the period at which it is near enough to uniformity for practical use. Mr. Edmonds seems to assume that during each whole period of its use, his constant q is absolutely fixed; but of this he does not give any evidence. For the logarithms of q, by the Carlisle Table, from 10 to 60, and from 60 to 100, Mr. Gompertz has 0·0126 and 0·0271. In Mr. Edmonds's book, the logarithms of p, from 12 to 55, and from 55 to the end of life, are 0·0128 and 0·0333. Mr. Edmonds calls his constants "now first discovered" (p. vi.).

I can find nothing in which Mr. Edmonds went beyond Mr. Gompertz. It is now for him to show that he not only went beyond Mr. Gompertz, but so far beyond that all which was due to Mr. Gompertz was comprised in the statement that the whole of what I have quoted gives the *single common point* that a^{b^x} has *some connexion* with tables of mortality. Should he wander from this point—which is *the* point—it will be my part to request insertion of a short exposure of irrelevancy. I will conclude by observing that I do not care to inquire whether or no what Mr. Edmonds published, as above quoted, was done independently of Mr. Gompertz, as asserted. On this point it will be well to suspend opinion until it is seen what Mr. Edmonds can say in justification of the suppressive mention which, I submit, has been fully established.

On some Considerations suggested by the Annual Reports of the Registrar-General, being an Inquiry into the Question as to how far the Inordinate Mortality in this Country, exhibited by those Reports, is controllable by Human Agency. (Part II.) By H. W. PORTER, ESQ., B.A., *Assistant Actuary to the Alliance Assurance Company, Fellow of the Institute of Actuaries and of the Statistical Society.*

[Read before the Institute the 27th February, 1860.]

WITH respect to the inhalation of woolly matter, Mr. Leigh, a surgeon at Manchester, and Registrar of Deaths for the Deans-

gate sub-district of that city, in which cotton-mills are so numerous, states, in one of his Reports to the Registrar-General,
that in certain trades in Manchester almost every member above
40 is affected with emphysema of the lungs; and that persons
who work in dust—as fustian-cutters, cotton-carders, and the like
—are almost all affected with it, after working at their occupations
for a few years.

This disease, which is also called "pneumatosis," is not often
to be found recorded as a cause of death in the Registrar's returns,
probably because the disease usually turns to bronchitis, and the
deaths, which are caused primarily by emphysema, come to be
registered under the head of "bronchitis."

Considerable injury also is said to be caused by the inhalation
of fuliginous particles, with which the atmosphere of large towns is
so generally impregnated.

It is common to find, after death, Mr. Leigh reports, the
bronchial glands and pulmonary tissue perfectly black, from the
presence of carbonaceous matter.

The Acts lately passed, providing for the compulsory consumption of the smoke produced in manufactories—the application of
which will in course of time, no doubt, be extended all over the
country, but which do not, I believe, at present apply to Manchester, or other large manufacturing towns—will gradually tend
to remove this source of disease.

Some advantage is considered to have been derived by workmen engaged in occupations from which they are exposed to the
inhalation of dust and other obnoxious matter, by the practice,
which has lately become very general, of allowing the beard and
moustache to grow, at the suggestion of a physician in Edinburgh,
the late Professor Alison.

The movement was commenced, I believe, by the stone-cutters
in Scotland. In many occupations, where the inhalation of metallic
particles is to be apprehended, wire masks have been introduced,
with the object of arresting the progress of particles which might
enter the mouth, nostrils, or eyes; and magnetic guards, when
the metal that is being worked is steel or iron: the intended object
is, of course, to some extent effected, but the mortality among
workmen engaged in these processes of manufacture shows, too
clearly, that the effect is but partial, and the use of such contrivances is far from general.

I am aware that it is popularly considered that the entry, by
inhalation, of particles of foreign matter into the bronchial tubes

and lungs, or, at any rate, their remaining there, is rendered impossible by reason of the ciliary action.

I am induced to refer to this matter, because I find the error perpetuated in a recently-published Magazine, edited by Mr. Thackeray, whose name carries so much weight with the public that it may be as well to enter a little into the question.

The writer of the article, which is entitled " Studies in Animal Life," after describing the nature of the cilia, states as follows, that " while the direction in which the cilia propel fluids and particles is generally towards the interior of the organism, it is sometimes *reversed;* and, instead of beating the particles inwards, the cilia energetically beat them back, if they attempt to enter. Fatal results would ensue if this were not so. Our air-passages would no longer protect the lungs from particles of sand, coal-dust, and filings, flying about the atmosphere; on the contrary, the lashing hairs which cover the surface of these passages would catch up every particle, and drive it onward into the lungs. Fortunately for us, the direction of the cilia is reversed, and they act as vigilant janitors, driving back all vagrant particles." " In vain," it is stated, owing to the direction of the cilia being reversed, as it is in man, " does the whirlwind dash a column of dust in our faces— in vain does the air, darkened with coal-dust, impetuously rush up the nostrils: the air is allowed to pass on, but the dust is inexorably driven back. Were it not so, how could miners, millers, ironworkers, and all the modern Tubal Cains contrive to live in their loaded atmospheres ? In a week, their lungs would be choked up." " Perhaps," continues the writer, " you will tell me that this *is* the case : that manufacturers of iron and steel are very subject to consumption ; and that there is a peculiar discoloration of the lungs which has often been observed in coal-miners, examined after death."

I shall certainly tell him that the highest medical authorities have no doubt whatever as to the possibility of the entry by inhalation of particles of foreign matter into the respiratory apparatus. It is a matter of fact that carbonaceous matter is found in the lungs after death. Mr. Queckett, the Conservator of the Museum of the Royal College of Surgeons, and Professor of Histology there —and a more competent authority I can hardly, I conceive, bring forward—has repeatedly found carbonaceous matter as the nucleus of tubercle, examples of which he has been kind enough to show me under the microscope, in which inhaled matter of this description formed clearly the nucleus of the tuberculous matter; and there is said to be little reason to doubt that such matter not only forms the nucleus of tubercle, but that it is often the exciting cause of tuberculous deposit.

The writer of the article to which I have just alluded does not, I think, very clearly describe the action of the cilia, which is not of a nature to *prevent* the entry of foreign particles into the lungs, but, by means of the reversed action to which allusion has been made, to *expel* them from the lungs after they have reached those organs ; and this, to a certain extent, no doubt they do.

Dr. Carpenter, one of the first authorities on the question of animal physiology, thus describes the cilia :—" The larynx, trachea, bronchial tubes, and air-cells of the lungs, in all air-breathing vertebrata," he says, " are lined by a mucous membrane, which is continued from the back of the mouth ; and this membrane, like the gills of aquatic animals, is covered with cilia, which are in continual vibration." He explains that the purpose of this ciliary movement must be here different from that which is fulfilled by the same action on the surface of the gills of polypes, and which is stated to be to cause a constant change in the water that is in contact with their surface, and so to secure a constant supply of the air contained in the water.

Dr. Carpenter describes the cilia to be " minute vibrating hair-like filaments ;" and the purpose of the ciliary movement "probably serves," he explains, " to get rid of the secretion which is being continually poured out from the surface of the mucous membrane, and which, if allowed to accumulate there, would clog up the air-cells, and in time produce suffocation."

The constant vibration of the cilia always in one direction, towards the outlet, no doubt serves to get rid of foreign particles which have effected an entrance into the lungs—not, as stated by the Magazine writer, to *prevent* their entrance. It serves to get rid of them to a certain extent, but the power of this action is probably quite insufficient to cause the ejection of the whole of the very great quantity of different dusts, and of the stony and metallic particles, which the constant exposure of persons subjected to such noxious influences might cause to enter the respiratory apparatus ; and we may draw this conclusion because we know, from ocular observation, that the power of the cilia is insufficient to expel from the lungs the carbonaceous matter, which, being of a lower specific gravity, is far more likely to be driven out than the stony and metallic particles, of the presence of which in the lungs it is difficult to bring forward actual proof. Seeing, therefore, that the cilia have not the power to eject the carbonaceous matter, we may not unfairly conclude that if the particles of the higher specific gravity alluded to are not to be found after death, their absence may be ascribed to some other cause than to that of the ciliary movement. The assumption, therefore, of the author of the Magazine article, that "although consumption may be frequent among the Sheffield workmen, the cause is not to be sought in their breathing filings, but in the sedentary and unwholesome confinement incidental to their occupation," I look upon as wholly untenable.

With reference to the peculiar discoloration of the lungs of coal-miners, which has been observed, the writer says, in some, but not in all or in many, of those engaged, and which has been observed as well in persons who were not miners, and who were not exposed to the influence of any unusual amount of coal-dust, I may refer to the reports of Mr. Leigh, the Registrar of the Deansgate sub-district of the city of Manchester, from which I have already quoted, and I would ask, whether those persons referred to as not being exposed to the influence of coal-dust to any unusual extent, were not, equally with all the inhabitants of large towns, exposed to the influence of the fuliginous particles with which the atmosphere is constantly impregnated, and evidence of the inhalation of which has already been adduced? On the other hand, it is true that evidence of the deposit of pigment in the cells of the lungs, as well as of an entire change of colour of other organs of the body, is to be found in our collections of morbid anatomy.

The ciliary action may be seen to advantage in the living oyster and mussel—in the latter particularly—in a portion taken from the gills, and viewed with a magnifying power of 200 diameters or upwards. An obvious agitation of the water, caused by the unceasing movement of the cilia, is clearly to be seen. Mr. Queckett has discovered that the cilia have a feathering action, like an oar, the object of which seems to be to prevent their motion producing two currents in opposite directions, the result of which would be to counteract altogether the effect intended to be produced. This he has observed in the mussel, and it is fair to assume that it is the same in the human subject, in which it is not easy to find the opportunity of witnessing the experiment, as the ciliary action is not visible for more than about 12 hours after death.

The ciliary motion of the human mucous membrane can, however, be seen for a much longer period on the surface of a recently extracted nasal polypus.

The ciliary action, which must be seen to be appreciated, forms altogether one of those wonderful provisions of nature of which the microscope is now every day opening up a fresh store, to delight, instruct, and astonish us.

The great and concluding argument, however, against the possibility of the inhalation of foreign matter, the writer of the article under consideration tells us, is found in an experiment by Claude Bernard on a rabbit, over the mouth`of which a bladder containing powdered charcoal was tied, so that whenever the animal breathed the powder would be likely to be inhaled. The bladder was. kept on constantly, except at feeding time, for many days, and yet, it is stated, when the rabbit was killed no charcoal was found in the lungs or bronchial tubes—the cilia being considered to have acted as a strainer, and to have kept the particles of the dust from the air tubes.

Now, from what has been said, it will appear that no such thing took place. If the cilia did act in the matter—which, possibly, they did to some extent—it was by expelling the noxious particles after they had entered the air tubes; but, I look upon it, other reasons could be adduced to account for the absence of the particles of charcoal from the lungs; for instance—

1st. Because the process was probably not continued sufficiently long to counteract the effect which the cilia, it is admitted, have, to some extent, in expelling from the respiratory organs particles of low specific gravity.

2nd. Because, owing to the low specific gravity of charcoal (which is, when in powder, 1·5 only, while that of stone, marble, granite, and slate, I may mention for comparison, varies from 1·7 to 2·85, and that of iron and steel is respectively 7·788 and 7·81), it would be very easily affected by the ciliary movement.

And 3rd. Because there is a peculiarity in the construction of the lungs of rabbits which might have influenced the experiment— the minuteness of the subdivisions of the lungs of those animals into small cavities or air cells being just of a nature to favour the experiment of Claude Bernard.

I may mention another fact which favours the opposite conclusion as to the inhalation of foreign bodies into the respiratory organs—namely, that experiments have shown that animals can be made phthisical by the introduction of particles of metallic mercury into the lungs.

The elaborate investigations, however, of Dr. Holland, physician to the Sheffield General Infirmary, entirely set at rest the question of the injurious effect on the lungs of grinding in all its branches.

I shall, hereafter, more particularly allude to the inquiries of Dr. Holland.

M. Morni, in a communication to the Academy of Sciences in Paris, stated that the sword-makers on the continent die before attaining the age of 40 or 45 years. Deslandes, an eminent French writer, mentions, in his *Dictionnaire de Médecine et de Chirurgie Pratiques,* that polishers of steel scarcely reach their 36th or 40th year, and that those are considered to be old who attain the age of 45.

It is necessary to be most careful how we receive as positive facts statements put forth in popular articles on scientific subjects, lest we may be led into error. Professor Babington, with reference to the examinations in natural history at Cambridge, speaks of the inaccuracies contained in popular books on the natural sciences, on which point he thinks it necessary to caution the students of the University.

There is no doubt that many popular errors are perpetuated by means of the elementary handbooks on scientific subjects so much in vogue at the present day ; and among the retarding influences with which sanitary reformers have to contend, are the prejudices founded upon errors to which those who hold them often most obstinately cling.

The evidence of the inhalation of stony and metallic substances into the respiratory organs that I have been enabled to obtain is but of a negative character.

With the exception of the carbonaceous matter that I have referred to, I have not been able to *see* any foreign particles that had been received into the lungs by inhalation.

I have taken some trouble to ascertain if any preparation existed showing the fact, and have ascertained that there is no such preparation in the Museum of the College of Surgeons in London. I found several specimens of calcareous and earthy matter—carbonate and phosphate of lime principally—which had been taken from the lungs of oxen and other animals, but this was the result of the ossification of tubercle.

I have not been able to ascertain, upon inquiry of several medical men in London, of the highest standing and experience, that they have ever actually found any stony or metallic particles in the lungs, as the result of inhalation, though no doubt seems to exist as to the fact that such bodies *do* effect an entrance, of which, indeed, the presence of carbonaceous matter, which is repeatedly found, is sufficient evidence.

One case of a mason has been mentioned to me, in which such a discovery is stated to have been made after death; but my informant had not the facts of the case at his command, and it is very possible that the deposit in this case was the result of disease, and not of inhalation.

The absence of such appearances at the autopsy are accounted for pathologically from the fact that the irritation of their presence induces inflammation, and the particles, in consequence, become detached and are expectorated. This, it is easy to understand, may be the case to some extent; but it is scarcely conceivable that the very large quantity of foreign particles that are likely to be inhaled by persons engaged in certain occupations should be so effectually ejected in this manner as to leave no traces after death of their former existence.

It seems reasonable to suppose that, unless a *post mortem* examination were made with the special object of detecting evidences of the inhalation of foreign particles, they might, from their minuteness, escape observation; and yet, minute though they be, they might have quite sufficient power to cause most serious mischief, by the mere mechanical irritation produced by their contact with the delicate surface of the mucous membrane, in the same way that in poisoning by strychnine, though the actual presence of the poison in the nervous tissue might escape detection from the extreme minuteness of the crystals, still death by that poison may, nevertheless, be the result of the mechanical irritation of the nerves of the body by the sharp, angular, and indissoluble crystals of the strychnine. I am now assuming the truth of a theory which has been propounded to account for the way in which death from this poison is produced—viz., by the actual mechanical action of the sharp-pointed angles which result from the crystallization of strychnine, and which work their way into the substance of the nervous fibre, it being a characteristic of this poison that the crystals are of a peculiarly indissoluble nature. In fact, it is only in microscopical preparations of portions of the lungs that such particles, as from their minuteness have the power of entrance, are likely to be observed. A particle of stone or metal sufficiently minute to be inhaled, would generally occupy far less space than a particle of carbonaceous matter; and yet such a portion of the latter as is likely to be found forming the nucleus of tubercles, when magnified 250 times, appears only about the size of a moderate-sized pin's head.

Thackrah mentions that Diemerbrock relates that he had found, in dissecting the bodies of masons, heaps of sand in the lungs; and that, in dividing the pulmonary substance, he seemed to be cutting a sandy body.   Thackrah, however, suggests whether the sand thus found was inhaled, or whether it was not rather a calcareous deposit formed by disease.

From what I have said as to the specimens of calcareous and earthy deposits preserved in the museum of the College of Surgeons, which were found in the lungs of animals, and which arose, in fact, from the ossification of tuberculous matter, it is not at all unlikely that the idea of Thackrah was correct.

The expectoration of calcareous matter, moreover, is not uncommon in some of the stages of phthisis, showing clearly the existence of such matter in the lungs, as the product of disease.

It is further stated, in the Magazine to which I have already referred, that "although consumption may be frequent among the Sheffield workmen, the cause is not to be sought in their breathing filings, but in the sedentary and unwholesome confinement incidental to their occupation."

Surely the writer of this article can hardly be aware of the extensive medical investigations that have been instituted into this subject!   I have before referred to Dr. Holland, whose official connection with the Infirmary at Sheffield—the head-quarters of the grinding trade—afforded him the fullest opportunities of satisfactorily determining the effects of this occupation on the health of the operatives, while the fact of this physician having studied under the celebrated Laennec, rendered him most peculiarly competent to undertake such an investigation.

. Dr. Holland has minutely inquired into the effect produced on the respiratory organs, by the inhalation of metallic particles, upon workmen employed in all the different branches of this trade— viz., upon scissor grinders, upon fork, needle, razor, penknife, table-knife, saw, file, and scythe grinders; and has shown, beyond the possibility of question, that such occupations are absolutely lethal in their nature.   Thoracic diseases, Dr. Holland tells us, are prevalent to a great extent in Sheffield, and are more owing to the effect of dry grinding than to that which is done on wet stones.

It is in the grinding of small articles, such as needles, penknives, &c., that the greatest amount of dust is produced; and the evils entailed in these branches are accordingly greater than those resulting from the grinding of saws and scythes, which are ground on the wet stone only.   The want of proper ventilation to carry off the dust is one great cause of the sufferings of the grinders,

who are constantly working in clouds of dust. Without quoting
the pathological reasons which lead to the conclusion, I may
mention that it seems to be considered that the mucous membrane of the air passages is the seat of the disease from which the
grinders suffer—grinders' asthma—the tendency of which is to
terminate in tracheal consumption.

Tubercles, however, are found in numerous cases in the lungs
of grinders, and can be traced to the inhalation of dust; and many
—the delicate in constitution particularly—die very early of a
disease more nearly approaching to tuberculous phthisis.

The principal mortality among these operatives takes place
between the ages of 21 and 35, and not many appear to survive
much beyond the latter period, while the more delicate in organi-
zation die long before attaining the age of 35.

There seems to be no reason to doubt that the disease of which
grinders die, whether it be bronchial or tubercular consumption, is
produced from the mechanical irritation of the hard metallic particles upon the delicate surface of the mucous membrane. In the later
stages of the disease, hard, black, and gritty masses are not unfrequently expectorated, which are stated to appear to be accretions of
dust, as large as a pea, and even larger. Some grinders expecto-
rate these masses for years; and the fact that extensive expectoration of this nature goes on throughout the progress of the disease,
would appear to account for the very rare discovery of the accumulation after death; another reason for the non-discovery being,
that the workman is not exposed to the inhalation of dust for
some time previous to his death, as he is incapacitated from following his occupation for months prior to the fatal issue. The position
of the grinder while at work has, no doubt, a most injurious effect,
the body being bent nearly at right angles, so that the lungs
have not the free play so necessary to keep them in a healthy
state. The remuneration of this class of workmen is low, notwithstanding the deadly nature of their employment; and they
are stated to be reckless and dissipated, so that the fatal results
which ensue from their occupation are, in many cases, unnecessarily hastened.

I have previously spoken of the want of knowledge of the
labouring classes as operating against any speedy amendment of
their sanitary position. Dr. Holland refers to the ignorance
of the Sheffield operatives as tending, in the first place, to induce them entirely to disregard the premonitory symptoms of
their illness; and, in the second place, as tending to make them

consider that such evil results are necessarily inherent in their calling.

We see from these, and many similar considerations, the importance of introducing some instruction in human physiology into the ordinary routine of education, as the only means of teaching the results that must necessarily ensue if the laws of nature are disregarded, and of showing the labouring community how much it rests with themselves to guard against many of the evils from which they now suffer, and which are, to so great an extent, capable of being remedied.

A peculiar appearance is mentioned as being often observed in *post mortem* examinations of the respiratory organs of grinders: this is, a change of structure of the bronchial glands, at the bifurcation of the trachea. They seem to be converted into a black, gritty substance, the composition of which is unknown. This abnormal condition seems to be the result of the more destructive branches of the trade, such as fork-grinding. These gritty bodies are also found in the lungs, and are usually about the size of currants, but they are found much larger, and, at the bifurcation of the trachea, even as large as a hazel nut. More than a quarter of a century ago an invention was introduced by Mr. Abraham, which, it was hoped, would tend to diminish the fatal results caused by the inhalation of metallic particles. This was, a magnetic guard for the mouth; the object of which was to attract the particles of metal, and so prevent their entrance into the respiratory organs. The invention, however, does not appear to have been generally adopted, and it has long since fallen into disuse.

The dust produced in grinding is stated to consist of gritty as well as of metallic particles; the latter alone would be arrested in their course by the magnets, and the gritty particles are said to be no less injurious than the metallic. Great prejudices, however, exist among the workmen against all inventions of this kind, and it is very difficult to make them see the advantage of them, to whatever point of perfection they may be brought, and however fully they may be shown to answer the end in view.

It became, accordingly, a desideratum to introduce some arrangement, the success of which should not depend upon the men, and a contrivance was invented with this view. This was, a wooden funnel, from 10 to 12 inches square, placed above the surface of the revolving stone, on the side the farthest from the grinder, terminating in a channel immediately under the surface of the floor. Each of the grinders working in the same room has his own funnel and channel, all of which terminate in a common channel close to the external wall, where a fan is placed, like those used for winnowing corn, so arranged as to cause a current of air to flow from the mouth of the funnels, the effect of which is to carry away all the particles thrown off from the wheels. The principle is similar to that which I have before described as having been introduced into Mr. Evans' flour mills.

This arrangement is still in use in the different works in Sheffield, but no legislative enactment compels its adoption. Notwithstanding the continued use of the invention, however, the most recent inquiries into the mortality of the manufacturers of cutlery in Sheffield and elsewhere show that the high mortality among this class of operatives still continues. Dr. Greenhow, to whom reference has previously been made, states that "the greatest excess of death-rate is in Birmingham, Ecclesall Bierlow, and Sheffield, where, especially in the two last-mentioned places, the finer kinds of metallic manufactures are made." "The pernicious influence on health of certain operations connected with the manufacture of cutlery," he adds, "has long been recognized, and it is here rendered evident by the high pulmonary death-loss among the males of Ecclesall Bierlow and Sheffield."

If, then, we suppose that the apparatus in use to carry off the dust is really and entirely effective, we must come to the conclusion that the high pulmonary death-rate is due to other causes—in a great measure, perhaps, to the bodily position of the workmen employed in the process of grinding, and to which reference has already been made. For this there seems, at present, to be no remedy, as the men explain that they cannot, in any other posture of the body than that which they habitually adopt, obtain the necessary purchase for keeping the article they are grinding in its due position on the wheel. It seems unlikely, however, that the inventive art of this country would be unequal to the production of some apparatus to prevent the injury thus accruing, should it be clearly shown to exist. To give an idea of the excessive mortality among the operatives employed in the Sheffield grinding trade, I may mention, that, in the fork-grinding branch—which is stated to be the most destructive—Dr. Holland found that, out of 1,000 deaths occurring among persons between the ages of 20 and 30, while the proportion in England and Wales was 160, among the Sheffield fork-grinders it was 475. In the next decade of ages, 30 to 40, a similar disparity was observed, the proportion in England and Wales being 136, and among the fork-grinders 410.

The mortality in this branch of the trade was, therefore, very nearly three times as great as that among the general population of the country. Carrying the inquiry a step further, we find that, between the ages of 40 and 50, the proportion for England and Wales was 126, and among the fork-grinders 115.

The sum of these three numbers, which indicate the deaths of the Sheffield workmen, it will be observed, is exactly 1,000, so that before reaching the age of 50, every operative engaged in this perilous branch of the Sheffield trade has actually been killed off, while the proportion of deaths per 1,000 among the population of England and Wales amounted only to 422. These investigations were made some years ago, but, as I have before observed, the inquiries of Dr. Greenhow show that the mortality of this class of operatives is, at the present time, most excessive, as is evidenced by the high pulmonary death-rate in Sheffield and in other places in which the manufacture of metals is carried on—the death-rate from pulmonary affections, per 100,000 males, being, for England and Wales, 569, while for Sheffield it was as high as 839, and nearly the same for Birmingham—viz., 838. In Ecclesall Bierlow it was 736.

In Sheffield the operatives are chiefly engaged as cutlers, file-makers, grinders, and other workers in steel, and as goldsmiths; in Birmingham, as brassfounders, button makers, goldsmiths, gunsmiths, iron manufacturers, tool makers, nail makers, and glass makers, and in Ecclesall Bierlow, chiefly cutlers, file makers, grinders, and other workers in steel. A very large proportion of the inhabitants of Sheffield and Ecclesall Bierlow are employed in the injurious occupation of steel working—the proportion of adult males engaged in metal manufactures in the former place being 40·9, and in the latter 43·6 per cent. Notwithstanding the improvements that have been introduced into the workshops of the Sheffield grinders to prevent the evil consequences which have been shown to result from the inhalation of the products of the grinding of metals, and particularly of the dry-grinding process, there is abundant evidence that the evil is diminished only, and not entirely prevented; and the mortality that, at the present time, still takes place annually among grinders at the principal seats of manufacture of metals is clearly too high to be the result of the "sedentary and unwholesome confinement incidental to their occupation," which the writer of the article in the Magazine to which I have referred seems to consider the cause of the frequency of phthisis among the operatives of this class. It will be evident, from the

satisfactory action of the inventions for remedying the fatal effects
that result from this branch of trade, that the means of diminishing the high mortality that, at the present time, still obtains, are
in our own hands; but the influence of the Government to compel
the due adoption of the means at command seems to be impera-
tively necessary.

To show to what extent proper attention to check the inhalation of dust may tend to diminish the mortality from pulmonary
complaints, which exposure to its influence creates, I may refer to
some statistics of the Pentonville prison, the discipline of which
enforces the separate system. Several trades are carried on by the
prisoners, some of which, such as mat and rug making, produce
great quantities of dust, the inhalation of which by the prisoners,
it was considered, might tend to the production of the high mortality from pulmonary disease which the records of the prison
showed. An investigation into the sanitary condition of the
prisoners was accordingly instituted by Dr. G. Owen Rees, the
principal medical officer, in whose Report on the subject, appended
to the Report of the Commissioners of the prison, the result of his
inquiries will be found recorded. The prisoners in the Pentonville prison, it is stated, were more subject to injury from the
inhalation of dust than those in prisons where the men do not
sleep in their work-rooms. The investigation into this question
was made on the 14th March, 1846, when the prison was full.

The proportion of men employed in dusty, as distinguished
from those engaged in non-dusty occupations, was, at that date, as
1 to 2·38—so that, if the chances of death from consumption were
equal for men employed in both ways, it would have been expected
that more than twice as many cases would have occurred among
men employed in non-dusty trades; but 8 men employed in dusty
trades appeared among the casualties, which, were the chances of
immunity equal, would give 19·04 for the number of deaths and
free pardons among the non-dusty trades.

It is clearly necessary to include the free pardons on medical
grounds as deaths from pulmonary disease, in all these comparisons,
as, had not the pardons been granted, it is quite clear that death
from such disease would have happened within the prison.

The result of the inquiry of Dr. Rees, however, he reports, did
not show this to be the case, for only 9 of such men suffered,
showing that the chance of escape from the disease in the prison
was twice as great for those employed in the non-dusty trades.

Acting on the information thus obtained, means were taken to lessen the evils shown to result from dusty occupations. The wind-ing of worsted was prohibited, the coir used in mat-making was ordered to be picked and beaten before being given to the prisoners to work up; and, probably, other measures, not detailed in the Report, were adopted to diminish the evils that were shown to have been the consequence of the inhalation of dust; and the result became apparent by the end of the year 1846, when the deaths by consumption scarcely exceeded the returns for the general popula-tion at the same ages—being, including free pardons, ·47 per cent. It therefore appeared that consumption in the Pentonville prison was an accidental, rather than a necessary, accompaniment of the system of discipline carried on there. Dr. Rees reports that the result of his investigation satisfied him that the admission of dust, of an irritating character, into the minute structure of the lungs, and particularly in occupations that require workmen to remain in an atmosphere of dust, leads to an excess of mortality by con-sumption; and that such is the case whether the dust created consists of metallic or of vegetable particles.

The result of this inquiry is important in two ways:—

1st. As showing that prison discipline, even when carried on upon the separate system, is not necessarily the cause of the tuber-cular cachexia, which has always been considered to be one of the evils of confinement in prisons; the improvements adopted with reference to the prevention of the inhalation of dust by the men, and the great attention devoted to ventilation, drainage, and the sanitary arrangements of the prison generally, under the able and zealous directions of Dr. Rees, having had the effect of so reducing the mortality of the prison, that, including all free pardons as deaths, the total mortality came but little to exceed that of the general population—the excess being 1·73 per cent. per annum—showing most satisfactorily the efficacy of the introduction of the sanitary measures recommended:

And 2nd. As evidencing how much we hold in our hands the means of diminishing the mortality from phthisis and other cognate diseases, which cut off, in the very prime of life, so large a pro-portion of the population of this country.

In our inquiries into the subject of pulmonary disease, we must not fail to consider the high mortality from this cause in the British army on home service. This question has recently begun to attract attention.

The disclosures in the Report of the Commissioners appointed in May, 1857, to investigate the sanitary condition of the army, were more startling in their nature than the public were prepared to expect.

It appears that one of the greatest enemies the soldier has to fear, far more than the sword or the rifle of the enemy, is the disease we are considering—phthisis.

It will be impossible for me, within the limits of this paper, to do more than give a slight outline of the causes tending to produce the effect observed.

Notwithstanding the circumstance of the soldier being, in fact, a selected life on entering the army, and that he is proved to be then free from any bodily defect or infirmity—notwithstanding that he is placed in a much better position than the rest of the community of the same social standing with himself, as to food, lodging, and clothing; that he has the best medical attendance upon the slightest suspicion of illness; and that, altogether, his position is one which might naturally be imagined must conduce to a high state of bodily health and to a lengthened existence— notwithstanding all this, the result of the inquiry is stated to be that the mortality in the army is very much greater than among any class of civilians.

At a meeting held lately at the Hanover Square Rooms, on the subject of the sanitary condition of the army, it was stated, by Dr. Gourlay, that the life led by the Guards in town unfits them entirely for campaigning; and that, upon the occasion of the Crimean war, after one night in the trenches, a great proportion of the newly-arrived soldiers were on the sick list.

It was stated further, that while the death-rate in the Prussian army was only 6 in 1,000, that in the British army was no less than 33 in 1,000, or nearly six times as great, and that 70 per 1,000 were invalided—a severe loss, even in a pecuniary point of view, to the country, when it is considered that each soldier, after he has learnt his duties, has cost the nation the sum of £100.

I am not aware what authority Dr. Gourlay had for this statement; as it appears, from the evidence of Colonel Tulloch before the Commission, that the mortality in the Grenadier Guards was 21·5 per 1,000—an enormously high rate as compared with that of the civil population, which was stated to be only 9·6 per 1,000 at the same ages—viz., 20 to 25.

The principal causes for this high rate appear, from the evidence before the Commission, to be—

1st. Defective arrangements connected with night duty.

2nd. Want of exercise and suitable employment.

3rd. Intemperance and vice.

4th. Bad barrack accommodation.

5th. Improper food and cooking.

Now, without going into any discussion of these five heads, there can be no doubt that the evil results traceable from such causes as these are within our power to remedy, and it is, accordingly, very satisfactory to learn that the Army Medical Department is now being reorganized by the Director-General, with the object of carrying out the recommendations contained in the Report of the Royal Commissioners for the improvement of the sanitary condition of the army.

A new statistical nomenclature, we are told, has been adopted for the army returns—viz., that which was drawn up at the Statistical Congress held in Paris in 1855, and which is identical with that of the Registrar-General; so that, in future, we shall be in a better condition to compare the mortality of our army with that of the armies of foreign powers, as well as with that of the civil population of this country.

One great cause of the maintenance of the high rate of mortality from phthisis, is, no doubt, the intermarriage of persons in whom the hereditary taint of this disease exists.

There is a theory put forth by Walker, that if one parent be afflicted with pulmonic disease, it is an even chance that any one child that may be the result of the marriage of two persons so situated should derive the organs of the chest from the parent so afflicted; and it has been suggested that this theory might account for the very capricious selection that this disease makes in the different members of the same family for its attacks.

Be this as it may, we know perfectly well that the hereditary transmission of this malady is an undoubted fact; and Assurance Companies, whose experience is so large on such points, very generally feel called upon to reject applicants for life assurance, whose apparent health may be undoubted, when several cases of phthisis have occurred among the members of the family.

Some Companies, indeed, think that it is safer to reject the case when the mother or father, or even one brother or sister, has died from this disease, if the applicant be young—say under 45.

In Walker's work on " Intermarriage," another theory is broached—viz., that, under certain restrictions, the male gives to the progeny the external or locomotive organs, and the female the internal or vital organs.

Arguing from the analogy of breeding among animals, this is

more than a mere theory; the fact is well known to breeders of stock. So much, indeed, do certain known laws with respect to propagation prevail, and so thoroughly have they been made the subject of scientific investigation, that a cattle-breeder can produce, within certain limits, almost any class of animal he desires.

If we recognise the truth of this second theory, the necessity for extreme caution, on the part of Life Assurance Companies, in accepting the lives of individuals whose mothers have died from any hereditary disease, becomes clearly apparent. Still, arguing from what has been proved to be the case with respect to animals, Assurance Companies might, with reason, carry their inquiries into the family history of lives proposed for assurance much further than they have yet thought of doing.

It would, perhaps, startle a person desirous of assuring his life, to ask him whether his mother had been married previously to her marriage with his father, and if so, whether her former husband died of phthisis; and yet this would not be by any means a ridiculous inquiry.

It is a fact quite beyond dispute, and one to which public attention was first drawn by Lord Morton in the *Philosophical Transactions,* that the female retains traces of the influence of a prior impregnation, and transmits the same to a second, third, and even fourth progeny.

This has been clearly proved to be the case with horses, mules, sheep, dogs, pigs, and fowls; and what is most extraordinary in the matter is, that it is not necessary that a prior impregnation, in order to influence a subsequent one, should have resulted in the production of progeny.

In illustration of this, I may mention two authentic cases communicated to the Newcastle Farmers' Club by Mr. Reginald Orton, a surgeon at Sunderland; the first being the case just referred to as having been brought before the public by Lord Morton, who allowed a quagga he possessed to breed with a thorough-bred chestnut mare. The produce was a quagga mule, having the stripes, and, in many respects, the characteristics of the male parent. The next season the same mare was allowed to breed with a black Arab horse, and the result was, to the astonishment of the owner, a foal bearing strongly the marks of the quagga; and this effect continued to be visible through three successive foals produced by horses.

The second case is recorded in the *Transactions of the Royal Society,* and is perfectly analogous to the first. It had reference to a mare, the property of Sir Gore Ousley, which was allowed to breed with a zebra. The result was an animal bearing the zebra stripes. The mare next bred with a blood-horse, and produced a striped foal. The following year the same thing happened again.

A precisely similar result has been obtained in the case of other animals.

I mentioned that it had been shown, by experiments with animals, that a prior impregnation need not necessarily have resulted in the production of progeny, in order that evidences of it may be transmitted to a subsequent generation. Of course, with human beings, if there were any children of a first marriage, and phthisis showed itself in them, the case would be stronger; but the result, according to this theory, would be the same in the case of a *primipara*, as of a widow who had borne children.

With respect to animals, the fact under consideration is undoubted; and if we may fairly argue by analogy that the same holds for the human species, there is, probably, some reason to believe that we may, in this way, account for some of those sad but inscrutable cases which we all of us see happen occasionally among our friends and relations, in which phthisis comes on suddenly and unexpectedly, without our being able to trace either any hereditary tendency to the disease, or to find any reason to account satisfactorily for the origin of the affliction.

Should our reasoning be correct, the desirableness of marriage with a widow whose husband has died of phthisis or other hereditary disease, may, at least, be questioned; and, according to what has been said, whether she had borne children to her first husband or not; and possibly the attention of those engaged in the conduct of life assurance business may not inappropriately be directed to the subject.

With the view to confirm the truth of some of the foregoing opinions as to the effect of occupation on phthisis, I applied to the authorities of the Hospital for Consumption and Diseases of the Chest, at Brompton, for some statistics on the subject; and, through the kindness of the medical committee of the hospital, and of the resident medical officer, Mr. Vertue Edwards, I have been furnished with a list of the occupations of 557 male and 318 female in-patients who died of phthisis, or some disease of the chest, in the hospital in the ten years from January, 1850, to December, 1859. It is necessary to confine our attention to the in-patients, as the results of the cases of the out-patients cannot be ascertained. Although the sum of these numbers, which amounts to 875, forms but a small portion of the numbers recorded under similar heads in the Registrar's Reports, the list is sufficiently extensive to show which are the conditions of life in London that tend to the production of the particular class of diseases treated in this hospital.

In considering any records of this nature, we must take into account the situation, apart from the cause of the disease, in which the persons who are likely to become inmates of such an institution are placed.

We are at first struck by the number of domestic servants, both male and female—the latter particularly—who have died in the hospital, and we are not aware of any particular predisposing cause to phthisis in their case ; indeed, in comparison with almost all the other classes in the list, the condition of domestic servants, as a class, contrasts very favourably ; yet we find this class contributing the greatest number of deaths—viz., males 78, and females 192, or 14 and 60 per cent. respectively of the whole cases under observation.

On reflection, however, we see that this is the class, more than all others, likely to be sent to the hospital. Their illness comes on under the very eyes of their employers, who are, for the most part, willing, as well as able, to gain admission for their servants.

We may, therefore, consider that the highest classes on this list are those of labourers, among males, and of needlewomen, milliners and dressmakers, and other similar workers for shops, among females.

These classes contribute respectively 14 and 20 per cent. of the whole number of cases; next in order follow governesses and teachers.

This is just what we might naturally expect, that those who fare the hardest and have the greatest struggle to gain a livelihood, are among the first to succumb to the great malady of the country; and, further, in proportion to their poverty will they be the most likely classes to be left to die in a hospital, while the classes a little above them in the social scale will be more likely to be removed by their friends when their cases are found to be incurable.

The other unhealthy occupations as tested by these returns, arranged in order of unhealthiness, are, among males, those of clerks, carpenters, shoemakers, tailors, porters, drapers' assistants, painters, gardeners, soldiers, sailors, bricklayers, engineers, cabinet makers, bakers, policemen, and compositors; and, among females, governesses, teachers and schoolmistresses, laundresses and ironers.

With some slight exceptions, these, too, follow very much the order that we might expect.

We are not surprised to find clerks, shoemakers, tailors, and drapers' assistants, high in the list, seeing that their employment is of a sedentary nature; that these classes have very little opportunity for active exercise in the open air, and that the three first classes are not only engaged in sedentary employment, but such as, in addition, necessitates stooping very much over their work—the clerks over their desks, against which they most unwisely press their chests, to the very serious injury of the thoracic viscera, and the tailors and shoemakers over their stitching; the injury to the latter, besides, is increased by the pressure of the last against the chest, to which I have before referred.

The long hours during which the drapers' assistants work—the close, hot, crowded, gas-laden shops in which their lives are passed —the constant standing posture, in combination with the almost total impossibility of taking out-door exercise in the daytime— render this class particularly prone to phthisis. The mortality from this cause, in the large metropolitan drapers' shops, is said to be very great. The want of daylight in the work-rooms in which a considerable portion of the labouring classes perform their daily avocations, is stated to conduce to this disease.

It is satisfactory to see that there is just beginning to be some slight improvement in respect to the long hours during which these shops are open; and some large establishments have, in a very praiseworthy manner, instituted libraries and reading-rooms on their premises for the use of those employed.

The number of deaths recorded in the hospital returns, particularly among persons following other callings than the four classes I have enumerated, is too small for us to deduce much from it, even by comparison with the number of persons employed in the different trades, and which the Census returns of 1851 give us the means of knowing.

By combining together, however, the numbers of deaths stated to have happened in the hospital from phthisis, among persons working in metals—such as engineers, watchmakers, smiths of all kinds, cutlers, optical instrument makers, and the like—we find the total number proportionately very high; in fact, this class compares, for unhealthiness of occupation, with clerks, who stand almost at the head of the list—each class producing, in fact, $5\frac{1}{2}$ per cent. of the whole mortality from this disease; and thus we are enabled, to some extent, to confirm, from the statistics of the hospital, the opinion as to the injurious effect upon the lungs of the inhalation of metallic particles.

No report has yet been issued by the authorities of the hospital of the result of the last ten years' experience to which I have just been alluding, in respect of which certain information has been furnished to me.

An elaborate medical report for the first six years, however—viz., from the establishment of the hospital in September, 1842, to the 31st of December, 1848—has been issued, and to this I have been kindly allowed access.

About the same number of in-patients are included in this period as in that embracing the last ten years; the exact number was 888, and consisted of 542 males and 346 females.

The out-patients numbered in all 10,051, but those treated for phthisis amounted only to 3,470—viz., males 2,137, and females 1,333.

The result of the medical inquiry based upon the facts recorded in the above cases, showed that there was a greater liability to phthisis in males than in females. I mention this, as the result is stated to be not in accordance with the opinion entertained on this subject by writers of authority in this country or on the continent.

The fact that the female population in London, during the period under observation, exceeded that of the males, was not lost sight of in the investigation.

The intensity of the disease, judging from the hospital statistics, appears to be between the ages of 25 and 35 for males, and at ages a little younger for females. The unhealthy occupations seem to follow very much the same order as those in the more recent period, and the results tend to confirm, to a very great extent, our preconceived views on the subject; and the more satisfactorily, as the number of cases under observation which were classified under the heads of the occupations of the patients that died of phthisis embraced 4,358 cases—viz., males 2,679, and females 1,679—or about five times as many as those that I have considered previously. Among the males we find the class of labourers, as before, at the head of the list; next come clerks, warehousemen and shopmen, mechanics, servants, tailors, carpenters, shoemakers, coachmen and cabmen, printers and compositors, painters and glaziers, weavers and glovers, bakers, and butchers; and among the females we find, as before, a very large and probably undue proportion of servants; then follow, as before, needlewomen, and those engaged in similar occupations, laundresses and governesses. Of course, we cannot accurately compare the relative mortality from any particular disease among persons following different occupations, unless we know the respective numbers engaged in those occupations.

An investigation of this nature would lead us further into the details of the question than is contemplated in this paper.

The question of the contagious nature, or otherwise, of phthisis, as bearing upon the production of this disease, is not an unimportant one. The difference of opinion, however, on this point appears to be very great, and the evidence on the subject to be very contradictory. As long, however, as the question admits of a doubt, it would certainly be well to be on the safe side, and, at all events, to avoid needless exposure to contagion.

The medical report of the hospital contains the results of a most interesting investigation into the question of hereditary predisposition to phthisis.

The information which is the result of such inquiries as these cannot but be of great interest to those connected with the business of life assurance, and of the highest importance to the Companies themselves, if they wish, not as a matter of course, to reject applicants for life assurance in whom an hereditary taint is suspected; in fact, if they desire to act upon principles which are the result of actual inquiries, instead of treating the cases brought before them in an arbitrary and unscientific manner, as they are now very much obliged to do.

With reference to the transmission of the disease, we learn from the report that it has been observed that one or more of several children of the same individuals may exhibit traces of the disease of their parents, and yet the other children of the family, though not apparently diseased themselves, may, on becoming parents, transmit to their offspring the elements of the disease, which only requires some exciting cause for its development. Thus a diseased parent in one generation will have offspring to all appearance healthy, and in the next generation an apparently healthy parent will be found to have diseased children.

The result of the inquiry of the medical officers of the Brompton hospital, as to the proportion of cases on the books in which hereditary taint could be traced, showed that out of 1,010 cases, comprising 669 males and 341 females, 122 males and 124 females, forming 18 and 36 per cent. respectively of the whole, or $24\frac{1}{2}$ per cent. of males and females combined, were born of phthisical parents; or, in other words, 1 in every 4 patients, nearly, of the 1,010 cases under observation was traced to be hereditarily predisposed to the disease. It appears, too, that females are more likely to inherit the disease than males, in the proportion of 2 to 1. With the view to show how very large is the comparative proportion who appear to inherit phthisis from their parents—and, accordingly, to show how peculiarly severe the power of the hereditary transmission of this disease is—a comparison was instituted with the similar power in cases of insanity, which is also known to be, unhappily, too frequently transmitted through successive generations; and the result was, that 11·9 per cent. males and 13·4 per cent. females—or, combined, $12\frac{1}{2}$ per cent. of the cases under observation—were born of insane parents. The probability, therefore, of the hereditary transmission of phthisis, as compared with that of insanity, is, taking both sexes together, as 2 to 1; and while the probability of inheriting phthisis, as just shown, is twice as great among females as among males, that of a similar transmission of insanity is only $1\frac{1}{2}$ per cent. greater.

(To be continued.)

CORRESPONDENCE.

ON A METHOD OF USING THE "TABLE OF QUARTER SQUARES."

To the Editor of the Assurance Magazine.

Sir,—Allow me to call your attention, and that of your readers, to a method of using the " Table of Quarter Squares" computed by me and published in 1855,* which, although not pointed out in the introduction to that work, is one which will extend the usefulness of the table, and, in certain cases, materially diminish labour.

A constant factor is of frequent occurrence in some of the calculations of an actuary; for my present purpose it will be sufficient to instance the expression for the value of a policy when the premium is just due and not paid, viz.—

$$1 - \frac{1 + a_{m+n}}{1 + a_m} = 1 - (1 + a_{m+n})\left(\frac{1}{1 + a_m}\right),$$

in which the term $\frac{1}{1 + a_m}$ is constant, whatever may be the value of the other factor. Supposing a computer to be about to form a table of the values of a policy for a series of years—say from 1 to n years—it will manifestly be advantageous if he can save one reference to the table in each calculation.

---

* *Table of Quarter Squares of all Integer Numbers up to* 100,000, *by which the Product of Two Factors may be found by the aid of Addition and Subtraction alone.* C. & E. Layton, Fleet Street.

This saving can be effected, where one of the factors is a constant; as may be seen by reference to the equation [2], at p. xiv. of that work, where it is shown that

$$ab = 2\left(\frac{a^2}{4} + \frac{b^2}{4} - \frac{(a-b)^2}{4}\right) \quad \cdots \cdots \cdots \quad [2]$$

But this expression requires the tabular results in every case to be multiplied by 2; and the object of the present communication is to show that where a constant factor occurs, as in the instance above quoted, the trouble of doubling the tabular results may be saved. For, transforming the above expression [2], by doubling the factor a, we obtain

$$(2a+b)^2 = 4a^2 + 4ab + b^2 \quad \cdots \cdots \cdots \quad [A]$$
$$(2a-b)^2 = 4a^2 - 4ab + b^2 \quad \cdots \cdots \cdots \quad [B]$$

By addition of [A] and [B], we obtain

$$4a^2 + 4ab + b^2 + (2a-b)^2 = 8a^2 + 2b^2;$$

$$\text{hence } 4ab = 4a^2 + b^2 - (2a-b)^2,$$

$$\text{and } ab = a^2 + \frac{b^2}{4} - \frac{(2a-b)^2}{4};$$

and since $a^2 = \dfrac{(2a)^2}{4}$, the expression may be further transformed into

$$ab = \frac{(2a)^2}{4} + \frac{b^2}{4} - \frac{(2a-b)^2}{4}.$$

It will, therefore, be obvious, that if, before entering the table of quarter squares, we double the factor a, we shall obtain precisely the same result as by the use of the expression [2].

I am aware that, in an isolated case, it is immaterial which of these methods be adopted; in general, in such a case, it will be preferable to use the other formula stated in the work, viz.—

$$ab = \frac{(a+b)^2}{4} - \frac{(a-b)^2}{4}. \quad \cdots \cdots \quad [1]$$

It may be said that, by the process here recommended, three tabular entries are necessary, while by the use of the formula [1] two tabular entries only are needed; but it must be borne in mind, that in the operation by the former of these modes we have only to take the difference of the factor, while in the latter both the sum and difference of the factors must be found. In isolated cases, the amount of work by either formula is about equal; but with a series of quantities to be multiplied into a constant factor, the computer will, I think, find it more convenient to adopt the method now suggested. It may further be remarked, that the process here pointed out is quite as simple, if not more so, than if done by logarithms. The amount of work is not greater, and we obtain a result directly in natural numbers.

In illustration of the above remarks, I append an example of the values of a policy taken out at age 25, at the end of 1, 2, 3, &c., years, the premium being just due and not paid (Carl. 3 per cent.). For the sake of comparison, I place in juxtaposition the logarithmic process. The constant is supposed to be written on a card, and moved onward for each operation, and for this reason the constant figures do not appear except at the head of each column.

Here the constant factor is $\dfrac{1}{1+a_{25}} = \dfrac{1}{21 \cdot 666} = \cdot 0461575$, and this quantity being doubled, for the reason before pointed out, $= \cdot 092315$.

$2 \cdot \left(\dfrac{1}{1+a_{25}} \right)$	(1) Constant factor, $\cdot 092315$.	Quarter square of (1), 213052.	Colog. $(1+a_{25})$, $8 \cdot 66424$.
$1+a_{26}$ Difference	21·442 70873	11494 Ar. Co. $\overline{1}874425$ ·98971	1·33127 $\overline{9}$·99551$= \cdot 98971$
$1+a_{27} =$	21·212 71103	11249 $\overline{1}873609$ ·97910	1·32658 9·99082$= \cdot 97909$
$1+a_{28} =$	20·981 71334	11005 $\overline{1}872787$ ·96844	1·32183 9·98607$= \cdot 96844$
$1+a_{29} =$	20·761 71554	10776 $\overline{1}872001$ ·95829	1·31725 9·98149$= \cdot 95828$
$1+a_{30} =$	20·556 71759	10564 $\overline{1}871266$ ·94882	1·31294 9·97718$= \cdot 94882$
$1+a_{31} =$	20·348 71967	10351 $\overline{1}870519$ ·93922	1·30852 9·97276$= \cdot 93921$
$1+a_{32} =$	20·134 72181	10134 $\overline{1}869748$ ·92934	1·30393 9·96817$= \cdot 92933$
$1+a_{33} =$	19·910 72405	9910 $\overline{1}868938$ ·91900	1·29907 9·96331$= \cdot 91900$
$1+a_{34} =$	19·675 72640	9678 $\overline{1}868086$ ·90816	1·29392 9·95816$= \cdot 90816$
$1+a_{35} =$	19·433 72882	9441 $\overline{1}867205$ ·89698	1·28854 9·95278$= \cdot 89698$

The foregoing results, subtracted from unity, give the value of a policy at the end of 1, 2, 3, &c., years.

I have the honour to be, Sir,

, Your obedient Servant,

Eagle Life Office,   SAMUEL L. LAUNDY.
   26th March, 1860.

INSTITUTE OF ACTUARIES.

PROCEEDINGS OF THE INSTITUTE.

Fifth Ordinary Meeting, Session 1859-60.—*Monday,* 26th March, 1860.

CHARLES JELLICOE, Vice-President, in the Chair.

The minutes of the last ordinary meeting were read and confirmed.

The Secretary announced various donations to the library.

It was announced to the meeting that the following Report had been received and adopted by the Council, and that the Syllabus appended would be in future acted upon :—

" The Examiners of the Institute have the honour to report to the Council, that they have for some time past, in the discharge of the duties entrusted to them, been sensible that the existing Syllabus required modification, in consequence partly of the time which has elapsed since it was formed, and partly of the experience gained whilst it has been in operation; that they have accordingly had several meetings to discuss such modifications as appeared to be desirable; and, having decided upon them, beg leave to submit a copy of the Syllabus in its amended form, with a recommendation that the same be adopted and published."

A. H. BAILEY.   H. W. PORTER, B.A.
SAMUEL BROWN.   T. B. SPRAGUE, M.A.   } *Examiners.*
W. B. HODGE.   ROBERT TUCKER.
C. JELLICOE.

SYLLABUS.

MATRICULATION EXAMINATION.

	Questions.
Vulgar fractions	
Decimal fractions	
Logarithms	*l*
Evolution	
Equations, simple and quadratic	4
Series, arithmetical and geometrical	
Permutations and combinations	8
Binomial theorem	
Finite differences	2
Geometry—First four books of Euclid	4
Total	25

Maximum number of marks 500, of which 250 must be obtained to enable the Candidate to pass.

and for this reason the cons
of each column.

Here the constant factor is
quantity being doubled, for the

appear except at the head

= ·0461575, and this

ted out, = ·092315.

> Colog. $(1+a_{25})$,
> 8·66124.

2.
$1+a_{25}$

$1+a_{26}$	21·442		1·33127
Difference	70873		$\bar{9}$ 99551 = ·98971
$1+a_{27}$	21·21?		1·32658
	̖1103		$\bar{9}$·99082 = ·97909
$1+a_{28}$	20·981		1·32183
	71331	1872787	$\bar{9}$·98607 = ·96844
		·96844	
$1+a_{29}$	20·761	10776	1·31725
	71551	1872001	$\bar{9}$·98149 = ·95828
		·95829	
$1+a_{30}$	20·556	10564	1·31294
	71759	1871266	$\bar{9}$·97718 = ·94882
		·94882	
$1+a_{31}$	20·348	10351	1·30852
	71967	1870519	$\bar{9}$·97276 = ·93921
		·93922	
$1+a_{32}$	20·134	10134	1·30393
	72181	1869748	$\bar{9}$·96817 = ·92933
		·92934	
$1+a_{33} =$	19·910	9910	1·29907
	72405	1868938	$\bar{9}$·96331 = ·91900
		·91900	
$1+a_{34} =$	19·675	9678	1·29392
	72640	1868086	$\bar{9}$·95816 = ·90816
		·90816	
$1+a_{35} =$	19·433	9441	1·28854
	72882	1867205	$\bar{9}$·95278 = ·89698
		·89698	

The foregoing results, ... the value of a policy at the end of 1, 2, 3, &c., years.

I have the honor to be, &c.,

SAMUEL L. LAUNDY.

Eagle Life Office,
29th March, 1860.

INSTITUTE OF ACTUARIES

PROCEEDINGS OF THE COUNCIL

Fifth Ordinary Meeting, Session 1859-60,—Monday, 26th March, 1860.

CHARLES JELLICOE, Vice-President, in the Chair.

The minutes of the last ordinary meeting were read and confirmed.

The Secretary announced various donations to the Library.

It was announced to the meeting that the following Report had been received and adopted by the Council, and that the Syllabus appended would be in future acted upon:—

"The Examiners of the Institute beg leave to report to the Council, that they have for some time past, in the discharge of the duties entrusted to them, been sensible that the existing Syllabus needed modification, in consequence partly of the time which has elapsed since it was framed, and partly of the experience gained whilst it has been in operation; that they have accordingly had several meetings to discuss such modifications as appeared to be desirable; and, having decided upon them, beg leave to submit a copy of the Syllabus in its amended form, with a recommendation that the same be adopted and published."

A. H. BAILEY.	H. W. .... B.A.	
SAMUEL BROWN.	.... SMITH, M.A.	*Examiners.*
W. B. HODGE.	....	
C. JELLICOE.		

SYLLABUS.

MATRICULATION EXAMINATION.

	Sections
Vulgar fractions	
Decimal fractions	
Logarithms	2
Evolution	
Equations,	4
Series, and	
Permutations	8
	9

.... Candidate

SECOND YEAR'S EXAMINATION.

Questions.

Theory of logarithms ⎫
Elements of the theory of probabilities ⎬ . . . 6
Compound interest and annuities certain ⎭

Tables of mortality . . . . *

Construction of auxiliary tables ⎫
Annuities and assurances on lives ⎪
Annuities and assurances on survivorships ⎬ . . . 15
Miscellaneous questions ⎭

Total 25

Maximum number of marks 750, of which 375 must be obtained to enable the Candidate to pass.

THIRD YEAR'S EXAMINATION.

Questions.

Life Assurance Finance.—Construction and graduation of tables of mortality; existing tables of mortality—the mode of their construction and their respective merits; methods of determining the surplus in an Assurance Company and of distributing it amongst the assured 5

Legal Principles.—Acts of Parliament; Charters of Incorporation; Deeds of Settlement; Partnerships, limited and unlimited—powers and duties of persons constituting them; Policy considered as a Contract; Probates and Letters of Administration; Assignments; personal representatives; Bankruptcies . . 5

Statistics.—Methods for the arrangement and collection of data; tests of accuracy; preparation of abstracts and reports; general system of the country's finance; funded and unfunded debt, and fiscal arrangements; taxation 5

Currency, Banking, and Investments.—Currency, metallic and paper; nature of banking—Bank of England, Private and Joint Stock Banks; Bills of Exchange; comparative value of securities; high and low prices; fluctuations in the rate of interest . . . 5

Miscellaneous.—Book-keeping; auditing; valuation of marketable securities; and approximate calculations 5

25

Maximum number of marks 1000, of which 500 must be obtained to enable the Candidate to pass.

A paper was read, " On the Life Assurance Companies of Germany; their business and position in 1858." By Herr Rath G. Hopf, of Gotha, Corresponding Member.

Thanks having been voted to Herr Hopf, the meeting adjourned to 30th April, 1860.

———

Sixth Ordinary Meeting, Session 1859-60.—*Monday, 30th April,* 1860.

WM. BARWICK HODGE, Vice-President, in the Chair.

The minutes of the last ordinary meeting were read and confirmed.

The Honorary Secretary read a letter from Mr. A. G. Finlaison, announcing the decease of his father, Mr. John Finlaison, the President of the Institute; and Mr. Hodge, after expressing the regret with which that intelligence had been received, reported that Mr. Jellicoe had been unanimously elected by the Council to fill the office until the annual general meeting in June. Mr. Jellicoe briefly expressed his acknowledgments for the honour conferred upon him, and the business of the meeting was then proceeded with.

The Secretary announced various donations to the library.

Frederick Bigg, Esq., duly nominated at the last ordinary meeting, was unanimously elected a Fellow of the Institute.

Mr. Hodge then read Part 4 of his paper, " On the rate of interest for the use of money in ancient and modern times."

A vote of thanks was very cordially given to Mr. Hodge, and the meeting adjourned to the 26th November, 1860.

Annual General Meeting, Saturday, 2nd June, 1860.

CHARLES JELLICOE, President of the Institute, in the Chair.

The circular convening the meeting having been read,

The minutes of the last ordinary meeting were read and confirmed.

The Report of the Council was then read by Mr. J. Hill Williams, one of the honorary secretaries, and was as follows :—

" The Council have the satisfaction of reporting, that, notwithstanding the great changes which have taken place of late as regards the various assurance institutions established in the metropolis, the number of members on the roll of the Institute suffers no decrease of importance. At the present time it appears there are 46 Fellows, 20 Official Associates, and 81 Associates—147 members in all—a number differing but little from that reported of late years as constituting the whole body.

" The receipts during the year amount to £612. 12s. 7d., and the payments to £392. 2s. 8d.; a further small purchase of Consols has been made, and the assets now consist of £220. 10s. 11d. in cash, and £192. 8s. 6d. invested in (£198. 16s. 2d.) Stock, making together the sum of £412. 19s. 5d. This includes the donation so obligingly made by Mr. Gompertz towards an extension of the library.

" The following papers have been read during the session, viz., ' On the rationale of certain actuarial estimates,' by Charles Jellicoe; ' On a formula for calculating the value of a survivorship assurance,' by M. Réboul; ' On the purchase of life assurance policies as an investment,' by Archibald Day; ' On some considerations suggested by the Reports of the Registrar-General, being an inquiry into the question how far the inordinate mortality in this country, exhibited by these Reports, is controllable by human agency,' by H. W. Porter, B.A.; ' On the Life Assurance Companies of Germany, their business and position in the year 1858,' by Herr Rath G. Hopf; ' On the rate of interest for the use of money in ancient and modern times' (Part IV.), by William Barwick Hodge, Vice-President.

" Some of these have already appeared, and the remainder will shortly be published, as usual, in the *Journal* of the Institute.

" The members have been informed that the prize offered last year by the Institute for an essay on the methods of distributing the surplus among the persons assured in a Life Assurance Company, has not been awarded. The Council have therefore determined on repeating their offer of a similar prize, and again invite the associates to compete for it. The subject of the essay and the regulations to be the same as before.

" The members have also been made aware of the heavy loss which the Institute has sustained in the death of its President, the late Mr. Finlaison. The office of president was tendered to that gentleman at the formation of the Institute, in accordance with the unanimously-expressed wishes of its founders, and was accepted by him, without hesitation, at a time when its capabilities were unknown and its success uncertain.

" Throughout its career the late President manifested a lively interest in its proceedings, and never failed, when taking part in them, to exhibit the utmost kindness and courtesy to its members. The Council forbear to touch upon the events of a life of great activity and usefulness, whether regarded from a public or a private point of view. They trust that a more fitting occasion for doing so will shortly present itself, and meanwhile content themselves with this brief record of their respectful remembrance and unfeigned regret.

"In accordance with Article 28 of the Laws and Constitution of the Institute, it was necessary to appoint a successor to Mr. Finlaison, for the period intervening between the time of the vacancy occurring and that of the annual general meeting; and Mr. Jellicoe having, at the request of a majority of the members of the Council, allowed himself to be put in nomination for the office, was balloted for at a special meeting of the Council on the 28th April last, and found to be unanimously elected. Mr. Jellicoe has, accordingly, exercised the functions of the presidentship until the present time, and, as the members will observe, is now recommended for re-election."

The abstract of the receipts and payments of the Institute, for the financial year ended 31st March last, was then read (*see* p. 120).

On the motion of the Chairman, the Report was unanimously adopted.

The election of a President, Vice-Presidents, and Officers, for the year ensuing, was then proceeded with.

Mr. Bailey and Mr. Cutliffe were appointed scrutineers.

On the result of the ballot being obtained, the following was declared to be the list, viz. :—

President.
CHARLES JELLICOE.

Vice Presidents.

SAMUEL BROWN.	PETER HARDY, F.R.S.
WILLIAM BARWICK HODGE.	ROBERT TUCKER.

Treasurer.
JOHN LAURENCE.

Honorary Secretaries.

JOHN REDDISH.	J. HILL WILLIAMS.

The following gentlemen were also unanimously elected Auditors of the Institute :—Edward Cutbush, John Coles, and James Terry.

It was resolved unanimously—"That the best thanks of the meeting be given to the President and Council for their services during the past year."

Mr. Jellicoe said—" Gentlemen, on behalf of the Council I have to express our acknowledgments for your kindness in adopting so readily the resolution which has just been passed ; and on behalf of the Vice-Presidents, the other officers and myself, to thank you for the honour you have done us in this day's election, and to assure you that while we are sensible of the compliment you have paid us, we are not forgetful of the responsibilities which it involves. We have laboured now for some years under the auspices of our late President, whose loss we so much regret. In the language of our laws, we have striven 'to elevate the attainments and to promote the efficiency of those engaged in our pursuits; and we have sought to extend and improve the data and methods of that science which has its origin in the application of the doctrine of probabilities to the affairs of life.' That our efforts have not been wholly unavailing I think any impartial person must acknowledge, who regards for a moment the number and variety of the subjects discussed in our *Journal*, and who can appreciate the clear light which has there been thrown upon many questions heretofore hidden in the deepest obscurity. The mathematical theories applicable to our science have been developed and extended, and the ideas of almost every writer upon them collected and examined. The manifold processes required in our daily practice have been in like manner reviewed—improvements introduced into them where improvement was needed, and new methods substituted where the old ones were found to be radically defective—and thus have been brought about a precision and uniformity of practice which, it must be admitted, were a few years ago altogether unknown. I will not dwell upon the advantages to be derived from the bringing into one focus such a mass of information on the subject of assurance generally, both at home and abroad, and from the collection of so much valuable and original data having reference to it, nor upon the utility of the discussions which serve to throw no little light upon many questions of public interest and importance; neither will I now detain the meeting by any lengthened allusion to the very satisfactory progress which the educational section of the Institute has exhibited, and to the wholesome and purifying influence which it is now exercising, and which it evidently promises to exercise in time to come—an influence more needed, as it would seem, in our pursuit than in almost any other. Suffice it to say, that the general result of our efforts is such as fully to justify them, and to encourage us to persevere with increased energy in the course upon which we have entered. I need hardly say there is much to be done : we have to inquire into the desirableness of adopting the modification so ingeniously suggested by Mr. Farren, and so ably illustrated by Mr. Younger, of what we may, perhaps, without impropriety, refer to as our peculiar ' calculus,' and to determine whether the advantages accruing are sufficient to justify an alteration, which, while it undoubtedly calls into exercise a much more refined method of determining our averages, will still leave them to be dealt with as are now those found in the more ordinary way, and which will apparently render it necessary to recalculate almost the whole of the immense collection of our tabulated values. We have also to ascertain whether any material change will appear in our mortality tables after they shall embody the results of a more prolonged experience, or whether they will remain unaffected by the events which have yet to be registered. It is surmised, as I need hardly mention, that the rates which they now exhibit may be more or less modified by the character of the withdrawals occurring in later years. It is, too, becoming more and more important to ascertain whether the rate of human mortality in the United States and elsewhere, amongst persons assured, corresponds with that found to obtain in this country : a vast amount of risk is being incurred on the assumption that it does, but there exists at present little or no data to warrant such an assumption. I am glad, however, to be able to state that an investigation, with a view to the elucidation of this matter, is now going forward in New York, under the able superintendence of gentlemen connected with the native Companies in the States. Some difference of opinion still exists as to the most accurate mode of dividing the surplus amongst the persons assured in a Life Assurance Company. It is very desirable that any doubts on this subject should be set at rest, and that, as far as possible, an uniformity of practice should obtain in regard to it. Amongst the purely statistical investigations which we have already touched upon are those connected with the questions of the decimal system of weights, measures, and coinage, and direct taxation. The arguments in relation to the former seem to be nearly exhausted; nevertheless, the somewhat unreasonable opposition made to the proposed reform calls for further efforts, and it is to be hoped that additional advocates, or at least expounders, will not be wanting amongst the members of our Society. As regards the question of taxation, it is very desirable that further investigations should be made, to determine whether it be true or not that trade is paralysed by the imposition of such duties as those of the customs and excise, and whether it be true or not that a vast means of employment for the population is thereby destroyed, and a needless poverty created and enforced. On the other hand, there remains abundant room for discussing what is the true measure of liability in a system of direct taxation, and for illustrating, with the aid of statistical records, the merits or demerits by which such a system may be found to be characterised. Such are a few of the subjects which may hereafter profitably engage our attention, and I trust I may be allowed to express the hope that the younger members of the Institute will assist in the investigation of them. I would ask them to bear in mind that unless the habits which these labours require are formed in early life, they become irksome and distasteful in later years ; that if they desire to exercise their best faculties then, it is needful that they exert some energy now ; and that the Institute must of necessity look to them at no very distant day to maintain and improve whatever reputation or distinction it may then have acquired from the persevering labours of their predecessors."

The meeting then adjourned.

INSTITUTE OF ACTUARIES.

Abstract of Receipts and Payments for the Year ending 31st March, 1860.

Dr.

RECEIPTS.

			£	s.	d.
March 31, 1859.					
To balance brought forward			64	10	8
Special subscriptions, viz.—25 at £3 3 0 ..... £78 15 0					
1 " 1 1 0 ..... 1 1 0					
1 " 1 0 0 ..... 1 0 0			80	16	0
Subscriptions for 1858-9 (arrears)			5	5	0
Subscriptions due for 1859-60—					
46 Fellows.........at £3 3 0	£107	2	0		
12 Country " 2 2 0	25	4	0		
20 Official Associates..18 Town.. " 3 3 0	56	14	0		
2 Country " 2 2 0	4	4	0		
82 Associates........64 Town.. " 2 2 0	134	8	0		
18 Country " 1 1 0	18	18	0		
			346	10	0	
Less Subscriptions of Members not paid, viz.—					
4 Associates, Town ...at £2 2 0.. £8 8 0					
2 " Country " 1 1 0.. 2 2 0		10	10	0	
			336	0	0
Guarantee Society			90	8	1
Examination Fee			5	5	0
Dividends on Messenger Legacy			5	4	0
Sundries			25	3	10
			£612	12	7
March 31, 1860. To Balance brought down			£220	9	11

Cr.

PAYMENTS.

	£	s.	d.	£	s.	d.
1859.						
Rent	75	0	0			
Salaries	98	8	7			
Journal (5 quarters)	119	3	0			
Library	8	6	10			
Stationery and Printing ...	16	19	0			
Postage and Receipt Stamps	6	15	0			
Firing and Lighting	12	17	7			
the Meetings..............	14	0	8			
Advertising Examinations (2 years)	4	18	0			
Miscellaneous	18	1	4			
Total of general				374	10	0
Purchase of £18. 12s. 3d. 3 per cent. Consols (Dividends from Messenger Legacy)				17	12	8
March 31, 1860.						
Balance carried forward....				220	9	11
				£612	12	7

Note.—The Assets of the Institute, on the 31st March, 1860, consisted of £198. 16s. 2d., 3 per cent. Consols, say 179 10 1

Cash.................. 220 9 11

Books in Library, say.............. 300 0 0

Total

Examined and approved :—

12, St. James's Square, London,
18th *May*, 1860.

C. Child,
Edward Cutbush, } *Auditors.*
Chas. Watkins,

THE

ASSURANCE MAGAZINE,

AND

JOURNAL

OF THE

INSTITUTE OF ACTUARIES.

On the Construction of Life Tables, illustrated by a new Life Table of the Healthy Districts of England. By W. FARR, ESQ., M.D., F.R.S.[*]

THE *Transactions of the Royal Society* contain the first Life Table. It was constructed by Halley, who discovered its remarkable properties, and illustrated some of its applications. The Breslau observations did not supply Halley with the data to frame an accurate table, for reasons which will be immediately apparent; but the conception is full of ingenuity, and the form is one of the great inventions which adorn the annals of the Royal Society.

Tables have since been made correctly representing the vitality of certain classes of the population; and the form has been extended so as to facilitate the solution of various questions.

In deducing the English Life Tables from the national returns, I have had occasion to try various methods of construction; and I now propose to describe briefly the nature of the Life Table, to lay down a simple method of construction, to describe an extension of its form, and to illustrate this by a new table representing the vitality of the healthiest part of the population of England.

The Life Table is an instrument of investigation; it may be called a *biometer*, for it gives the exact measure of the duration of life under given circumstances. Such a table has to be constructed for each district and for each profession, to determine their degrees of salubrity. To multiply these constructions, then, it is necessary

[*] Extracted, with permission, from the *Philosophical Transactions*, 1860.

to lay down rules, which, while they involve a minimum amount of arithmetical labour, will yield results as correct as can be obtained in the present state of our observations.

I. GENERAL DESCRIPTION OF A LIFE TABLE. (*See* Table C.)*

A Life Table represents a *generation of men* passing through *time ;* and time under this aspect, dating from birth, is called age. In the first column of a Life Table, *age* is expressed in *years,* commencing at 0 (birth), and proceeding to 100 or 110 years, the extreme limit of observed lifetime.

If we could trace a given number of children, say 100,000, from the date of birth, and write the numbers down that die in the first year, living therefore less than one year, against 0 in the Table, and on succeeding lines the numbers that die in the second, third, and every subsequent year of age until the whole generation had passed away, these numbers would form a *Table of Mortality,* showing at what ages 100,000 lives become extinct.

Again, if the 100,000 children were followed, and the numbers living on the first, on the second, and on every subsequent birthday until none was left, the column of numbers would constitute a *Table of Survivorship.* So if, of 100,000 children born at a given point of time, the numbers dying (d_x) in each subsequent year were written in one column, and the numbers surviving (l_x) at the end of each year in another column, the two primary columns of the Life Table would be formed.

It is evident that if one of these columns is known, the other may be immediately deduced from it; for if, of 100,000 children born, 10,295 die in the first year of age, 3,005 in the second year of age, it follows that the numbers living at the end of one year must be 89,705, at the end of two years 86,700. Upon adding the column (d_x) from the bottom up to the number against any age (x), the sum will represent the whole of the numbers *dying after that age;* and, consequently, the numbers *living at that age,* as shown in the collateral column (l_x).

The 100,000 children born at the same moment, and counted *annually* to determine the numbers *living* at *the end of every year,* would, by our Table, completely pass away in less than 107 years. If another generation of 100,000, born a year afterwards, were followed, the numbers dying in the various years of age would not be very different, the circumstances remaining the same; and the numbers of those entering each year of age would vary inconsider-

* The tables and plates referred to in this and the following pages will be found, with the concluding portion of the paper, in the next number.

ably from those of the first series.   If 100,000 children again were born at annual intervals, and were subject to an invariable law of mortality, they would form a community of which the numbers living at each age would be represented by the successive numbers (l_x) in the Life Table.   The sum of these numbers, by the new Table of Healthy Districts, would be 4,951,908.   The births are here assumed to take place simultaneously, at annual intervals; immediately before the births, therefore, in such a community, its population would be 4,851,908, to which it would fall progressively from 4,951,908 by 100,000 successive deaths in the year.   The average number constantly living would be some number between 4,951,908 and 4,851,908; and it would be very nearly the mean of these limiting numbers.

In the ordinary course of nature, the births in a community take place in remittent succession; and if it is assumed that the 100,000 births occur at equal intervals over every year, it is evident that, at any given date, a certain number will be found living at all the intermediate points of age between 0 to 1 year, 1 to 2, 2 to 3, and all the remaining years of age.   The population in the above instance would be found, by enumeration, to be nearly 4,899,665.

The annual *births* would be 100,000 in such a community. The annual deaths would also be 100,000; and by taking out the deaths at each year of age, from the parish registers of a single year, the second column (d_x) of the Life Table would be found. By adding this column of deaths up, and entering the sum of the numbers year by year against every year of age (x), the third column (l_x) of the Life Table would be obtained; for it has been already shown that the numbers attaining any age x are equal to the numbers dying at that age and all the subsequent ages.   From the registers of the deaths, a table of the numbers of the *population living* in a parish *so constituted* could be immediately determined without any enumeration.   Its deviations from the truth would be accidental; and they would be set right by taking the mean of many years.   So, also, from a simultaneous enumeration of the *numbers living in each year of age,* the two columns d_x and l_x of the life could be constructed without reference to any registry of the deaths at different ages.

The *mean age at death* in such a community would express the mean lifetime, or the expectation of life at birth; and the product of the number expressing the annual births multiplied into the mean age at death would give the numbers of the population.

The facts which a Life Table expresses in numbers may be

represented by the lines of a figure; age (x) being indicated by the abscissas measured from 0, the *numbers living* (l) at each age by the ordinates of a curve line, and the numbers living between any two ages by the plane surface within the two ordinates, the curve line, and the corresponding portion of the abscissa. The relative numbers living at the ages 20 and 21 are seen in the two lines of Plate XLII., fig. 1, over the ages 20 and 21; if the deaths in the intervening year all occurred immediately after the age 20 was attained, the numbers living would also be represented by the parallelogram having its two sides equal to the ordinate over 21, and for its base the portion of the abscissa between 20 and 21; but if all the deaths occurred only the instant before the age 21 was attained, the height of the parallelogram would be represented by the ordinate over the age of 20. The deaths occur at intervals between the two ages, so the numbers living, and the *lifetime* which is passed between the two ages, are correctly represented by the curvilinear area.

The deaths in each year of age are called the *decrements of life.* They are represented by the differences in the lengths of the successive ordinates. Thus, by cutting off a small portion of the ordinate at the age 20, the ordinate at the age 21 is obtained; this small portion, shown in Plate XLII., represents the decrement of life in that year of age. It will be observed that the decrements vary at every year of age; and this is more evident when they are exhibited on the larger scale of Plate XLII., fig. 2. The decrement in the first year is large; in the first five years the decrements of life are considerable; at the age of 10 to 15 they fall to their minimum, slowly increase to the age of 56, increase more rapidly until the maximum is attained at the age of 75, then decline gradually to 85, and, after that, more rapidly, until every life is extinct at the age 107 by this table.

II. PRINCIPLES OF CONSTRUCTION—THE FUNDAMENTAL COLUMN l_x.

The conditions of the hypothesis upon which the preceding reasoning rests are never precisely realised in nature—in the first place, the number of births fluctuates, increases, or decreases, from year to year; and the deaths fluctuate still more, rarely equalling the births in number. Immigration and emigration interfere. Under these circumstances, tables such as those which Halley, Price, and others, made from the observations on the *deaths alone* are never accurate, and require correction to give approximate

results. If it be assumed that the law of mortality remains invariable, and that migration does not interfere, then the nature of
the correction to be applied to a table framed from the deaths
alone will become immediately apparent by an example. The
births increase in England. Let the annual births in a portion
of the community be doubled in sixty years—thus, be 50,000 in
1796, and 100,000 in 1856; then the deaths of persons of the
age of 60 in 1856 must be doubled to obtain the deaths which
would have happened at that age if the annual births sixty years
before these deaths had been 100,000. If the births have been
accurately registered, formulæ for correcting the ordinary table
drawn up from the deaths at different ages will be suggested by
the above considerations.

I now proceed to describe another method which has been
adopted in framing the Table C, and is applicable wherever (1) the
number of annual births, (2) the numbers of the population living
at definite periods of age, (3) the deaths at the corresponding ages
during a certain number of years, in any community, are ascertained by observation. This method is not open to the previous
objections.

The aim is to obtain equations which will describe the curve
lines (Plate XLII., fig. 1) of the Life Table in the most direct
way; and these equations may be deduced from the determined
rate of mortality at certain intervals of age.

The relative numbers living at two ages, 20 and 21, can evidently be found from an equation which expresses the relation of
the average numbers living and dying between those ages during a
given time. This can be determined very nearly; for, although
the ages of the living are not ascertained with exact precision at
the census, still, by taking all the numbers living at the ages 15,
16, 17 years, up to 24 and under 25, together, the aggregate
represents very nearly the numbers living in that decenniad of
life. The deaths at the same ages are obtained with at least equal
accuracy from the registers of deaths. By this process, and by
extending the observations over five or more years, a number of
facts is obtained sufficiently great to yield average results; and it
may be assumed that the ratio of the living at the ages 15–25* to
the dying in a year at the same ages, 15–25, represents the annual
rate of mortality at the exact age 20. So also the mortality rate
at the ages 30, 40, 50, and other ages, may be determined. As
observations grow more exact, and the facts are multiplied, the.

* By this, 15 and *under* 25 years of age is understood, and so in all similar cases.

intervals of age may be diminished to 5 years, and ultimately to 1 year.

In determining the *rate* of *mortality*, a given number of persons living a year is considered equivalent to twice that number living half a year, or to half the number living two years.

Thus, if nd represent the deaths in n years out of a number amounting, on an average, to P during the same years, then $\frac{nd}{nP} = m =$ the rate of mortality, or the proportions of death in a *year* (always taken as the unit of time) out of *one year* of *lifetime*. It is found, from all the observations hitherto made on a large scale, that the rate of mortality varies at every interval of age; but at the same age it may, for the present purpose, be considered invariable under similar circumstances.

m_x therefore varies in every moment of age; but I have employed it to express the mean annual rate of mortality during the year following the year of age x, $\therefore \frac{d_x}{P_x} = m_x$, where d_x indicates the deaths, P_x the year of lifetime, after the year of age x. The m_x is the expression of the force of the causes that induce death, of the death-force, *vis mortalis*; and its reciprocal $\frac{1}{m_x} = u_x$ measures the forces that sustain life, the *vis vitalis*.

The vital force, under natural circumstances, may, by one hypothesis, be sufficient to sustain a whole generation alive for seventy or eighty years, and then suddenly collapse. The Life Table, if this hypothesis were true, would be represented by the *parallelogram* in which the curve of the Life Table is inscribed (Plate XLII., fig. 1).

By the hypothesis of Demoivre,[*] the rate of mortality is such, that, at the age of 20, 1 in 66 living at the beginning dies before the end of the year, leaving 65, 64, 63, 62, 61, to enter on each year of age until, at the age of 86, all are dead.

Upon this hypothesis the relative numbers living up to the age 86 form an arithmetical progression; and the deaths in the equal times are equal out of the diminishing numbers living. The rate of mortality increases on this hypothesis, as age advances, in the same ratio as $n - \frac{1}{2} : 1$; where n is the difference between the actual age x and 86. It is called the complement of life. The Life Table upon this hypothesis has equal decrements, and might be represented on Plate XLII., fig. 1, by drawing a diagonal line

---

[*] See *Treatise of Annuities on Lives*; Preface to 2nd edition.

through the parallelogram. Its deviation from the true curve on this scale is evident; but it is also evident that a series of straight lines, which would nearly represent the true curve, may be drawn from point to point of all the ordinates.

If the causes of death act with equal intensity at all ages, they may be represented by any simple external cause destroying an equal *proportion* of the numbers living in equal intervals of time. Thus, if 1,600 men were distributed equally over ground where they were exposed to certain dangers, represented by successive discharges of musketry, which, at every discharge, shot down one-half of the numbers remaining, they would be reduced successively from 1,600 to 800, to 400, to 200, to 100, to 50, and so on *ad infinitum*, if a fraction of a living man could be conceived; the numbers living at each year of age in a Life Table would not decrease at *these rates*, but they *would decrease* at a constant rate if the dangers at every stage of life remained *constant* and equally *great*. The numbers of the living at successive ages would be in geometrical progression, and would be represented by the ordinates of the logarithmic curve.

The law of mortality can only be derived from observation, and it is found to be less simple than either of these hypotheses implies. It can, however, be represented nearly by equations at different periods of age. Upon inspecting Table A, it will be seen that, at the age 55–65, which may be represented by the exact age 60, the mortality is such, that 2,162 women die in a year out of a number equal to 100,000 living a year; and the mortality, which is the ratio of the dying to the living in a unit of time, here set down as a year, is, therefore, $m = \cdot02162$. Again: the mortality at the age of 70 is $\cdot04992$, at the age of 80 it is $\cdot11866$, and at the age of 90 it is $\cdot26711$. The mortality increases rapidly, and is more than doubled every ten years. The four numbers differ little from the terms of a geometrical progression, the logarithms of which have a constant difference. Let the rate at which the mortality increases be r, and $r^{10} = 2\cdot3116$, and the first term (m) be $\cdot02177$; then a series of numbers will be formed differing little from those which express the value of m at decennial intervals of age.

Values of m at the precise age x.—*Females.*

Age (x).	60.	70.	80.	90.
By observation .	$\cdot02162$	$\cdot04992$	$\cdot11866$	$\cdot26711$
By hypothesis ,	$\cdot02177$	$\cdot05033$	$\cdot11633$	$\cdot26891$

Note.—It may be assumed that m at 60 is the mean value of m in its range from $m_{59\frac{1}{2}}$ to $m_{60\frac{1}{2}}$; and so in other cases.

The *annual rate* of the increase of m from the age of 55 to 95 is $r = 1\cdot0874$; and if m is the mortality at any age after 55, then $m_z = mr^z =$ the mortality at z years after the age at which m is taken. The common logarithm of r is $= \lambda r = \cdot03639$.

The mortality (m) of males at corresponding ages is higher than the mortality of females; but the rate of increase as age advances is nearly the same.

The value of m for females at the age of 20 is $\cdot00765$, and the mortality increases at the rate of nearly one-seventh part every ten years. The exact value of r is $1\cdot0149$, and $\lambda r = \cdot006423$.

Values of m.—Females.

Age.	20.	30.	40.	50.
By observation	·00765	·00894	·00998	·01192
By hypothesis	·00760	·00882	·01022	·01185

By these observations, in the healthy districts the mortality (m) of men at the ages 15 to 45 is lower than the mortality of women at the same ages; yet, during that period, the rate of increase r is nearly the same for the two sexes. From the age of 40 to 50, and 50 to 60, the mortality of males increases at a rate intermediate between the rates of manhood and mature age.

Limits of Ages.   *Females.*

15 to 55 or 20 to 50	$r = 1\cdot0149$	$\lambda r = \cdot00642$
55 to 95 or 60 to 90	$r = 1\cdot0874$	$\lambda r = \cdot03639$

Males.

15 to 45 or 20 to 40	$r = 1\cdot0148$	$\lambda r = \cdot00640$
55 to 95 or 60 to 90	$r = 1\cdot0874$	$\lambda r = \cdot03640$

The subjoined Table exhibits the series of values for m derived from the hypothesis of two constant rates, and from direct observation. The values of r for females may be evidently applied to males in every period, except in the ten years of age, 40 to 50.

Mortality (m) of males and females, (1) derived from observation, and (2) from the hypothesis that m *increases at the preceding rates.*

Precise Age.	ANNUAL MORTALITY TO 100 CONSTANTLY LIVING AT EACH AGE (m).			
	Males.		Females.	
	By observation.	By hypothesis.	By observation.	By hypothesis.
20	·691	·696	·765	·760
30	·818	·807	·894	·882
40	·928	·935	·998	1·022
50	1·273	1·083	1·192	1·185
60	2·294	2·329	2·162	2·177
70	5·486	5·385	4·992	5·033
80	12·817	12·451	11·866	11·633
90	28·350	28·785	26·711	26·891
100	40·000?	66·550?	45·000?	62·160?

The observations on the numbers living and dying of the age of 95 and upwards are exceedingly uncertain; and it is probable that many of the persons believed to be 100, &c., are really persons five or ten years younger; so that these values of m_x, by the hypothetical method, are probably as correct as the direct numbers.

I shall now notice briefly the application of this hypothesis, first suggested by Mr. Gompertz, and applied by him to the interpolation of the Northampton and other Tables.* Mr. Edmonds, in 1832, extended the "Theory," and applied it to the construction of three Life Tables.† He gave an elegant formula, similar in principle to that of Mr. Gompertz, from which the curve of a Life Table can be deduced, upon the above hypothesis.

In the equation $\frac{s}{t} = v$, where s indicates space, t time, v velocity, the units of measure must be fixed before numbers can be inserted in the general expression; and then v will express, in the measure that has been applied to space, the number of such units of space described in *one* unit of time. Here v is a ratio—it is the rate at which the body moves; and, in the same manner, m, in the equation $\frac{d}{l} = m$, is the *rate of dying*—that is, as I shall express it, the *mortality;* or it is the ratio of the dying to the living in a given unit of time, the time during which the deaths occur being of precisely the same duration as the time during which the living are under observation—

l (living during 1 year) : d (dying during a year) :: 1 (year of life) : m.

If for l the number 100,000 is substituted, it is assumed that immediately a death occurs another life is substituted; and as the time is a year, then 760 will represent the value of d at the age 20, according to the preceding Table; $\therefore m = \cdot 00760$. If the *time,* instead of *one year*, be the *thousandth part* of one year, then $m = \cdot 0000076$; and if the time be infinitely short, m will be infinitely small: m is a ratio; the quantity of life existing during the time is represented by 1, and the quantity of life destroyed by a fraction, m. Whether the life inheres in the first organic molecule after conception, in the infant, or in the man, the vital action has a certain force of continuance, which is constantly varying; and the amount of this *force* that is *extinguished* at a given instant of time will be represented by the force of mortality—namely, by

---

* *Philosophical Transactions*, 1825; paper by B. Gompertz, Esq., F.R.S.
† *Life Tables founded on the Discovery of a Numerical Law regulating the Existence of every Human Being, &c.* By T. R. Edmonds, B.A., 1832.

m at that instant.   Then let the age $x=z+a$, where a represents the number of years up to the age at which a given rate (r) of increase of m begins, then $z=x-a$; and the mortality at any instant of age, in an instant of time at the end of z years or parts of years, will be mr^z.   Now, let y represent the living at that precise age, then the decrement of y in an infinitely short time will be $-dy=ymr^zdx$; the dy being negative, as it is taken in a direction opposite to that in which the ordinate y of the curve is assumed to be drawn.   Transferring y to the other side of the equation, this becomes $-\dfrac{dy}{y}=mr^zdz$; and integrating both sides, we have ($\lambda_\epsilon y$ being put for the hyperbolic logarithm of y, and $\lambda_\epsilon c$ for the difference between the constants of the two integrals)—

$$\lambda_\epsilon c-\lambda_\epsilon y=\lambda_\epsilon \frac{c}{y}=\frac{mr^z}{\lambda_\epsilon r}; \quad \cdots \cdots \quad (1)$$

$$\therefore \ \lambda_\epsilon y=y_\epsilon c-\frac{mr^z}{\lambda_\epsilon r}, \quad \cdots \cdots \cdots \quad (2)$$

and

$$\lambda_\epsilon c=\lambda_\epsilon y+\frac{mr^z}{\lambda_\epsilon r}. \quad \cdots \cdots \cdots \quad (3)$$

When z is made zero, let $y=1$; then $\lambda_\epsilon y$ will also disappear, and $\lambda_\epsilon c=\dfrac{m}{\lambda_\epsilon r}$.   Upon substituting this value of $\lambda_\epsilon c$ in equation (2), it becomes—

$$\lambda_\epsilon y=\frac{m}{\lambda_\epsilon r}-\frac{mr^z}{\lambda_\epsilon r}=\frac{m}{\lambda_\epsilon r}(1-r^z). \quad \cdots \quad (4)$$

Upon passing to the numbers, equation (4) becomes

$$y=\epsilon^{\frac{m}{\lambda_\epsilon r}(1-r^z)}$$

$=$ the value of y (taken as 1 at the origin) at the end of z years.

Let λ denote the common logarithm with the base 10, then $\lambda_\epsilon y=\dfrac{\lambda y}{k}$, where k is the modulus of the common system of logarithms; as also

$$\lambda_\epsilon c=\frac{km}{\lambda r}, \quad \text{and} \quad \frac{mr^z}{\lambda_\epsilon r}=\frac{kmr^z}{\lambda r}.$$

Equation (2) becomes, after the required substitutions,

$$\frac{\lambda y}{k}=\frac{km}{\lambda r}-\frac{kmr^z}{\lambda r}$$

and

$$\lambda y=\frac{k^2m}{\lambda r}(1-r^z); \quad \cdots \cdots \quad (5)$$

so the equation becomes finally

$$y = 10^{\frac{k^2 m}{\lambda r}(1 - r^z)} \quad \ldots \ldots \ldots \quad (6)$$

This is the form given by Mr. Edmonds, and is convenient for use.

By making z successively 1, 2, 3, ... up to any number less than the number of years of age within which r remains constant, the number l_x being known, the number living at any other age within that range will be obtained by multiplying l_x by the corresponding value of y. Thus, if y_{10} is the value of y when $z = 10$ in equation (6), then, putting l_{20} for the numbers living at the age 20, the living at the age 30 will be $y_{10} \times l_{20} = l_{30}$.

This hypothesis does not express the facts deduced from the observations exactly. If m_z could be expressed exactly over more than 20 years by $m_z = m_0 r^z$, the first differences (δ^1) of the logarithms in the series following would, in a certain number of cases, be equal.

Females in Healthy Districts of England.

Precise Age.	Annual Rate of Mortality.	Logarithms of the Annual Mortality.	First Decennial Differences of λm_x.	Second Decennial Differences of λm_x.
x.	m.*	λm.	δ^1.	δ^2.
20	·00765	$\bar{3}$·8835	·0677	$-$·0197
30	·00894	$\bar{3}$·9512	·0480	·0290
40	·00998	$\bar{3}$·9992	·0770	·1817
50	·01192	$\bar{2}$·0762	·2587	·1047
60	·02162	$\bar{2}$·3349	·3634	·0126
70	·04992	$\bar{2}$·6983	·3760	$-$·0236
80	·11866	$\bar{1}$·0743	·3524	$-$·1259
90	·26711	$\bar{1}$·4267	·2265	
100	·45000	$\bar{1}$·6532		

The inequalities in the second differences vary in every separate class of observations, but there is generally a tendency in the first and in the second differences to increase, over a certain extent of the series. The error of the hypothesis is slight if the rate of increase (r), of which λ·00677 is the logarithm in the case in hand, is only assumed to remain uniform for the ten years 20 to 30, or for the one year 20 to 21. Now, let the number living at the age

* Here, at the age 20, m is the mean mortality that rules over the age 19¼ to 20¼ years of exact time.

20 be represented by l_{20}, and the number living at the age 21 by l_{21}; then put $\dfrac{l_{21}}{l_{20}} = p_{20}$. Here it is evident that if l_{20} and p_{20} be known, l_{21} is determined immediately by the equation $l_{21} = l_{20} \times p_{20}$. But p_{20} is the value of y in the equation $y_1 = 10^{\frac{k^2 m}{\lambda r}(1 - r z)}$, when z is put $= 1$. Taking the numbers from Table A, we have $m = {\cdot}00765$ at the precise age $20 = (19\frac{1}{2} + 20\frac{1}{2})\frac{1}{2}$; and $\lambda m = \bar{3}{\cdot}8835130$, $\lambda r = {\cdot}0067728$, and $\therefore r = 1{\cdot}015717$; k is put for the modulus of the common logarithms, $\therefore \lambda k^2 = \bar{1}{\cdot}2755686$; $k(\lambda r)$ is the complement of the *logarithm* of (λr).

λk^2	$\bar{1}{\cdot}2755686$
λm	$\bar{3}{\cdot}8835130$
$k(\lambda r)$	$2{\cdot}1692317$
$\lambda(1 - r)$	$\bar{2}{\cdot}1963697$
$-{\cdot}0033472$	$\bar{3}{\cdot}5246830$
$\bar{1}{\cdot}9966528$	

As the factor $(1 - r)$ is negative, it makes the exponent of 10 negative; and, upon taking the complement of this, the logarithm of y is found to be $\bar{1}{\cdot}9966528$. This is also the logarithm of $p_{20} = {\cdot}99232$, and it enables us to pass, in the construction of a Life Table, from the living at the age of 20 to the living at 21. If we obtain the several values p_x at every year of age, the whole of the Life Table can be constructed.

It will be found that p_x is always a fraction, and it does not differ very much from $1 - m_x$; but while m_x* shows the *deaths* in a year out of a *unit* of *life* (which may consist of any *number* of individual *lives* constantly kept up), p_x shows how much out of a *unit of the same life* at the beginning of a year, the dead not being replaced, *survives a year* after the age x; and $1 - p_x$ is the amount of loss which occurs in the same year out of a unit of life at its commencement. Thus, as $p_{20} = {\cdot}99232$, it follows that $1 - p_{20} = {\cdot}00768$. In the same year of age, 20 to 21, the mortality is $m_{20} = {\cdot}00771$, or ${\cdot}00003$ more than $(1 - p_{20})$. If the unit of life is made 100,000 living at the age 20, then 99232 will survive, and 768 will die in the ensuing year of age. But if it is assumed that the deaths take place at equal intervals, it may also be assumed that the number of lives (100,000) being constantly sustained, the accession of 768 new lives takes place at equal intervals, conse-

---

* m serves to indicate the mean mortality in the year following the exact age x.

quently that they are under observation half a year on an average,

giving the equivalent of $\dfrac{768}{2} = 384$ years of lifetime at the age 20

to 21. Now, out of this number (384), at that age *three* die when the mortality is m_{20}. This accounts for the difference of ·00768 and ·00771, the former occurring in a year out of a unit of life of which the waste is not replaced.

From these considerations, it may be inferred that, if m_x is known, p_x may be deduced from it, upon the hypothesis of equal

decrements through the year, by the formula $p_x = \dfrac{1-\frac{1}{2}m_x}{1+\frac{1}{2}m_x} = \dfrac{2-m_x}{2+m_x}.$

Thus, m_{20} being ·0077072, we have $\dfrac{·9961464}{1·0038536} = ·99232,$* as be-

fore. The λp_{20}, by the previous method, is $\bar{1}·9966528$, and by this method it is the same. By either of the methods the value of p_x may be deduced for the subsequent ages, and p_{20}, p_{30}, $p_{40} \ldots \ldots p_{90}$, p_{100} will be obtained. These values are here given, and it will be seen that the results by the two methods are nearly identical at all ages, except the last two, when the observations themselves become less exact.

Females.

Age (x).	$\lambda p_x = \lambda y_1 = 10^{\frac{\kappa^2 m(1-r)}{\lambda r}}$.	$\lambda p_x = \lambda\left(\dfrac{1-\frac{1}{2}m}{1+\frac{1}{2}m}\right).$
20	$\bar{1}·9966528$	$\bar{1}·9966527$
30	·9960967	·9960967
40	·9956263	·9956264
50	·9946669	·9946676
60	·9902049	·9902073
70	·9773538	·9773557
80	·9463182	·9462643
90	·8809176	·8801776

It will be observed that the fraction $p = \dfrac{1-\frac{1}{2}m}{1+\frac{1}{2}m}$ approximates

to $1-m$ as m becomes less; for, upon developing it into a series, $p = 1 - m + \frac{1}{2}m^2 - \frac{1}{4}m^3 + \frac{1}{8}m^4 \ldots$. And taking m infinitely small, the terms after the first two may be neglected.

---

* m at the precise age 20 is nearly ·00765. The increase in this mortality from the age 20 to 20$\frac{1}{2}$, the middle of the year of age 20 to 21, is obtained by adding $\frac{1}{2}\lambda r$, as above given, to $\lambda m_{19\frac{1}{2}}$, that is, to the log. of $(m_{19\frac{1}{2}} + m_{20\frac{1}{2}})\frac{1}{2}$; ∴ $m_{20} = ·0077072$

$$\lambda m_{19\frac{1}{2}} \quad 3·8835130$$
$$\tfrac{1}{2}\lambda r \quad 0·0033864$$

$$\lambda m_{20} \quad \overline{3·8868994}$$

The values of m_0, m_1 .... m_5 may be obtained by the method already described; but it rarely happens that the population living at each year of age is accurately enumerated at the census; and, besides inaccuracies of statement, the numbers living at each of the early years of age fluctuate considerably, so that the numbers of children living of each year of age in 1851 do not represent the average numbers living of those ages in the five years 1849 to 1853 for instance.

The following method is less exceptionable. It may be assumed *for this purpose*, (1) that the births registered in the year 1848 represent the births in that year; (2) that the births are equally distributed over the years in which they occur; and, consequently, (3) that the *mean date* of all *the births* in the two years 1848, 1849, was immediately before January 1, 1849. The *half* of the births in those two years will consequently represent pretty accurately the number of births out of which the deaths of children *under one year* of age happened in the year 1849; and the deaths and survivors can be followed by this method year by year, as is evident in the annexed scheme:—

Age.

$$0 \begin{cases} \tfrac{1}{2}(\text{births } 1848, 1849) = \text{mean annual births of which the mean date} \\ \qquad\qquad\qquad\qquad\qquad\text{is January 1, 1849.} \\ minus \text{ deaths under age 1 in 1849} \end{cases}$$

1   =surviving on January 1, 1850.
   minus deaths age (1 to 2) in 1850
2   =surviving on January 1, 1851.
   minus deaths age (2 to 3) in 1851
3   =surviving on January 1, 1852.
   minus deaths age (3 to 4) in 1852
4   =surviving on January 1, 1853.
   minus deaths age (4 to 5) in 1853
5   =surviving on January 1, 1854.

By commencing with the mean number of births in the years 1849, 1850, and deducting the deaths, a similar series may be obtained; and thus a succession of similar series may be deduced, the mean of which will supply the ordinary series l_0, l_1, l_2, l_3, l_4, l_5 of a Life Table.

These series are liable to various disturbances. If all the births are not registered, the *rate* of mortality is overstated; if all the deaths are not registered, or if the children are carried off as emigrants, the decrements of life are understated. The annual number of births fluctuates, and now increases, in England; they are in excess also in the early months of the year. Several of the disturbances are slight, and some of them are in opposite directions.

The results can also be, and have been, checked by the results of the other method. The values of m_7 and m_{12} are deduced by dividing the annual deaths at the ages 5 to 10 and 10 to 15 by the mean population at those ages. The interpolation of the series λp_x from λp_3 to λp_{20} succeeds, taking λp_3, λp_7, λp_{12}, and λp_{20}, as the fixed points of the series, and λp_{12} being adjusted to allow for the turn of the curve.

The Tables A, B, and C supply the data from which the Life Table of Healthy English Districts was deduced. One or two arithmetical examples of the application of the method adopted in the earlier ages are also supplied.

III. INTERPOLATION.

We have therefore determined the values of λp_x at certain ages. The values of λp_x at the intervening ages may be determined by changing the value of r and making z successively $1, 2 \ldots . 10$ in the formula (p. 131). They may also be interpolated for every year of age by the method of finite differences; and, upon the whole, this method is preferable to any other. The logarithms of p_x are required, and to them it will be convenient to apply the interpolation directly. Any number of differences beyond four becomes cumbersome, and it will be therefore sufficient to give the general formula, which can be employed in deriving the first of either four or three orders of differences.

Investigation of Formulæ—Intervals equal.

Let any numbers of a series be so related that u_n, the nth from the first, u_0, is determined by the equation (1)—

$$u_n = u_0 \; \begin{matrix} + \\ - \end{matrix} \frac{n}{1} \delta^1 \; \begin{matrix} + \\ + \end{matrix} \frac{n(n-1)}{1.2} \delta^2 \; \begin{matrix} + \\ - \end{matrix} \frac{n(n-1)(n-2)}{1.2.3} \delta^3 \; \begin{matrix} + \\ + \end{matrix} \frac{n(n-1)(n-2)(n-3)}{1.2.3.4} \delta^4. \quad . \quad (1)$$

δ^1, δ^2, δ^3, and δ^4,* the first differences of the four orders, are unknown; they can all be determined from any five values of u_n. Now, let n be successively $1x, 2x, 3x, 4x$; then the coefficients of u_0, u_{1x}, u_{2x}, u_{3x}, u_{4x}, can be found to give the values of δ^1, δ^2, δ^3, and δ^4, in four equations. But when x is ten or more, the coefficients become large and the numerical calculation laborious; it is, therefore, well to obtain the numerical values of δ^4, δ^3, δ^2, δ^1, in succession. Thus, if the series is ascending or descending, the following are convenient forms—the upper rows of signs are used in the *ascending*, the lower rows in the *descending*, series :—

* It will be borne in mind that these imply first differences, or $\delta^1 u_0$, $\delta^2 u_0$, $\delta^3 u_0$, $\delta^4 u_0$.

$$\delta^4 = \frac{\overset{+}{+}\, u_{4x} \overset{-}{-}\, 4u_{3x} \overset{+}{+}\, 6u_{2x} \overset{-}{-}\, 4u_x \overset{+}{+}\, u_0}{x^4}. \qquad \dots \qquad (2)$$

$$\delta^3 = \frac{\overset{+}{-}\, u_{3x} \overset{-}{+}\, 3u_{2x} \overset{+}{-}\, 3u_x \overset{-}{+}\, u_0}{x^3} \overset{-}{+}\, \tfrac{3}{2}(x-1)\delta^4. \qquad \dots \qquad (3)$$

$$\delta^2 = \frac{\overset{+}{+}\, u_{2x} \overset{-}{-}\, 2u_x \overset{+}{+}\, u_0}{x^2} \overset{-}{+}\, (x-1)\delta^3 \overset{-}{-}\, \frac{(7x^2-18x+11)}{12}\delta^4. \qquad \dots \qquad (4)$$

$$\delta^1 = \frac{\overset{+}{-}\, u_x \overset{-}{+}\, u_0}{x} \overset{-}{+}\, \frac{x-1}{2}\delta^2 \overset{-}{-}\, \frac{(x^2-3x+2)}{6}\delta^3 \overset{-}{+}\, \frac{(x^3-6x^2+11x-6)}{24}\delta^4. \qquad (5)$$

It is necessary to be careful in deducing the successive values of δ from the values preceding; and, before commencing their use, their accuracy should be tested by inserting them in the checking equation—

$$u_{4x} = u_0 \overset{+}{-}\, \frac{4x}{1}\delta^1 \overset{+}{+}\, \frac{4x(4x-1)}{1.2}\delta^2 \overset{+}{-}\, \frac{4x(4x-1)(4x-2)}{1.2.3}\delta^3$$
$$\overset{+}{+}\, \frac{4x(4x-1)(4x-2)(4x-3)}{1.2.3.4}\delta^4. \qquad \dots \qquad (6)$$

x may be any number. If only four terms are given, δ^3 is assumed to be constant; and δ^4 being 0, all the terms into which it enters disappear. The above formulæ, if this is borne in mind, are applicable when δ^4, δ^3, or δ^2, are assumed to be constant, and serve, therefore, to supply the differences, when there are one, two, three, or four orders, by the most expeditious method.

In constructing the Life Table, x was made 10 from the age of 20, and on inserting the numbers, the equations (2, 3, 4, 5, 6) became

$$\delta^4 = \frac{\overset{+}{+}\, u_{40} \overset{-}{-}\, 4u_{30} \overset{+}{+}\, 6u_{20} \overset{-}{-}\, 4u_{10} \overset{+}{+}\, u_0}{10,000} \qquad \dots \qquad (7)$$

$$\delta^3 = \frac{\overset{+}{-}\, u_{30} \overset{-}{+}\, 3u_{20} \overset{+}{-}\, 3u_{10} \overset{-}{+}\, u_0}{1,000} \overset{-}{+}\, 13\tfrac{1}{2}\delta^4. \qquad \dots \qquad (8)$$

$$\delta^2 = \frac{\overset{+}{+}\, u_{20} \overset{-}{-}\, 2u_{10} \overset{+}{+}\, u_0}{100} \overset{-}{+}\, 9\delta^3 \overset{-}{-}\, 44\tfrac{1}{4}\delta^4. \qquad \dots \qquad (9)$$

$$\delta^1 = \frac{\overset{+}{-}\, u_{10} \overset{-}{+}\, u_0}{10} \overset{-}{+}\, 4\tfrac{1}{2}\delta^2 \overset{-}{-}\, 12\delta^3 \overset{-}{+}\, 21\delta^4. \qquad \dots \qquad (10)$$

The checking equation is—

$$u_{40} = {}^+_+ u_0 {}^+_- 40\delta^1 {}^+_+ 780\delta^2 {}^+_- 9880\delta^3 {}^+_+ 91390\delta^4. \quad . \quad . \quad (11)$$

If three orders of differences are used, the checking equation is—

$$u_{30} = {}^+_+ u_0 {}^+_- 30\delta^1 {}^+_+ 435\delta^2 {}^+_- 4060\delta^3. \quad . \quad . \quad . \quad . \quad . \quad (12)$$

After adding or subtracting any constant to or from a series of numbers, the differences remain the same; and if consecutive terms are multiplied or divided by the same factor, the differences are multiplied or divided by that factor. Thus $(b+a)-(c+a)=b-c$, and $ab-ac=a(b-c)$. Advantage is taken of these properties to reduce any one of the terms in the equations to *zero*.

Thus, let the logarithms to be interpolated be the following— values of p_{20}, p_{30}, p_{40}, and p_{50}, taken from the column headed *males*, Table B, then they may, among other ways, be interpolated as follows :—

As $\bar{1}\cdot 9969724$ is the contracted expression of $(\cdot 9969724-1)$, we have—

Age.
20 $\bar{1}\cdot 9969724 = -\cdot 0030276$ (1) Multiplying each term by 10,000,000—that is, striking out the decimal point and the two adjoining ciphers — and (2) then subtracting from each 30,276, the values of $u_x = \lambda p_x$ to be operated on become

30 $\bar{1}\cdot 9964260 = -\cdot 0035740$

40 $\bar{1}\cdot 9959051 = -\cdot 0040949$

50 $\bar{1}\cdot 9943048 = -\cdot 0056952$

$u^0 = -00000$

$u_{10} = -5464$

$u_{20} = -10673$

$u_{30} = -26676$

By inserting these values with their negative signs in the equations, and taking the upper signs, the three differences are found— that is,

$$\delta^3 = -11\cdot 049; \quad \delta^2 = 101\cdot 991; \quad \text{and } \delta^1 = -872\cdot 7715.$$

The differences are now divided by 10,000,000—that is, ciphers are added to their left-hand side, so that the above decimal point may be moved seven places in that direction—and the operation may be thus commenced. By adding the differences successively to each other and to $\lambda p_{20} = \bar{1}\cdot 9969724$, the successive values are found of λp_{21}, λp_{22}, λp_{23} . . . . λp_{50}, up to and including λp_{58}, for males, where the series joins naturally the subsequent series, commencing at λp_{59}.

δ^3.	δ^2.	δ^1.	λp_x.
—·000,0011,0490	·000,0101,9910	—·000,0872,7715	$\bar{1}$·996,9724,0000
(constant)	·000,0090,9420	—·000,0770,7805	$\bar{1}$·996,8951,2285
		—·000,0679,8385	$\bar{1}$·996,8180,4480
			$\bar{1}$·996,7500,6095

In the actual operation the δ^3 is *subtracted* from δ^2, δ^2 from δ^1, and δ^1 from λp_x; it is, therefore, convenient to substitute for their present values the complements of δ^3 and δ^1, as thus all the series become additive.

As $\lambda l_{20} + \lambda p_{20} = \lambda l_{21}$, and $\lambda l_{21} + \lambda p_{21} = \lambda l_{22}$, and, generally, $\lambda l_x + \lambda p_x = \lambda l_{x+1}$, it is evident that the λp_x is the *first difference* of the series λl_x; and the whole series, λl_x, from λl_{20} to λl_{58}, may be formed as in the subjoined example, where δ^3 becomes δ^4, δ^2 becomes δ^3, and so on.

Healthy Districts—Males.

δ^4 (constant)

9·999,9988,9510

Age.	δ^3.	δ^2.	$\delta^1 = \lambda \dot{p}_x$.	$u_x = \lambda l_x$.
20	0·000,0101,9910	9·999,9127,2285	9·996,9724,0000	4·584,1951,2769
21	0·000,0090,9420	9·999,9229,2195	9·996,8851,2285	4·581,1675,2769
22	0·000,0079,8930	9·999,9320,1615	9·996,8080,4480	4·578,0526,5054
23			9·996,7400,6095	4·574,8606,9534
24				4·571,6007,5629

Note.—The four last figures in the decimal portion of the series λp_x and in λl_x may, in practice, be omitted.

The corresponding values of λp_x, in the column headed " Females," Table B, are interpolated in the same way; and the λp_{60}, λp_{70}, λp_{80}, and λp_{90}, are interpolated by the same methods, the series being continued backwards to λp_{57} and forwards to λp_{105}; the actual observations of age after the age of 90 furnishing results less liable than those thus obtained, which bring a generation of 100,000 to their last end in 107 years. The successive values of λp_x, in the period from the age of 3 to the age of 19 inclusive, are derived from λp_3, λp_7, λp_{12}, and λp_{20}, which represent u_0, u_4, u_9, and u_{17}. As the terms of the series are here at unequal distances, the first differences cannot be derived from the preceding formulæ. The δ can in this and similar cases be derived from the proper equations by substituting figures for letters. But three literal equations supply formulæ for finding the three first differences from any four terms of series of the kind which have been discussed; u_0, which has a troublesome coefficient, can always be reduced to *zero*, and is, therefore, omitted. The first given term being u_0, let the second, u_x, be the xth from u_0, and u_y be the yth, u_z the zth from u_0; here $x < y < z$; then the following equations give the differences*:—

$$\delta^3 = \frac{6\left\{(y-x)\dfrac{u_z}{z} - (z-x)\dfrac{u_y}{y} + (z-y)\dfrac{u_x}{x}\right\}}{(y-x)\{(z-1)(z-2)-(y-1)(y-2)\}-(z-y)\{(y-1)(y-2)-(x-1)(x-2)\}} . \quad (13$$

$$\delta^2 = \frac{2}{y-x}\left\{\frac{u_y}{y} - \frac{u_x}{x} - \{(y-1)(y-2)-(x-1)(x-2)\}\frac{\delta^3}{6}\right\} \quad \ldots \ldots \quad (14$$

$$\delta^1 = \frac{u_x}{x} - (x-1)\frac{\delta^2}{2} - (x-1)(x-2)\frac{\delta^3}{6} \quad \ldots \ldots \ldots \ldots \quad (15$$

By making $y=2x$, and $z=3x$, these equations assume the same forms as equations (3), (4), (5), with the term δ^4 struck out.

Putting $x=4$, $y=9$, and $z=17$, the three preceding equations become those which were actually used in constructing the series p_3 to p_{19}: u_0 is reduced to zero and is not used.

$$\delta^3 = \frac{45u_{17} - 221u_9 + 306u_4}{13260} \quad \ldots \ldots \ldots \ldots \ldots \quad (16)$$

$$\delta^2 = \frac{4u_9 - 9u_4 - 300\delta^3}{90} \quad \ldots \ldots \ldots \ldots \ldots \quad (17)$$

$$\delta^1 = \frac{u_4 - 6\delta^2 - 4\delta^3}{4} \quad \ldots \ldots \ldots \ldots \ldots \quad (18)$$

Checking equation.

$$u_{17} = u_0 + 17\delta^1 + 136\delta^2 + 680\delta^3 \quad \ldots \ldots \ldots \ldots \quad (19)$$

* *A useful Table in applying the above formulæ.*

x.	$(x-1)(x-2)$.	x.	$(x-1)(x-2)$.	x.	$(x-1)(x-2)$.
20	342	34	1056	47	2070
21	380	35	1122	48	2162
22	420	36	1190	49	2256
23	462	37	1260	50	2352
24	506	38	1332	51	2450
25	552	39	1406	52	2550
26	600	40	1482	53	2652
27	650	41	1560	54	2756
28	702	42	1640	55	2862
29	756	43	1722	56	2970
30	812	44	1806	57	3080
31	870	45	1892	58	3192
32	930	46	1980	59	3306
33	992				

Table of first differences in the *Life Table* of *Healthy Districts* of *England.*

Age x.	$\lambda l_x.$	$\lambda p_x = \delta^1.$	$\delta^2.$	$\delta^3.$	$\delta^4.$
		MALES.			
3	4·631,5849,0000	9·993,2422,0000	0·001,2416,1260,934	9·999,8012,4393,666	0·000,0141,9648,567
20	4·584,1951,2769	9·996,9724,0000	9·999,9127,2285	0·000,0101,9910	9·999,9988,9510
59	4·403,7768,0454	9·990,6137,0980	9·998,9756,9020	9·999,9704,0800	9·999,9843,4520
60	4·394,3905,1434	9·989,5894,0000	9·998,9460,9820	9·999,9547,5320	9·999,9843,4520

Note.—The last series p_x was carried backwards from λp_{60} to λp_{59}.

Age x.	$\lambda l_x.$	$\lambda p_x = \delta^1.$	$\delta^2.$	$\delta^3.$	$\delta^4.$
		FEMALES.			
3	4·623,2586,0000	9·993,2928,0000	0·001,2164,1598,794	9·999,7874,2556,561	0·000,0170,4566,365
20	4·570,6868,3846	9·996,6528,0000	9·999,9241,5455	0·000,0060,2930	9·999,9994,2530
57	4·405,2189,6826	9·992,9332,3725	9·999,0836,2675	0·000,0123,2100	9·999,9838,1950
60	4·381,2818,8126	9·990,2049,0000	9·999,0720,4825	9·999,9637,7950	9·999,9838,1950

Note.—The last series p_x was carried backwards from λp_{60} to λp_{57}.

(20)

A series of the form $v^x l_x + v^{x+1} l_{x+1} + v^{x+2} l_{x+2}$ is required in rendering the Life Table applicable to the solution of questions in annuities and life insurance.

The logarithms of the series are obtained by making the first term of the new series $\lambda(v^x l_x)$, and the first term of the first order of differences $\lambda(v p_x) = \lambda v + \lambda p_x = \delta^1$; the δ^2, δ^3, and δ^4 of the original series remaining unchanged. Taking the interest of money at 3 per cent., $v = \frac{1}{1·03}$; and $\lambda v = \overline{1}·9871627,753$.

The derivation of the new series from this value of λv, and from the above Table (males), is shown in the annexed example. Any value of v^x may be introduced in the same way.

$$\delta^4 = 9·9999988,951$$

Age.	$\delta^3.$	$\delta^2.$	$\lambda(v p_x) = \delta^1.$	$u_0 = \lambda(l_x v^x).$
20	0·0000101,991	9·9999127,2285	9·9841351,7530	4·3274506
	·0000090,942	·9999229,2195	·9840478,9815	·3115858
		·9999320,1615	·9839708,2010	·2956337
			·9839028,3625	·2796045
				·2635074

In describing the first English Life Table, I ventured to express the belief that the chances of life may ultimately be calculated by Mr. Babbage's machine.* Mr. Babbage's conception has been realized in the original and ingeniously-constructed machine of the Messrs. Scheutz, which was favourably reported upon by a com-

* Letter to the Registrar-General, in Appendix (p. 352) to his *Fifth Annual Report*, year 1843.

mittee of the Royal Society. The first differences to be inserted in the machine can be immediately deduced from those given above; and we may hope ere long to see the logarithms of Life Tables, for single and for joint lives, printed from types cast in moulds stamped by the machine now in the course of construction by the Messrs. Donkin, for Her Majesty's Government, at the instance of the Registrar-General.

(*To be continued.*)

On the Clearing of the London Bankers. By SIR JOHN W. LUBBOCK, BART., F.R.S., *formerly Treasurer of the Royal Society, and Vice-Chancellor of the University of London.**

> Atque equidem, extremo nî jam sub fine laborum
> Vela traham, et terris festinem advertere proram;
> Forsitan et pingues hortos quæ cura colendi
> Ornaret, canerem, biferíque rosaria Pæsti:
> Quóque modo potis gauderent intyba rivis;
> Et virides apio ripæ, tortúsque per herbam
> Cresceret in ventrem cucumis: nec sera comantem
> Narcissum, aut flexi tacuissem vimen acanthi,
> Pallentésque hederas, et amantes litora myrtos.

THE operation of the clearing has the effect of enabling all the payments from one bank to another to be performed without the passing of bank-notes; and the result is, that if any bank has to receive from the clearing, say, £50,000, the account of that bank at the Bank of England is better by £50,000 at 9 o'clock the next morning; and if a bank has to pay into the clearing £50,000, the amount of that bank is worse by £50,000.

The following description of the mode of conducting the clearing is taken from Mr. Babbage's *Treatise on the Economy of Machinery and Manufactures* (second edition, page 124) :—

"In London all checks paid in to bankers pass through what is technically called ' *The Clearing House.*' In a large room in Lombard-street, about thirty clerks from the several London bankers take their stations, in alphabetical order, at desks placed round the room; each having a small open box by his side, and the name of the firm to which he belongs in large characters on the wall above his head. From time to time, other clerks from every house enter the room, and, passing along, drop into the box the checks due by that firm to the house from which this distributor is sent. The clerk at the table enters the amount of the several checks in a book previously prepared, under the name of each bank to which it is due.

* The object of this paper is one in the attainment of which the readers of this journal are probably little interested; but the paper itself affords so remarkable an instance of the application of the doctrine of-probabilities to the ordinary affairs of life, that we have not hesitated to insert it.—ED. *A. M.*

" Four o'clock in the afternoon is the latest hour to which the boxes are open to receive checks; and at a few minutes before that time some signs of increased activity begin to appear in this previously quiet and business-like scene.   Numerous clerks then arrive, anxious to distribute, up to the latest possible moment, the checks which have been paid into the houses of their employers.

" At four o'clock all the boxes are removed, and each clerk adds up the amount of the checks put into his box and due from his own to other houses. He also receives another book from his own house, containing the amounts of the checks which their distributing clerk has put into the box of every other banker.   Having compared these, he writes out the balances due to or from his own house, opposite the names of each of the other banks; and having verified this by a comparison with the similar list made by the clerks of those houses, he sends to his own bank the general balance resulting from this sheet, the amount of which, if it is due from that to other houses, is sent back in bank-notes.

" At five o'clock the *Inspector* takes his seat, when each clerk, who has upon the result of all the transactions a balance to pay to various other houses, pays it to the inspector, who gives a ticket for the amount.   The clerks of those houses to whom money is due, then receive the several sums from the inspector, who takes from them a ticket for the amount.   Thus the whole of these payments are made by a double system of balance, a very small quantity of bank-notes passing from hand to hand, and scarcely any coin.

" It is difficult to form a satisfactory estimate of the sums which daily pass through this operation: they fluctuate from two millions to, perhaps, fifteen.   About two millions and a half may possibly be considered as something like an average, requiring for its adjustment perhaps £200,000 in bank-notes and £20 in specie.   By an agreement between the different bankers, all checks which have the name of any firm written across them must pass through the Clearing House; consequently, if any such check should be lost, the firm on which it is drawn would refuse to pay it at the counter; a circumstance which adds greatly to the convenience of commerce.

" The advantage of this system is such, that two meetings a day have been recently established—one at twelve, the other at three o'clock; but the payment of balances takes place once only, at five o'clock.

" If all the private banks kept accounts with the Bank of England, it would be possible to carry on the whole of these transactions with a still smaller quantity of circulating medium."

Since this was written, the system of paying in bank-notes has been completely done away with, and the amounts are settled at the Bank of England, by an entry to the debit or credit of each banker, as the case may be.   There are at present 33 clearing bankers.

The details given by Mr. Babbage are not quite accurate.   It is the duty of the inspector to be always present during the hours of business, to maintain order.

To calculate the probability of the concurrence of any number of independent events, the probabilities of each separately being given, is the most elementary question in that science which is called the

Theory of Probability, and is treated of in all books upon that subject.—*See* Bethune and Lubbock, *on Probability*, p. 9.*   De Morgan's *Essay on Probabilities*, p. 30.

It is only necessary, in order to understand what follows, to bear in mind—

1. That the probability of the concurrence of any number of independent events is equal to the product of the probabilities of each considered separately.

2. When the number of trials is infinite, the number of times an event happens is to the number of times another independent event happens, in the ratio of their simple probabilities.

If any event in the clearing happens n times in m trials or days, I shall consider the probability of the event to be properly represented by $\frac{n}{m}$, so if I find that my house has to receive from the clearing 200 times in 400 clearings, I shall estimate the probability of that event happening on any day in future to be $\frac{200}{400}$ or $\frac{1}{2}$, and so in any other case.   Strictly speaking, this probability being unknown, and only to be derived from experience, belongs to what De Morgan calls inverse probabilities; such are the probabilities which occur in questions relating to insurances on lives.

I do not pretend in these pages to give a treatise on probability, but only to show the application of the most elementary principles of that science to questions which arise in the clearing.

If a be the probability of any bank having to receive on a given day, and b the probability of that bank having to pay in, as one of these events must happen, $a+b=1$.

This is the only principle which can be predicated with certainty without recourse to observation, but it is upon the numerical values of these quantities a and b that this inquiry is based.

I think it will be admitted that $a=b$, and therefore $a=\frac{1}{2}$; at any rate, I find this to be about the case at my own bank, and I have no doubt the same obtains with all other banks, but any banker can verify it in his own case at once by reference to his pass-book.

The probability of any number of bankers having to receive on a given day is, therefore, nearly the same as the probability if thirty

* This treatise was published by the Society for the Diffusion of Useful Knowledge. When this work was written, no work on " Probability" had appeared since the time of Simpson and De Moivre.   If anyone shall pretend that this work was written by De Morgan, I can produce the letter of my lamented friend with which he furnished our manuscript to Mr. Coates.

shillings are tossed that the same number of them comes heads. But as two events are equally impossible, namely, that all thirty banks either bring in or take out, these events must be excluded and the other numbers must be increased in the ratio of 1 to $1-(\frac{1}{2})^{29}$.

The next principle will, I think, also be admitted; and, at any rate, I find it in my own clearing that the probability of having to receive any given sum decreases with enormous rapidity; so that, for example, if the chance, in case of having to pay in or receive any sum between nothing and £50,000 is $\frac{1}{2}$, the chance of having to pay in or take out any sum between £50,000 and £100,000 is probably less than $\frac{1}{4}$, and so on in a rapidly decreasing ratio.

It is probable, further, that the probability decreases on each side zero in the same ratio, and that for a given bank the chance of paying in £5,000, or any given sum, is precisely the same as that of taking it out.

The next question which arises, and which is of great importance, is to obtain the accurate value of the probability of a banker paying in or taking out any particular sum. Any banker can obtain this without difficulty from his own pass-book with the Bank of England, and I think it may fairly be taken for granted that the probability for any sum within certain limits varies directly with the amount of the deposits. So that if the probability, that if a bank whose deposits are half a million takes out, it takes out a sum between 0 and £10,000, be $\frac{1}{2}$, the probability that if a bank whose deposits are a million takes out, it takes out a sum between 0 and £20,000, is also $\frac{1}{2}$, and so on. Thus this quantity, which may be called p, will vary for every bank, and as I know nothing of the phenomena which the clearing presents of any bank but my own, I am thrown upon the necessity of making the best conjectures I can from my own pass-book with the Bank of England.

The deposits of the joint-stock banks are not quite 50 millions; I conjecture the deposits of the private clearing banks to be about 50 millions. In order, therefore, to simplify the calculation, I will suppose the thirty banks to have each deposits to the amount of three millions each. I conjecture the value of p to be about $\frac{1}{2}$ for such a bank, and for any sum not exceeding £50,000.

The probability that any event will happen amongst these thirty banks is given very nearly by the corresponding term in the development of $(\frac{1}{2}+\frac{1}{2})^{30}$, so that the chance of

29 banks taking out is $30(\frac{1}{2})^{30}$ very nearly.

28   „   „ $\frac{30\ 29}{1\ 2}(\frac{1}{2})^{30}$ „

Upon the supposition, therefore, that the business of each bank is so conducted that it is an even chance on any day whether any given bank has to pay or to receive in the clearing, the following table shows the chance of any given event happening on any given day, or the number of times that the event will happen in 1,000,000,000,000 days, on the average of a great many years. The upper line is given to complete the table; but as one bank at least must bring in, it is necessary to exclude the two events of no bank bringing in and no bank taking out, and the numbers in the table should be increased in the ratio of 1 to $1 - (\frac{1}{2})^{29}$, that is, they should be multiplied by 1·000000002, by which, of course, they would not be sensibly altered.

No. of Bankers.	No. of Days, or Probability.		No. of Bankers.
	·0000000009313	0·9691000	
29	·0000000279397	2·4462213	29
28	·0000004051250	3·6075893	28
27	·0000037811700	4·5776260	27
26	·0000255229000	5·4069298	26
25	·0001327190000	6·1229331	25
24	·0005529960000	6·7427218	24
23	·0018959800000	7·2778350	23
22	·0054509600000	7·7364728	22
21	·0133246000000	8·1246530	21
20	·0279816000000	8·4468723	20
19	·0508756000000	8·7065096	19
18	·0805530000000	8·9060820	18
17	·1115350000000	9·0474111	17
16	·1354350000000	9·1317320	16
15	·1444640000000	9·1597609	15

The table reads thus—the extreme columns to the right and left contain the number of banks. The chance of fifteen banks having to pay in the clearing, on any given day, is ·14446· The third column gives the logarithms of the numbers in the second column.

The following table shows the number of days, in the year of 313 days, that any given event will happen, on an average :—

No. of Bankers.	No. of Days.	No. of Bankers.	No. of Days.	No. of Bankers.	No. of Days.
29		19	15·93	9	4·17
28		18	25·21	8	1·71
27		17	34·91	7	·59
26	·01	16	42·39	6	·17
25	·04	15	45·21	5	·04
24	·17	14	42·39	4	·01
23	·59	13	34·91	3	
22	1·71	12	25·21	2	
21	4·17	11	15·93	1	
20	8·76	10	8·76		

The table reads thus—in a year of 313 days, fifteen banks on an average have to pay on forty-five days, on the average of a great number of years. Of course, in order to ascertain what will take place in ten years, or 3,130 days, it is only necessary to move the decimal point one place to the right. So that in ten years seventeen banks on an average have to pay on 349 days.

I have already said, that I have little doubt that the value of a does not differ sensibly from $\frac{1}{2}$, and, if so, the values given in the table above cannot differ much from the truth. The records of the clearing would furnish the means of determining the exact value of a if there were any advantage in so doing.

The value of p is subject to greater difficulty; it may be different for different banks, and so may the value of a; but the conditions of the problem are such as to render it evident that if a is less than $\frac{1}{2}$ for one bank it is probably greater than $\frac{1}{2}$ for another, and unless its average value for all the thirty banks differs widely from $\frac{1}{2}$, my results will not be affected. The same remark applies also to my value of p; it may, and no doubt it does, differ very much for every bank, varying with their mode of conducting the business, and with the magnitude of their deposits. But it is the average value with which we are mainly concerned, and unless its value differs considerably on the average of all the banks from the value which I shall assign to it, my results will still hold good.

One thing I hold to be certain with regard to p, which is, that its value decreases rapidly; so that, if its value for any given bank and for any given b pounds is p, the value for the b following pounds is far less than p.

The greatest difficulty we have to contend with in determining the value of p, arises from the possibility of its varying with the state of the times, that is, with the state of the money-market; and it is, probably, some unknown function of the rate of discount. I do not think it can vary with the rate of discount; it may, perhaps, be higher in times of panic than at others, but I do not think so. It may, of course, be very much higher for days when the dividends are being paid, and when the clearing is affected by any large operation. It would be difficult to determine the value of p for any sum within small limits, and for any given bank, say, for £1,000, but, of course, the wider the limits the less danger of error.

I think, for a bank holding £3,000,000 deposits, and for a sum not exceeding £50,000, the value of p does not differ materially from $\frac{1}{2}$. Of course, if I knew the returns of the clearing-house I could tell its value, and for any given bank I could easily ascertain its value from their bank pass-book, but with only my own experi- ence to guide me in this difficult determination, I give this value with much diffidence. I wish the reader to bear in mind, that while I am groping my way in the dark, a few minutes' examination of the clearing-house returns, properly tabulated, would give a clear insight into this and all similar questions.

Upon this hypothesis, the chances of the Bank having to pay more than £50,000 is, of course, also $\frac{1}{2}$, and the chance of fifteen banks taking out each £50,000 is the probability given in the fol- lowing table, multiplied by $(\frac{1}{2})^{15}$. So that the probabilities given in the table are enormously reduced, and the probability of fifteen banks having to pay, which is

$$\cdot 14446$$

an event which would happen 14,446 times in 100,000 trials, upon the supposition that, if all the fifteen banks have to pay at all, they have to pay at least £50,000, becomes very much reduced.

Mr. Farley has kindly examined the pass-book of my bank with the Bank of England, and, from the results Mr. Farley obtained, I conjecture the following numbers to apply to a bank whose deposits are about £3,000,000 :—

	Probability.	Logarithm of Probability.	Probability.	Logarithm of Probability.	
£ 250,000	·01383	8·1408155	·01383	8·1408155	£ 250,000
200,000	·01330	8·1237822	·02713	8·4334498	200,000
150,000	·03298	8·5182338	·06011	8·7789467	150,000
100,000	·07074	8·8496938	·13085	9·1167737	100,000
50,000	·14841	9·1714464	·27926	9·4460037	50,000
0,000	·22074	9·3438903	..	..	0,000
0,000	·22074	9·3438903	..	..	0,000
50,000	·14841	9·1714464	·27926	9·4460037	50,000
100,000	·07074	8·8496938	·13085	9·1167737	100,000
150,000	·03298	8·5182338	·06011	8·7789467	150,000
200,000	·01330	8·1237822	·02713	8·4334498	200,000
250,000	·01383	8·1408155	·01383	8·1408155	250,000

The first column gives the sum in thousands.

The second column gives the number of times the event happens in 100,000 trials.

The third column gives the logarithm of the number in the second column.

The table reads thus :—A bank with £3,000,000 deposits has to receive between £50,000 and £100,000 14,841 times in 100,000 trials or clearings. The chance of a bank holding £3,000,000 deposits having to pay above £50,000 is ·27926, the logarithm of which is 9·4460037.

I apprehend this law or curvature of p will be similar for all banks, or nearly so; but, of course, for banks whose operations are greater, their deposits being also greater, the figures in the first column will be proportionably greater.

The numbers in the third column are nearly in a geometrical ratio, of which the common ratio is $\frac{1}{2}$, confirming the opinion given in p. 146, of the rapidity with which these numbers decrease on each side zero.

It is evident, that the chances of large sums being taken out of the clearing diminish, for two reasons: the one is, the great improbability of the number of bankers who have to receive greatly exceeding the number of those who have to pay; and secondly, that for each bank the chance of having a moderate sum to pay greatly exceeds that of having a large sum to pay.

Supposing the values of the probability to be approximately correct, it is easy to find the chance of the clearing amounting to any given sum on a given day.

At present the London bankers are compelled to keep very large sums of money unemployed, in order to provide for the possible results of each day's operations. It is, of course, impossible to ascertain the amount of notes kept in the tills of the several banks, but the returns presented to Parliament give, for each week, the total amounts of the balances kept by the bankers with the Bank of England, up to the end of 1857.

A large amount might, therefore, safely be employed by the bankers collectively, and a considerable profit obtained, if the clearing could be worked out of a common fund, so as to assimilate the position of the banks to what it would be if they were all united in one establishment.

In 1839, the actual transfer of money, or *difference of the sides*, was not, upon the average, more than £20,000, and now, probably, £500,000 would be sufficient to provide for this payment. Although, however, the average difference is so small, the uncertainty is so great, that the London bankers are compelled to keep at the Bank of England a balance which varies between £2,500,000 and £4,000,000, and, probably, never falls below the smaller amount.

It is somewhat remarkable, that during the panic at the end of 1857 the balances of the London bankers exceeded £6,000,000.

It is probable also, that independent of the actual profit, the collateral advantages of a combination of the bankers, and the work-ing the payments in the clearing out of a joint account, instead of thirty-three separate accounts at the Bank of England, would be by no means inconsiderable. Although the London bankers enjoy the confidence of the public, still they would, if united, be able more effectually to provide against any panic, and to prevent the recurrence of a state of events such as that which in November, 1857, endangered the whole of our monetary system.

As my career is drawing to a close, or, in the words of the poet, as my bark is nearing the shore, I must leave the details of such a plan to be worked out by younger heads.

On some Considerations suggested by the Annual Reports of the Registrar-General, being an Inquiry into the Question as to how far the Inordinate Mortality in this Country, exhibited by those Reports, is controllable by Human Agency. (Part II.) By H. W. PORTER, ESQ., B.A., Assistant Actuary to the Alliance Assurance Company, Fellow of the .Institute of Actuaries and of the Statistical Society.

(Concluded from page 112.)

AS regards the hereditary transmission of phthisis, another investi-gation of the statistics of the Hospital for Consumption at Brompton showed, that where one parent only was affected with pulmonary disease, the fathers being so affected transmitted the disease to their sons in 63 out of 106 cases, being 59·4 per cent. of the whole number observed; and to their daughters in 47 cases only out of 108, being 43½ per cent.: while the mothers being phthisical transmitted the disease to their sons in 43 cases, being 40·6 per cent.; and to their daughters in 61 cases, being 56½ per cent. of the cases under obser-vation. Judging, therefore, from these figures, it would probably appear, if a large number of cases were registered, that the power of transmission of disease by phthisical fathers to their sons, and by phthisical mothers to their daughters, is about the same.

Very similar—in fact, almost identical—results are shown by a similar comparison of the statistics of insanity.

From the consideration of such facts as these, it appears that we hold in our hands the power of checking the increase of diseases of an hereditary character by discouraging marriages under cer-tain circumstances—the extreme inexpediency of which, in some cases, seems clearly apparent; and Life Assurance Companies, by attending closely to such subjects, may deduce valuable rules to enable them to deal with proposals for the assurance of lives in which a possible tendency may exist to phthisical or other hereditary disease.

With respect to the mortality from fever, without going very much into the question—the length of this paper having already, I fear, exceeded its fair limits—I may mention that there is abundant evidence to show that we hold in our own hands the means of checking, to a very great extent, the loss of life from this cause. Mr. Simon, in a recent report to the Board of Health, speaking of fever, quotes an expression of the late M. Baudens, one of the physicians of the French army in the Crimea, in support of this statement—" *On pourrait le faire naître et mourir à volonté.*"

In a former paper read before this Institute, and in some communications to the *Assurance Magazine,* in reference to the efficacy of sanitary improvements as respects the health of the labouring community, I had occasion to refer to the fact of its having been shown by Dr. Southwood Smith, that the result of such improvements as had been introduced and carried out by the Metropolitan Association for the Improvement of the Dwellings of the Industrious Classes actually exceeded belief.

The improvement of the sanitary condition of the districts under observation had been attended with marked results ; and it was shown that it was quite within our means to hold in check not only fever, but other contagious diseases.

Similar results were shown to have attended a like scheme originated by Lord Shaftesbury ; and the working of the Common Lodging-houses Act had been equally satisfactory.* In fact, all the evidence derived from the reports on these subjects tends to show how much we hold in our own hands the key to the improvement of the health and longevity, to say nothing of the comfort and happiness, of the working classes, which form the great bulk of the entire community.

It is painful to see how very little, comparatively, has been done in this respect ; and the great proportion of what has been effected is due almost entirely to the philanthropic exertions of a few individuals—what has been done by the Government having only been tardily conceded at the earnest and repeated solicitations of a few sanitary reformers, among whom Lord Shaftesbury stands pre-eminent.

Hygienic reform is now, no doubt, beginning to produce satis-

* *Assurance Magazine,* vol. iv., pp. 112, 260; vol. vi., p. 110.

factory results. In a recent report, Sir Benjamin Hall's Act is stated to be working with success. The medical inspectors, it is said, " have done quietly a great deal of good work, and it is probable have saved many lives and *prevented* much sickness."

The alarming excess in the returns of deaths from small-pox has lately given rise to much public discussion as to the cause of this increase.

Dr. Collinson and Mr. Norway, in a recent letter to the *Times*, stated that it was the conviction of a number of medical men that small-pox is likely again to appear as a scourge in this country unless efficient means are taken for its prevention; adding that the alarming increase of the disease was due to the inefficacy of vaccination as at present pursued, and attributing such inefficacy to the fact that the virus now used for vaccination is the same as that which was employed by Jenner at the time of his great discovery, and transmitted since through countless human bodies— that it has never been renewed from the cow since that period—and accordingly, that it is not now of sufficient efficacy to prevent the infection of small-pox. These gentlemen suggest that a fresh supply of virus from the cow itself should be obtained, as Jenner originally procured it. They consider, moreover, that in a number of the poor infant population lurks the poison of inherited syphilis, scrofula, and cancer, or other degenerated states of the constitution; and that, in addition to this, owing to bad drainage and ventilation, overcrowding in their dwellings, and the peculiar diet of the children of the poor, cutaneous diseases are common; and that, accordingly, the transmission of the cow-pox matter through such bodies as these must have had the effect of deteriorating the virus.

Now this seems plausible enough; but, unfortunately, as happens in all medical discussions, a large body of the profession take an opposite view, and maintain that there is no more reason why the strength of the virus should be diminished and its power to ward off infection lessened, than that the disease itself—or, in fact, any other disease—should wear itself out, or be less virulent than it used to be ages ago, by reason of its subsequent transmission through the bodies of whole generations of men. This is as far as regards the first part of our statement: with reference to the second, I will appeal to the gentlemen present whether many do not, in their own families or in those of their friends or acquaintances, know of cases where it is easy to trace disease and weakly constitution to the fact of the victim having been vaccinated from an unhealthy subject; and if this be the case, how much more likely—nay, how certain it is—that a constant transmission of the virus through generations of the sickly infant population of the poor must tend to the effect above referred to.

We must bear in mind, when we read in the Reports of the Registrar of the increase of any disease, that it is the impoverished lower classes that contribute the bulk of the cases reported in his department; and as a considerable prejudice against vaccination exists amongst this class, and as, owing to this and other causes, the Act regulating vaccination is very much evaded, the increase of this disease among the lower orders need not surprise us. Complaint is made that the public vaccinators are not so careful as they might be, and, it is added, that the difficulty of procuring lymph is much complained of. In Paris, a regular establishment—not a Government one—has for many years been kept up, from which a constantly fresh supply of lymph is always procurable, a number of calves affected with cow-pox being maintained, from which establishment, conducted by Dr. Mangeant, of No. 8, Faubourg St. Denis, a supply of pure lymph may always be procured; and it can be forwarded to all parts of the world, in hermetically sealed tubes, at 5 francs each.

It is a curious fact that it is found necessary to transmit the virus through seven or eight calves before it is safe to use it upon the human subject, otherwise the disease would be transmitted in so severe a form that the result would be highly dangerous, if not fatal.

The number of deaths recorded by the Registrar under the head of small-pox varies considerably from year to year.

In 1851 and 1852, for example, the deaths from this cause, in England, were 6,997 and 7,320, more than double the numbers recorded in the two following years, 1853 and 1854, which were 3,151 and 2,808. In the years 1855 and 1856, the numbers were still less—viz., 2,525 and 2,277 only. In the year 1857, the deaths from small-pox appear again to be inclined to increase, being 3,936. The complete returns for 1858 and 1859 are not yet issued.

In the last weekly report as to the mortality in London for the year ending 31st December, 1859, we are told that " the mortality from small-pox increases slowly but steadily"; and we hear that the disease has been raging as an epidemic in Paris, and has been observed to have the characteristic of attacking, not the young alone, but persons of advanced age who have not taken the precaution of being re-vaccinated.

It would seem, therefore, from these figures, that this disease may be an epidemic, as many other diseases are at particular times.

It is a curious fact, and a satisfactory thing to know, that, while a serious epidemic of any kind is raging, the deaths from other epidemic diseases are always observed to be less numerous. From the fact that there does not appear to be a regular constant increase in the mortality from small-pox, but that it varies in its intensity in particular years, as other epidemics do, it would seem that it is not the lymph that is in fault, but that the disease is epidemic at certain periods, in conformity with some law of which we are at present ignorant; and that the power of vaccination, as now practised, is not sufficient to overcome the disease.

There is no lack of scientific investigation constantly being made into this subject; but the pursuit of such inquiries has, until a certain point of information is attained, a tendency to confuse the mind, and probably to shake the public confidence, to some slight extent, as to the power of vaccination to destroy the disease entirely, which, theoretically, it is supposed it should do in the course of time, if satisfactorily and completely carried out.

Vaccination is, no doubt, often imperfect in its results. At times there is a great scarcity of lymph, so that many cases arise in which the system is not properly saturated with the disease, though the vaccinator may be enabled to certify that the operation had been duly performed. I say nothing of those cases in which the manual work has been carelessly or inefficiently done, as in such cases cow-pox is not produced; but there are, I believe, great causes of complaint in this respect, and possibly some more stringent legislative enactment on the subject, or, at least, some more effectual method of seeing that the present Act is duly carried out, may be the result of public attention being called to the subject.

The necessity is, happily, now beginning to be recognised for providing special carriages for the conveyance to the hospitals of small-pox and fever patients. Too many are still conveyed in the common street cabs, by which course these diseases are, no doubt, much disseminated.

The Board of Guardians of the Greenwich Union have lately established a conveyance of their own for the above purpose; and so recently as at the December Middlesex Sessions, the Assistant-Judge, Mr. Bodkin, called the attention of the Court to the practice of removing to the hospitals, in street cabs, prisoners labouring under small-pox. The Court immediately gave orders for the discontinuance of the practice. It is believed that the disease is propagated to an alarming extent by means of street cabs employed for the conveyance of patients.

In consequence of the recent increase in the returns of deaths from small-pox, the attention of the Government has been seriously called to the matter, and the necessity has been recognised of endeavouring to increase the efficacy of vaccination; the result of which has been the issue, by the Privy Council, of certain special instructions to the different parish authorities on the subject, embodying some minute directions as to the mode of performing the operation of vaccination; and medical inspectors under the Public Health Act have been appointed to inquire into the present state of vaccination, and to communicate the views of the Government to the proper authorities, for the safety of the public, to the end that unprotected persons may be immediately vaccinated; and generally to take stringent steps to enforce the Act, which has been heretofore so very much evaded.

The question of the necessity, or otherwise, of re-vaccination after the lapse of a certain number of years, is one that we may hope will now be set at rest; and if it be determined that the necessity exists, steps will, no doubt, be taken to render the performance of vaccination imperative after the lapse of a certain number of years.

In connection with this subject, I may refer to a fact bearing on the statistics of the results of vaccination in cattle, which has lately appeared in the newspapers. It seems, that of three Cattle Assurance Companies in Holland, one made a practice of having all the cattle they assured vaccinated, as a safeguard against pneumonia; another had vaccination performed as soon as the disease had broken out in the animals' stalls; and the last took no precaution of the kind. It is stated that the first Company in a certain period lost 6 per cent. of the assured cattle, the second 11 per cent., and the third 40 per cent.

It does not, of course, follow that the circumstances were identical in which the cattle assured by the different Companies were placed; but the numbers of the losses differ so widely, that, allowing a large margin for this reason, it is not too much to assert that the vaccination of the cattle must have had some prophylactic effect.

I am not aware that any general system of vaccinating cattle exists in this country, though it has long been the practice to vaccinate kennels of hounds, and other valuable dogs, against distemper; but I think it has been done in an amateur kind of way, and not from any positive knowledge of its efficacy.

We now come to the consideration of the mortality among the infant population.

In this class the sacrifice of life is fearful. It appears from the last published Annual Report of the Registrar-General, that, in the year 1857, no less than 57,285 male and 45,942 female children—making a total of 103,227, or about 1 in 6 of the whole number of children born alive in England, and comprising very nearly one quarter of the whole number of deaths registered in the year—died before attaining the age of a twelve-month.

I had intended entering into an investigation of the causes of this high mortality, but the subject is so extensive that I feel called upon, owing to the length of time I have already trespassed upon the attention of the members, to refrain from doing so on the present occasion. I shall only state that an inspection of the Registrar's Reports clearly shows that a large proportion of the deaths of infants is to be attributed to accidental causes quite under human control—probably one of the greatest of which is the absence of physical power in the mothers to supply the requisite nutriment to their offspring, owing, generally, to the want of proper food for themselves. Of course, the humbler classes cannot afford the expensive process of bringing up their children by hand, and so they die of debility and atrophy, the Registrar records—but of *starvation*, if we do not shrink from applying the correct term, to the reproach of the age in which we live.

I have before referred to the too early employment of children in factories as affecting the question of education; the consideration also arises as to its effect on their health and longevity.

There can be no doubt, I believe, that a great amount of disease and of undue mortality is created by the working of young children in factories before they are old and strong enough to bear the strain upon them of the long hours of labour imposed by the manufacturers, as well as by the unhealthiness engendered by the processes of manufacture—some evils arising from which, it is true, are necessarily inherent in the requirements attending on the law of supply and demand, but others, it is no less true, are remediable, if not entirely, at least partially; and it is by such means as these, amongst others, that the mortality, as recorded in the annual returns, is unduly swelled, to the discredit and opprobium of our legislative system.

With respect to the hours of labour, the Factory Act does not allow of young persons working after six o'clock p.m.; but the reports of convictions for infringements of the Act in this respect show that the law is systematically evaded.

There is some difficulty in directly tracing the deaths of factory operatives to the local circumstances by which they are really caused, and in ascertaining what proportion of such deaths may be due to influences which might be controlled, though it is perfectly well known that the number is far from insignificant; but there is no difficulty in ascertaining the loss of life from accidents connected with the machinery used in these factories, as we shall presently see.

Many classes of disease, the inordinate mortality from which, it is submitted, may be to a great extent lessened, are fostered and encouraged, nay, more, perpetuated to a fearful extent—for the hereditary nature of phthisis is now perfectly well known—by the unhappy position in which great masses of our labouring community are placed, and numerous fatal accidents are caused, owing, in some degree, to the apathy of employers, but partly, it must be admitted, to the want of forethought and knowledge of the workmen themselves; for while we find, on the one hand, that employers are reluctant, in many cases, to incur the trouble and expense of improvements in which their own interest is not immediately apparent, with the view to the prevention of accidents, and, what is of infinitely more importance, of disease, so we find, on the other hand, an equal degree of reluctance on the part of the operatives to avail themselves of those precautionary measures within their reach, which the humanity of the employers in many cases leads them to introduce for the advantage of their workmen.

It is painful to read the accounts of the accidents that are of constant recurrence in factories; some due to neglect on the part of the masters, and others to the inherent recklessness of the men, but which more stringent regulations might check, particularly if a proper system of.overlooking were adopted with a view to prevent the workmen, before it is too late, disobeying the rules devised for their advantage.

There is no lack of inspection to see that the hands are constantly duly attending to their work, but a very great want of the other sort of overlooking.

Probably, an inspector, in addition to the usual overlookers of the works, placed on each floor of a factory where machinery is in use, whose sole duty it should be to see that the men did not act in opposition to the regulations of the place—which the millowners are now content with merely having placarded on the walls—would fully answer the required end; and this can hardly be considered to be asking too much of the rich millowners, when we consider the sad detriment to life and limb that now unhappily arises daily under the present system.

The mistaken leniency of the unpaid magistracy is considered to tend to encourage this sort of neglect.

It is true, that clauses 10 and 71 of the Factory Act of 1844 preclude " the occupier of a factory, or the father, son, or brother of the occupier of a factory, being a justice of the peace," from deciding upon cases in which his own pecuniary interests are involved; but even of this most proper restriction, the National Association of Factory Occupiers—a combination organized for the express purpose of opposing the existing law—are said to be straining every effort to obtain a repeal : it being considered, by the parties more immediately interested, to imply " an unwarrantable suspicion upon the honourable conduct of that portion · of the magistracy who are engaged in manufactures."

The power of the rich millowners is immense ; and it is easy to see how, by a combination among them, they may avert from themselves for a time the inconvenience of what they term " meddling legislation."

This Association, some time ago, issued a pamphlet, written by Miss Martineau, in defence of the factory occupiers, and which, while purporting to be a justification of their conduct, was, in point of fact, a tirade against the Government, on the subject of the legislative enactments devised with the sole object of checking some of the evils which are, more or less, necessarily inherent in the conduct of manufactures upon a large scale.

The pamphlet in question, moreover, accuses Mr. Charles Dickens for unfairly calling the attention of the public, through the medium of his valuable weekly series, to the large number of these accidents.

Mr. Dickens would appear to be about the last person in the world to be so attacked.

He has evidenced throughout his works—all written, I conscientiously believe, in a purely philanthropic spirit—an anxious desire to benefit the working-classes; and such attacks as these, therefore, only make it more than ever apparent how weak the cause of the millowners is ; and we are reminded of the instructions to counsel, in the case of a trial in which the defendant had but a poor defence to make—" No case—bully the opposite counsel."

It should be remembered, that it is almost entirely by means of popular articles in such works as these, with which, fortunately, the present generation is well supplied at a cheap rate, that the general body of the people come to be cognizant of and to understand these questions, which so nearly affect their well-being.

The great mass of the people, within whose means these serial writings are, do not belong to the Statistical Society or to the Institute of Actuaries; and writers of eminence who take the trouble to get up the statistics of such subjects and to put them into a popular shape, as Mr. Dickens does so well, are most highly to be commended; and that this sort of writing is of rather an invidious nature, as tending to place the author at issue with some of the *Dives* of the community, the controversy to which I have alluded very clearly shows.

With reference to the alleged unfairness on the part of this writer in dealing with the statistics of the question, it is not difficult to reconcile the discrepancies that appeared to exist between his statements and those of the factory occupiers as to the number of accidents in a given time. Without troubling you with the figures, I may mention that the differences in question appear to have arisen from Mr. Dickens quoting the gross number of accidents that happened in the period of observation, while the opposite party referred to such accidents only as were caused by the shafts alone.

A few remarks on the subject of these accidents may not, perhaps, inappropriately be introduced here. The question is so far important, that when an erroneous opinion on any subject is believed to exist, it is desirable to remove it, if possible. Now, as in the education question, to which I have alluded, certain notions have been referred to as prevailing that will not bear the test of investigation : so in this matter, I believe, a very general opinion obtains, that the accidents, so very constantly occurring in factories, are almost invariably due to the fact that occupiers object to incurring the expense of securely fencing off their machinery, in conformity with the Act of Parliament—that they fail to provide sufficient means for the purpose generally of preventing the recurrence of such accidents as may fairly be considered to be controllable—and that to secure themselves an immunity from the pecuniary penalties to which they thereby lay themselves open, the National Association of Factory Occupiers has been instituted.

Now this is not altogether the case. A great difference of opinion prevails, even among scientific men, as to the fencing off the machinery used in mills; and this being so, the occupiers must not have all the onus of deciding the matter thrown upon them, and particularly as the Act of Parliament which regulates the question does not seem to be sufficiently stringent.

The terms of the Factory Act of 1844 require that all machinery in certain positions and under certain circumstances and conditions, shall be "securely fenced."

It seems difficult to determine in what the *secure fencing* of machinery actually consists, and on this point the millowners and the inspectors of factories have long been at issue.

The Act of Parliament, from its vagueness, is much at fault; and it will be the object of any new legislation on this subject to remedy the evil complained of in this respect.

The permanent casing of the machinery with wood or iron, insisted on in many cases by the inspectors, is considered by the millowners to be likely, in some cases, to *increase* the tendency to accidents; and this opinion is confirmed, it seems, by high engineering authority, Mr. Fairbairn having stated that "he did not see how it was possible to fence off the horizontal shafts of mills driving machinery, without incurring greater evils, and probably more danger, than at present exists from their being left entirely open."

Vertical as well as horizontal shafts should be fenced off when within seven feet of the floor, and the sources of danger from excess of fencing are said to arise from the necessity of suspending the fencing for shafts, which entails a risk of the suspenders giving way, and from the danger of the hands employed tampering with the shafts and pulleys.

Now it certainly seems strange, that, with all the mechanical skill at our command, we cannot obviate the first two of these difficulties; and the constant employment, as previously suggested, of an extra overlooker to act as inspector, would certainly prevent the last.

The Acts for the regulation of factories, as regards the safety of the operatives employed in them, were passed in 1833 and 1844, but were not enforced until the year 1853, when the great number of accidents constantly occurring was brought under the notice of the then Secretary of State, Lord Palmerston.

From the reports of the inspectors of factories for the year ending 31st October, 1859, it appears that the total number of accidents *arising from machinery* was 3,939, of which number 64 resulted in death, and 1,100 were of a serious character—a large proportion of these, viz., 545, rendering amputation of part of the body necessary.

Though these numbers may not appear large as compared with the total number of persons affected by the factory law, which has been estimated to be upwards of 500,000—and though the inspectors, who are, of course, competent judges of the matter, do not express themselves to be dissatisfied with these numbers—it is clear that a great sacrifice of life, and a terrible amount of bodily mutilation, takes place annually in factories from causes undoubtedly within human control; and the inspectors admit that " the number of serious accidents might be reduced if the excellent example set by many millowners, in fencing dangerous machinery whenever practicable, were more generally followed."

Mr. Fairbairn's opinion has been already quoted as against the safety of fencing horizontal shafts; but, on the other hand, it appears from the reports of the inspectors that in all the districts, except that of Lancashire, where the great opposition to the Factory Acts exists, a considerable amount of horizontal shafting, under 7 feet from the floor, *has been* properly cased; and that, in the west of England, " the experiment had been tried on a sufficiently large scale, and for a sufficiently long period, to prove the fallacy of the apprehensions that were expressed as to the practicability of fencing securely horizontal shafts."

That the factory operatives are far from being satisfied with the manner in which the Acts are administered, is evidenced from the circumstance, that, as recently as the end of last Session, a deputation from the operatives of Lancashire, in which district I mentioned the greatest opposition to the Acts of Parliament was to be found, had an interview with the Home Secretary, Sir G. C. Lewis.

The operatives did not seek, it was stated, any new legislation on the subject, but asked for more efficient inspection, to put a stop to the systematic violation of the law. It seems that there never was a time at which a better feeling existed between masters and men than exists at present, and that a large number even of the employers are anxious that the legal requirements of the Acts regulating factories should be strictly enforced. It was stated also, that the operatives themselves were not free from blame in conniving at violations of the law which they knew were going on. The violation was clearly on the increase, and this was said to be owing chiefly to the want of due inspection—the number of factories having increased 50 per cent. since the Government inspectors were first appointed. The number of these inspectors was originally only four, and of these one has recently retired.

It is stated that the Act under which their appointment was originally made contains no power to allow of vacancies arising in even this small number being filled up, unless such vacancies arise by death.

It seems clear that, if the existing Act is so faulty, that it can with impunity be evaded as it is at present, no more stringent method of carrying it into effect is likely to prevail; and that, therefore, some new and more satisfactory legislation on the subject is called for, to remedy, in the first place, the admitted defects, and to provide, in the second, for such additional amount of inspection as the great increase in the number of factories since the matter was last settled may seem to demand, as well as to bring under Government inspection any manufactories that are at present exempt from its operations, such as paper mills.

Sufficient, no doubt, has been said on this head to show that some, at least, of the mortality recorded in the Reports of the Registrar-General might be checked by proper legislative enactments stringently enforced.

But however much millowners and proprietors of manufactories, and of mines—for in these again the sacrifice of life and limb is fearful—may be to blame, and some amount of censure they will scarcely contrive to escape—and however cruel the deaths and permanent bodily injuries that arise from such accidents, whether preventable or not, may be—the loss of life and injury to bodily health from such causes as those to which I have referred—and I have said nothing about the injury caused by insufficient ventilation, or by the evils created by the processes of manufacture—the loss of life and injury to health, I repeat, from such causes is trifling in comparison with that arising from the neglect of sanitary measures, not in mills alone, but in all kinds of works and localities, for the prevention of the undue increase of mortality, and of the propagation of the most terrible bodily and mental diseases which are shown to be so universally prevalent; for, apart from the deaths from accidental causes, the early and laborious employment prematurely imposed upon young children injures their health, stints their growth, and renders them unfit for such labour as, when they had become adults, they should be properly able to endure.

A very great number of deaths from accidental causes takes place annually in mines.

The average annual number of deaths from accidents from collieries, for example, in England, is stated to be about 1,000, out of a population engaged in coal-mining of about 250,000, in addition to an enormous number of accidents. Of this number about one-fourth arise from explosion. Now a great proportion of these accidents—probably the whole number caused by explosions —may be considered to be preventable; and that this is the fact may be fairly assumed from the circumstance of the death-rate in different mines varying considerably, according to the precautions taken in each.

Surely, if this be the case, the proprietors of mines should be forced to adopt proper precautionary measures to protect the lives of the miners, and should, if they failed to do so, be held legally responsible for their neglect!

In the course of a paper on the subject of colliery accidents, read recently at the Society of Arts, by Mr. P. H. Holland, one of the Government Medical Inspectors under the Burial Acts, it was suggested, that, as a means of checking these accidents, no one should be allowed to work in a colliery without having his life assured at the expense of the proprietor of the mine, against death by accident, to a sufficient amount to secure his family from destitution in the event of his death.

The proprietor would be called upon to pay an extra premium should his mine be considered by the Accidental Death Office to be in a dangerous state, owing to neglect of due precautions; it would thus be the interest of the proprietor to bring his mine into the best possible condition, in point of safety. It is added that an increase in the price of coal would provide for the payment of the ordinary but not for the extra premium, as the one, of course, would be a uniform charge in all mines.

This is all very well on paper; but we know that, in practice, the premium for such an assurance would virtually come out of the wages of the men—and the premium would, moreover, probably, be so high, as to preclude the possibility of carrying out the views of Mr. Holland. It was suggested, whether an assurance scheme of this nature might not be made legally compulsory, though Mr. Holland stated that it was doubtful if assurances could be enforced by law.

Of course, any idea of the sort is not likely to be entertained in practice—it would be well for the Assurance Offices were it otherwise—but a law to render the actual proprietors of the mines liable to be tried for manslaughter, in case of a fatal accident caused by the neglect of those precautions which the law might require to be adopted, and a strict enforcement of the punishment for that crime, might probably tend to diminish the extreme mortality recorded in the reports of the inspectors.

It is scarcely necessary to remark, that a large proportion of the casualties registered as "accidents" eventuate in the death of the victims; but the deaths occurring some time after the accidents are accordingly not set down to the primary cause, but to the immediate cause of death, whatever that may chance to be.

The total number of deaths from violence recorded in the year 1857, was no less than 15,027, and even this large number does not include the deaths from drowning at sea.

I had intended to refer to some of the minor causes which tend unduly to increase the mortality of the country, but, owing to the length to which this paper has already extended, I feel called upon to defer carrying out my intention.

It may be as well for me to mention, that the statistics which I have quoted in the course of this paper have reference almost entirely to the year 1857 — the last Report of the Registrar-General, the 20th, not going beyond that date.

It is worthy of remark that while the deaths from one large class of diseases alone have been shown to be upwards of half a quarter of a million, those from "old age" are only 26,847. Probably, in a less artificial state of society than that in which we live, so manifest a disproportron would not exist; for in a natural state, I presume, there can be no doubt it is to "old age" that we should expect to find the great mass of the population owed their death. "Death by old age," Mr. Simon says, in a recent report to the Board of Health, "is, physiologically speaking, the only normal death of man."

Do we not at the present day—as the poet records, nearly two thousand years ago, the presumptuous race of man in primeval times had done—attempt to violate all natural laws, and so draw down upon ourselves consumption, and a new train of fevers before unknown, and thus accelerate the slow approaching necessity of death?

> " Macies et nova Febrium
> Terris incubuit cohors;
> Semotique prius tarda necessitas
> Leti corripuit gradum."

It would be a most interesting inquiry to follow the course taken by Thackrah, 30 years ago, in his round of inspection of the different manufacturing processes throughout the kingdom, with the view to see to what extent the sanitary state of the workmen employed in the different manufactures had been improved since his time.

Judging from the limited extent to which I have been able to carry my inquiries on this subject, I fear that but small progress has been made; and yet his work *ought* to have opened the eyes of the public to the matter.

An investigation of this nature could easily be carried out by Assurance Companies in combination with the visits of inspection of their agencies, which such Companies have in all parts of the kingdom. The expense of an enquiry of this nature would be, in fact, almost nothing, as it would, at most, but necessitate a few hours extra stay in the different towns which required to be visited; and I feel satisfied that such an investigation, even if conducted at some expense to the Office undertaking it, would more than repay the outlay, by the important information that would be derived as to the actual effect of certain occupations on health and longevity, and of which those engaged in life assurance operations have at present but a vague, and perhaps incorrect, idea.

In the foregoing remarks, I have endeavoured to show that, under the five principal heads, viz., phthisis, fever, small-pox, infantile diseases, and accidental death, an unnecessary amount of mortality takes place annually from causes greatly within human control.

These causes are as follows, viz.—

With reference to Phthisis and diseases of the Respiratory Organs.

1. The unhealthy nature of certain employments.

2. The bad arrangements, as respects ventilation, in manufactures and works of all kinds in which large bodies of the labouring classes are employed.

3. The quantity of drink taken by this class of people.

4. The apathy shown to the position in which they are placed as to sanitary matters, and the prejudices on the part of the operatives generally against all suggestions for the removal of the causes of the evils that injure them, in many conditions of life, particularly in the case of printers, millers, stonemasons, and persons employed in the manufacture of metals.

5. The defective sanitary arrangements as regards the Army.

6. The propagation of the disease by the marriage of those hereditarily affected with phthisis.

With reference to Fever.

I have referred to facts which prove that the means of checking fevers, and other contagious diseases, depend entirely upon sanitary arrangements, and that we, in fact, hold in our own hands the power of prevention of this class of diseases.

With reference to Small-pox.

I have endeavoured to show that this complaint does not, perhaps, differ from other epidemics; but that, probably, more stringent legislative enactments as to vaccination, or, at least, a more effectual method of carrying out the present Act, would have some effect in diminishing the number of deaths from this cause.

With reference to deaths from Infantile Diseases.

I have said enough to show that the mortality under this head is excessive, and that a considerable portion of it might be prevented.

With reference to deaths from Accidental Causes.

There is, I think, no reason to doubt that a great proportion of the very high mortality under this head arises from causes clearly within human control.

These causes are principally as follows, viz.—

1. Fractures and contusions.
2. Drowning.
3. Burns and scalds.
4. Hanging and suffocation.
5. Wounds and violent deaths, not classified.
6. Poison.

It will be admitted, I apprehend, without question, from the foregoing considerations, that an enormous unnecessary mortality takes place annually in this country. Without venturing myself to assess even an approximate number, I may state that the Registrar-General speaks of 100,000 deaths, in round numbers, as being of a preventable character.

Notwithstanding the evident unnecessary sacrifice of life in this country, there is no doubt that a gradual progressive increase in the mean duration of life has been maintained for some centuries past—with one exception, the 17th century—but it is probably only within our own time that any considerable increase in the longevity of the mass of the population will be apparent; and this will be owing as well to the improved habits of life of the people as to the reform in sanitary matters—to the improvements in the dwellings of the labouring classes—the greater attention that is now being paid to drainage—the abolition of intramural interment —the establishment of baths and wash-houses—to the shorter hours of labour that are now becoming daily more general, owing

in a great measure to the exertions of the "Early Closing Association," and of the Secretary of the society, Mr. Lilwall—to the
establishment of national play-grounds, and the encouragement to
indulge in manly exercises, of which the shorter hours of labour
now more readily admit—and to the greater care that is now devoted generally to the promotion of the well-being of that large
section of the community which the labouring classes form, a duty
which those above them in the social scale have heretofore too
much neglected, to the consequent undue increase of the deaths
recorded in the Registrar-General's annual returns, and the filling
of our prisons and reformatories.

It will probably be found that, in proportion as the energies of
the country are devoted to sanitary questions, so will our expenditure upon such institutions as these be diminished; and thus,
while we are promoting the good of the people, we shall be at the
same time economising the funds of the nation.

If we are not to begin to profit by the valuable information we
derive from the Registrar-General's department, which has been in
operation since the year 1837, and if we are not to endeavour to
adopt methods to check any unnecessary amount of mortality that
we thus have the means of ascertaining—what, allow me to ask, is
the advantage of the expensive machinery required for the due
carrying out of the elaborate system now in operation?

The very object of the introduction of the system of recording
the statistics of births, deaths, and marriages, of the state of
factories and mines, and of education, was that we might learn
where our system of social government was at fault, and that it
might be the means of directing our attention to the course necessary to be taken, with the view to remedy the evils that might be
observed to exist.

With a view to utilise the materials we possess, some general
national statistical system would, however, appear to be necessary.
Dr. Farr, in his Report on the International Statistical Congress,
held at Paris in 1855, proposed a scheme for the establishment
of a Statistical Board, for the purpose of digesting the national
statistics, which should consist, he was of opinion, of statistics of
finance, population and health, sickness, poor-law, friendly societies,
and charity; of learning, art, and science; of statistics of the
church and the law; of trade, manufactures, and agriculture; of
the army and navy; and of statistics of India and other foreign
countries.*

---

* *Vide* 16th Annual Report of the Registrar-General.

If a comprehensive scheme of this nature were carried out, we should, in the course of a few years, begin to derive some return from the capital which has been sunk in the purchase of the raw material for our subsequent operations, which has heretofore been, to a great extent, unproductive; for, as truly as the magnetic needle guides the mariner through the pathless ocean, so would the digests of the great masses of facts we possess direct the Government through the wide sea of legislation, and point out the course that must most surely be taken, if we wish to improve the resources of the country—if we wish to diminish sickness and to increase health and longevity—to lessen poverty and to promote learning—to take away the incentives that lead to the commission of crime—to ameliorate the condition of a great people—and to further their social and religious progress.

In concluding this paper, I will beg to express a hope that I may not be considered by the members of the Institute to have intruded upon them anything approaching to an *ex-cathedrâ* opinion. I have endeavoured to bring forward recognised authorities upon the different subjects on which I have touched, and I have taken the greatest pains to verify facts.

I may add, that one great motive I have had in coming forward on the present occasion, has been an earnest feeling that it is a duty incumbent upon every one to lend any assistance that lies in his power towards calling attention to matters affecting the health and longevity of the people; and there is no reason, I consider, because I am not myself equal to the task of doing justice to the question, that I should abstain from entering upon a subject which is only now, I may say, *beginning* to arrest public attention, and that I should forbear to open up the store of rich materials that we have at our command from which to derive instruction in such matters.

> " Ergo fungar vice cotis, acutum
> Reddere quæ ferrum valet, exsors ipsa secandi."

On the Composition for Leave to an Assured to reside Abroad.   By ROBERT CAMPBELL, M.A., *Advocate, Edinburgh, Fellow of Trinity Hall, Cambridge.*

IT is not an uncommon occurrence with certain classes of professional men who have settled themselves in this country, and perhaps assured their lives, without any intention of leaving it, to be called to a position which takes them to a residence permanently, or at least for a lengthened period, abroad. The principle on which an Office should increase their payments for assurance forms both a speculative and practical problem of some interest.

If the original contract of assurance be rigidly interpreted— that is, supposing it simply liable to forfeiture in the event of going to live out of the limits specified in the policy—the assured must then stipulate for the liberty by paying the difference between the value of a policy of £100 at death, according to the statistics of mortality at home and that according to those of the given place abroad—the premium payable in each case, during the remainder of life, being that of the original assurance. But it must be remembered that the liability to forfeiture on going abroad without a new contract, though nominally one of the original conditions of the policy, is not one for which, in the ordinary case, the assured receives or the Office gives any value; and therefore the assured, stipulating in the way above indicated, will really give more than the equitable consideration for his liberty.

Suppose the contingency of the party's going to live abroad had been contemplated from the first, let us see how the matter would have stood.

Suppose the party at the time of assuring—say in the year 1860—to be aged m years, and suppose he then stipulate for the additional payment required in the contingency of his being alive and wishing to go abroad at the end of n years:

Let A_m be the present value of a policy of £100 on a life aged m, according to the statistics of this country;

A'_m the same according to the statistics in the foreign place— the premiums payable during life being in both cases those fixed at first assuring.

If the party be alive in the year $1860+n$, and intend then immediately to go abroad, the value of the policy in that year, supposing him allowed to receive it without increase of premium, will be

$$A'_{m+} .$$

The value of the policy, supposing him to remain at home, would be

$$A_{m+n} .$$

The difference, $A'_{m+n} - A_{m+n}$, which would be the amount payable in the year $1860+n$ to obtain the liberty required on a strict interpretation of the conditions.

The value of that payment in the year 1860, at the time of

assuring, supposing it certain that, if alive, he would go abroad at the end of n years, would be

$$p_{m_1 n} r^n . (\text{A}'_{m+n} - \text{A}_{m+n})$$

($p_{m_1 n}$ being the probability of a life aged m surviving n years, r^n the present value of £1 due at the end of n years).[*]

Now, suppose, in the year $1860 + n$, the circumstances emerge that the party is alive and wants to go abroad. By the strict condition of the contract he will forfeit if he goes; but he has received no consideration for this condition. The only equitable conclusion is, that the original contract be amended, and it will stand thus: on the one hand he has a right to go, which we must suppose to have been constituted at first assuring; but, on the other hand, the Company have the right to the sum which he should have paid for the permission had it been part of the original contract—in other words, to the sum which he should have paid them for the permission, improved at compound interest during the n years when it has been owing.

The payment, therefore, that should be made, is—

$$p_{m_1 n}(\text{A}'_{m+n} - \text{A}_{m+n}).$$

The principle, therefore, of an equitable adjustment of such payments I take to be this, that when a party, having assured as for a home policy, wishes to take out a wider one, the additional payment for the permission should first be calculated according to the relative statistics of home and foreign mortality, *and then multiplied by the probability of a life of his age when first assured living for the number of years which he has remained in this country since his assurance.*

It should be noticed, as a very important point in the case of statistics of mortality abroad, that not only the ages, but the length of residence in the foreign country, should be looked at.

The suggestion may appear startling, practically, but the theory would apply equally to the case of assured persons dying having broken the conditions without stipulation, except that it is open for consideration whether the original conditions of the policy ought not, in this case, on grounds of general expediency, to be strictly interpreted. Suppose n to be the number of years intervening before breaking the condition, v the number of years after it, the ordinary premiums continuing to be paid up to death, the amount to be deducted from the policy would be

---

[*] Notation in Jones' *Annuities*, vol. i., p. 110.

$$\frac{p_{m,n}(A'_{m+n} - A_{m+n})}{r^v}.$$

Perhaps the most equitable conclusion would be this, that when conditions have been violated *without fraud,* the policy should be paid under the above deduction.

On the Discovery of the Law of Human Mortality, and on the antecedent partial Discoveries of Dr. Price and Mr. Gompertz. By T. R. EDMONDS, B.A., *formerly of Trinity College, Cambridge.*

IN the year 1832 there was published, in my name, an extensive collection of *Life Tables,* founded upon the discovery of the law which, in my belief, governs the mortality, according to age, of all nations and classes of men, from the earliest infancy to extreme old age. In these tables the numbers living or surviving at successive ages have been deduced from a simple formula expressing the proportion of survivors at any age in terms of the mortality. They are the first Tables of this kind ever published, and they have been used daily in the practice of life assurance, and in the valuation of life contingencies, for a period of thirty-two years, including four years before publication.

The following is the law of human mortality :—The whole duration of human life is divided into three well-marked stages, which belong to all animal life, and are—stages of growth, maturity, and decay. The mortality, in all three stages, increases or decreases, uniformly with the age, in geometrical progression, but in a different progression for each of the three stages. The constant ratios of progression belonging severally to the three stages have been ascertained, their values being the same for all populations at the same stages of life.

The three stages of human life may be conveniently designated as those of *infancy, florescence,* and *senescence.* The period of infancy most commonly ends at the age of 9 years; the period of senescence most commonly begins at the age of 55 years. The values of the annual constants of progression in the stages of infancy, florescence, and senescence respectively, are $\frac{1}{1\cdot479108}$, $1\cdot0299117$, and $1\cdot0796923$—the logarithms of these numbers being $-\cdot1700$, $+\cdot0128$, and $+\cdot0333$ respectively. If the force of mortality continued uniform for one year, at the beginning of any of the three stages of life, be represented by (a), and if the constant annual ratio of geometrical progression be represented by (p), we shall have, for any time or age (x) measured from the beginning of the stage, (ap^x) to represent the force of mortality at that time; the same being true for all values of (x), whether (x) is a whole number or a fraction. When the mortality has been ascertained by observation for any age, and when the exact length of the period of florescence is known, the mortality existing at every other age may be calculated from the formula (ap^x); the quantity (p) having one or other of the three values given above, according to the stage of life observed.

The limits of the periods of infancy and florescence vary slightly from the ages of 9 and 55 years respectively, as well in different populations as in the same population at different epochs. In every population the force of mortality, or the ratio of the dying to the living, is always least at the age or short period immediately preceding the commencement of puberty—say from the age of 9 to the age of 11 years, or from the age of 8 to the age of 12 years. In my principal theoretical tables, I have assumed the existence of a short period, from the age 8 to the age 12 years, wherein the mortality is constant and at a minimum. This assumption is in accordance with appearances presented by large numbers attaining severally their lowest mortality at different ages—near 9 years of age—although each individual may suddenly pass out of the first into the second stage without passing through the supposed intermediate stage.

The first intimation of the nature of the law of human mortality was conveyed to the public in an essay of Dr. PRICE, read at the Royal Society in April, 1769. This essay is printed in *Price's Observations on Reversionary Payments,* a work which was for many years the chief text-book used by students of the valuation of life contingencies. In this essay Dr. Price made the following remark (page 400, vol. ii., edition 7) on the mortuary register of the parish of Holy Cross, near Shrewsbury: — "This register exhibits, with remarkable regularity and consistency, the progress of human mortality from birth to old age—representing human life, in conformity to other observations, as particularly weak in the first month, and from that age as growing gradually stronger, till at 10 it acquires its greatest strength, which it afterwards loses ; but more slowly till 50, and after 50 more rapidly, till at 70 or 75 it is brought back to all the weakness of the first month."

In the above statement may be recognized the chief features of the true law of human mortality. The three stages of human life are designated, their limits are noted, and words are used descriptive of the effect of the different constants belonging to the three separate stages. Dr. Price, to perfect his discovery, would have had to make the following additional remark, viz.:—"That the rate of increase or decrease of the mortality was constant throughout each of the periods designated; that life growing *gradually stronger* was represented by the constant annual ratio of 1·479 to 1; that life losing strength *more slowly* was represented by the constant annual ratio of 1 to 1·0299; and that life losing strength *more rapidly* was represented by the constant annual ratio of 1 to 1·0797.

In the discovery of the law of human mortality, I was not assisted by any knowledge or recollection of the antecedent partial discoveries of Dr. Price or any other writer. My discovery `is founded on direct observation of the principal and best-established facts or statements of mortality on record. These facts, collected by me and expressed in their simplest form, denoted observed ratios of dying to living at successive equal intervals of age. At ages above 10 years, the mortality at successive quinquennial intervals of age was thus ascertained from an extensive variety of tables and observations. These successive quinquennial rates seldom indicated any regularity of increase; but, when combined so as to form rates of mortality for successive decennial intervals of age, the uniform rates of progressive increase became manifest, and the two numbers already indicated, as regulating the increase in the periods of florescence and senescence, were discovered. The constant ratio regulating the progressive decrease of the mortality during the period of infancy, was found by observations on the mortality exhibited at this period in a large variety of tables for successive annual and biennial intervals of age. The constant of the period of infancy was found to be the same in all tables and observations of mortality; so also the constant of the period of senescence was found to be the same in all observations. The constant of the period of florescence was found, however, occasionally to vary from its average magnitude, when small sections of a population were observed; but in large aggregates, consisting of mixed population, and in small sections, consisting of homogeneous population, the particular constant already indicated was manifested.

Immediately after the discovery of the law of human mortality, I constructed, for practical use, a table of mortality in which the numbers living or surviving at successive annual intervals of age were deduced by successive subtractions of the approximate annual decrements obtained from the formula $\Delta y = y \times ap^x \Delta x$. By this table was exhibited a series of annual decrements uniformly progressive with age, excepting at very advanced ages, when the uniformity was interrupted, although the defect was of no importance in practice. By reflection on this defect, manifested at advanced ages, I became convinced that uniformity of decrement according to age could not be obtained by means of an annual decrement of the form $\Delta y = y \times ap^x \Delta x$; and that such uniformity could only be secured by continually diminishing the intervals of age observed, from Δx to dx, or by calculating the decrement in infinitely small intervals of age by the formula $dy = -y \times ap^x dx$.

Until I had arrived at the conclusion that the decrement of the living at any age was of the form $dy = -y \times ap^x dx$, the idea had never occurred to me that there existed anything in common between the formula (g^{p^x}), used by Mr. GOMPERTZ to represent the number living, and my formula (ap^x)—with three permanent values of (p) determined, representing the decrement of life for every year of age. I had previously become acquainted with the formula of Mr. Gompertz by conversation with Mr. Gompertz himself; I had also tried the value of his formula by applying it to interpolation of the numbers living, in a table of mortality of my own selection, and found the result not to be in conformity with Mr. Gompertz's verbal statement. Not believing his theory to be well founded, I soon forgot everything relating to it, except that his formula was (g^{p^x}), and that he used the differential calculus in his investigation, either in descending from the integral to the differential, or in ascending from the differential to the integral. After the true law of human mortality had become known to me, on re-examination of Mr. Gompertz's theory, I became aware of the cause of my original failure to find any truth in his representations. The error was on the side of Mr. Gompertz, who, in his statement to me, made no mention of any limits of age circumscribing the application of his formula. In all probability, at my first examination of his formula, I chose for interpolation two numbers representing the survivors at two ages, on different sides of one of the limits. The result of the comparison would, in this case, evidently be contradictory to Mr. Gompertz's theory.

The statement of his theory made to me does not differ materially from the general statement given by Mr. Gompertz in his paper read at the Royal Society in June, 1825, for he there states it to be a mathematical consequence of his formula being found to be true by observation, "That the average exhaustions of a man's power to avoid death are such that, at the end of equal infinitely small intervals of time, he loses equal portions (or proportions) of his power to oppose destruction." As no limits of age are here mentioned, it might fairly be inferred from the above statement, that the vital force of man, measured by the ratio of the living to the dying, is in a constant state of decay from birth to the end of life, at one and the same uniform rate. Mr. Gompertz nowhere makes mention of the fact that the vital force increases at a high rate during the period of infancy; nor does he note the existence of any difference between the rate of the annual loss of vital force during the period of florescence, and the greater rate of annual loss of vital force during the period of senescence. All the examples in detail which he adduces in support of his theory are confined to specimens, taken from different tables, of numbers surviving at various ages, between 15 and 55 years, or comprehended in one stage of life only, that of florescence. With respect to ages above 55 years, Mr. Gompertz states the result of his examination of one table of mortality only, without entering into detail as in the other cases.

My knowledge of the law of human mortality is founded on direct observation of the ratios of the dying to the living, at successive intervals of age, in various populations. Mr. Gompertz's knowledge of the law has been obtained indirectly, being founded on compilations made by others, of ratios of dying to living. The laws of mortality in the last century of the populations of Sweden and of Carlisle are contained in the observed ratios of dying to living at various ages, collected and published by Wargentin and Heysham. Mr. Gompertz, in deducing the laws of mortality of these populations, relies upon tables of survivors, according to age, constructed by Price and Milne, who acted as compilers of the ratios of mortality supplied by Wargentin and Heysham. He accepts observations at second hand for his guide when the original ratios of mortality observed were easily accessible. Mr. Gompertz has done nothing without the aid of tables of survivors at successive ages previously constructed by others. Nearly all that he offers to show is, how *interpolations* may be made for intermediate ages when the number of survivors at the beginning, and the number of the survivors at the end, of a large interval of age are given. Mr. Gompertz never shows how a new table of mortality may be constructed independently of other previously-existing tables.

The paper of Mr. Gompertz read at the Royal Society in June, 1825, is entitled, " On the nature of the function expressive of the law of human mortality, and on a new mode of determining the value of life contingencies." This paper is divided into two chapters, the first containing 17 pages, and the second containing 58 pages. The subject of both chapters is *interpolation* for intermediate ages between the number of survivors at the beginning, and the number of survivors at the end, of intervals of age, of greater or less extent, given by different tables of mortality. Chapter I., or rather 12 out of its 17 pages, is occupied in showing that *interpolation* may be most correctly effected by the aid of the transcendental (g^{p^x}). The whole of Chapter II. is devoted to showing that interpolations for short intervals of age may be satisfactorily effected by means of a simple geometrical series, of which each term is of the form (g^x). As the two modes of interpolation are, in a certain degree, inconsistent with and opposed to one another, it is not easy to account for the publication of these two chapters side by side. The reader might fairly infer that the author could not have much faith in the virtues of his transcendental (g^{p^x}), when, in the inculcation of its truth and applicability, he had bestowed only one-fifth of the space and attention which he had bestowed on the simple and common geometrical series represented by (g^x).

If the two chapters are viewed together, the reader can hardly avoid coming to the conclusion that the patient labour expended in the production of Chapter II. was the source of the discovery exhibited in Chapter I., and that the ideas connected with the discovery have been presented to the public in an order which is the reverse of the order in which they occurred to the mind of Mr. Gompertz. If anyone acquainted with mathematical formulæ, and with the transcendental (g^{p^x}), had expended, with tolerable success, a vast amount of labour (as Mr. Gompertz apparently had done) in the interpolation of survivors according to age, by means of the geometrical form (g^x), it could hardly have failed to occur to his mind, in the course of his labour, that (g^{p^x}) being a form of much greater elasticity than (g^x), would yield numbers much more nearly coinciding with the facts intended to be represented. After he had, by examination, found his prognostication verified, he would not (at first, at least) regard the quantity (p) as a permanent quantity; for, since (g^{p^x}) is taken only as the substitute of (g^x), and as (g) in (g^x) has an endless number of values, the quantity (p) would also be considered as having an endless number of values. It would be supposed that (p) would change with every different interpolation, because (g) in (g^x) changes with every different interpolation.

As soon as Mr. Gompertz had discovered that the transcendental (g^{p^x}), when used for interpolation of survivors according to age, gave results nearly coincident with fact, his curiosity would be awakened as to the signification of the constituent parts of (g^{p^x}). The readiest means of examination would be afforded by differentiating the above quantity. The differential obtained would be found to be log. $g \times$ log. $p \times g^{p^x} p^x dx$, which is of the same form as $dy = a \times yp^x dx$, the differential arrived at by my investigation previously to any knowledge of its connexion with (g^{p^x}) as its integral. Thus Mr. Gompertz will have descended from his integral to my differential—there being no reason to suppose that he had ascended, from a differential like mine, to his integral. The former course is simple and obvious; but the latter course, though apparently simple, would not be obvious to a person dealing for the first time with quantities of a new and unknown character. It is often erroneously supposed that Mr. Gompertz has deduced his formula from the hypothesis of equal proportions of vital force being lost in equal intervals of age; but the reverse is the fact, the hypothesis having been arrived at as the mathematical or necessary consequence of the truth of the formula; for Mr. Gompertz states that "the hypothesis was derived from an analysis of the experience alluded to"—the experience alluded to consisting in (g^{p^x}) being found to yield correct interpolated numbers for the survivors according to age in many tables of mortality.

The formula (g^{p^x}) is a general formula, alike applicable to the expression of the true law of human mortality and to the imperfect law announced by Mr. Gompertz. The true law differs from the imperfect law only in respect of the value of (p) being determined in the former case, and being left undetermined in the latter case. According to the true law of human mortality, (p) has a fixed determined value for each of the three periods or stages into which human life is divided—growth, maturity, and decay—for the same and for different populations. According to the imperfect law of mortality declared by Mr. Gompertz, no limit is assigned to the variation of (p), whether in the same or different populations, at any period of life. The original idea entertained by Mr. Gompertz apparently was,.that by means of the formula (y^{p^x}), interpolations of survivors between any pair of ages, not very remote from one another, may be effected by assuming different values of (p). It is only by reference to examples of interpolation cited by Mr. Gompertz that the reader learns, indirectly, that, in the second period of life, or from the age 15 to 55 years, Mr. Gompertz regards (p) as constant in the same population at the same time, though variable for different populations and for the same population at different times. Mr. Gompertz says nothing of the value of (p) at any age under 15 years, and gives the value of (p) in one case only for the period of life above 55 years of age. In examples of the application of his formula to different populations, Mr. Gompertz always treats (p) as a fugitive quantity, and as of no value except for the particular interpolation then required to be made.

The view now taken of Mr. Gompertz's theory of mortality, with reference to the variability of the quantity (p), has evidently been also taken by the late Mr. Galloway, who, I believe, is the only person who has made a practical use of Mr. Gompertz's formula in the construction or adjustment of a new table of mortality. In the year 1841, whilst explaining the construction of tables which he had made of the mortality experienced by the Amicable Assurance Society, Mr. Galloway states that, in the adjustment of these tables, he "had recourse to the highly ingenious method (founded, indeed, on a hypothetical principle) proposed by Mr. Gompertz. .... In Table III. the same constants were used through the whole series, from age 45 to age 93 years. In Table IV., from age 24 to age 68, one set of constants was used ; and a second set from age 69 to age 93 years." Mr. Galloway does not state the value of the quantity (p) in either of the three cases, nor does he make any allusion to (p) as distinguished from other but less important constants used. He evidently regarded (p) as a fugitive quantity, of no value except for the use to which it had been applied, in smoothing the irregularities which would otherwise have appeared in the tables which he was then constructing. Both Mr. Gompertz and Mr. Galloway have treated (p) as an insignificant quantity, of accidental value, and of no use but to maintain the formula (g^{p^x}). The truth is, however, that (p), with its three determinate values, is independent of all formulæ, has existed as long as man has existed, and forms part of the foundations of the universe.

All my published tables of mortality have been deduced from the formula $y = 10^{\frac{k^2 \alpha}{\lambda p}(1 - p^x)}$. In arriving at this formula, I am not conscious of having received any assistance from the writings of

Mr. Gompertz, beyond the suggestion of the idea that the form which I had discovered for the decrement of human life at every age was one capable of integration by simple and well-known methods. I do not believe that this assistance was of any importance, except with regard to time, for I possessed all the materials for forming the series proposed for summation, and could hardly have failed to discover the law of that series as soon as my attention had been directed to, and concentrated on, the subject. The law of the series may be found as well without as with the aid of the differential calculus. My formula, as given above, is reduced to more simple terms than that given by Mr. Gompertz. The correctness of my formula, or of the process by which it has been obtained, has never been questioned. There is, however, ground for doubting the correctness of the process by which Mr. Gompertz has deduced his formula. To remove this doubt, the publication of his process in a more explicit form is requisite.

My formula, $y = 10^{\frac{k^2 a}{\lambda p}(1 - p^x)}$, gives the number living or surviving at any age (x) in any of the three periods of human life, expressed in terms of the mortality (ap^x) at that age, and known quantities; (k) being the modulus of the common system of logarithms and equal to $\cdot 4342945$, and (λp) being the common logarithm of (p), the constant ratio of the period. This formula may be written thus :—

$$y = 10^{c(1 - p^x)} = g^{1 - p^x} = \frac{g}{g^{p^x}}.$$

Mr. Gompertz, by changing the sign of (c) and introducing the superfluous quantity (d), makes his corresponding formula $y = dg^{p^x}$. If the change of sign had not been made, Mr. Gompertz would have stated the value of (y) to be $\frac{d}{g^{p^x}}$, instead of $\frac{g}{g^{p^x}}$, my value given above. According to my formula, $y = 1$ when $x = 0$, whilst Mr. Gompertz's formula gives $y = \frac{d}{g}$ when $x = 0$. The integration of Mr. Gompertz's formula requires to be performed between limits of $x = a$ and $x = a + n$, whilst in my formula the limits are of the simplest form between 0 and x. The defect in Mr. Gompertz's formula, caused by the addition of (d), is the same as that which would exist in a table of discount of money at compound interest if any other basis were adopted than the value of the sum of £1 receivable (x) years hence. It is only in cases of interpolation that the quantity (d) is of any utility.

It may be useful here to give an example of the application of my formula $y_x = \dfrac{g}{g^{p^x}}$; for this purpose we will assume (x) to be equal to 10 years, and to be measured from the beginning of the second and chief of the periods of human life—the period of florescence. In this period $p = 1\cdot0299117$, and the value of (a), or the minimum mortality, is, in most populations, on an average,

$= \cdot0063643$; hence $\dfrac{k^2 a}{\lambda p}$ or $c = \cdot0937799$, and 10^c or $g = 1\cdot241023$.

Since $p^{10} = 1\cdot342765$, we get $g^{p^{10}} = g^{1\cdot342765} = 1\cdot336363$;

whence $\qquad y_{10} = \dfrac{g}{g^p}10 = \dfrac{1\cdot241023}{1\cdot336363} = \dfrac{1}{1\cdot076823}$;

otherwise,

$$ y_{10} = g^{1-p^{10}} = g^{1-1\cdot342765} = g^{-\cdot342765} = \dfrac{1}{g^{\cdot342765}} = \dfrac{1}{1\cdot076823} , $$

as before.

The formula above given, when expressed in logarithms, becomes $\lambda y_x = \dfrac{k^2 a}{\lambda p}(1-p^x)$, whence may be derived the logarithms of the probability of living one year from the age (x), — quantities by the aid of which 'the construction of mortality tables is most facilitated. The logarithm of such probability is equal to the logarithm of the value of (y) at age $(x+1)$, less the logarithm of the value of (y) for the age (x).

Since—

$$ \lambda y_x = \dfrac{k^2 a}{\lambda p}(1-p^x), \text{ and } \lambda y_{x+1} = \dfrac{k^2 a}{\lambda p}(1-p^{x+1}), $$

we get, by subtraction,

$$ \lambda y_{x+1} - \lambda y_x = -\dfrac{k^2 a}{\lambda p}(p^{x+1}-p^x) = -\dfrac{k^2 a}{\lambda p} \times (p-1)p^x, $$

$$ \text{or } \lambda \dfrac{y_{x+1}}{y_x} = -mp^x ; \text{ putting } m = \dfrac{k^2 a}{\lambda p}(p-1). $$

That is to say, the logarithm (which is negative) of the probability of living one year from the age (x) is equal to (p^x) multiplied by the constant quantity $\dfrac{k^2 a}{\lambda p}(p-1)$.

If, instead of the common or decimal system of logarithms, having 10 for its base, we use the hyperbolic system, of which the base is $e = 2 \cdot 7182818$, we shall get, for the hyperbolic logarithm of the probability of living one year,

$$l \frac{y_{x+1}}{y_x} = -\frac{a}{\beta}(p-1)p^x,$$

$$= -\frac{a}{\beta}\left(\beta + \frac{\beta^2}{1.2} + \frac{\beta^3}{1.2.3} + \&\text{c.}\right)p^x,$$

$$= -a\left(1 + \frac{\beta}{2} + \frac{\beta^2}{1.2.3} + \&\text{c.}\right)p^x;$$

(β) representing the hyp. log. of (p), and (p) being developed in terms of its logarithm, (p) or e^β being $= 1 + \beta + \frac{\beta^2}{1.2} + \frac{\beta^3}{1.2.3} + \&\text{c.}$: also (β) bearing the same relation to $(p-1)$ that (a) bears to (a). The last quantity (a) measures the decrement in one year of age suffered in passing through that year by a given number $(1+a)$ of persons alive at the beginning of the year, when the mortality is constant or when $(p) = 1$.

The equation $y = \frac{g}{g^{p^x}} = g^{-p^x+1}$ having been obtained by integration, it may be useful to reverse the process and show the steps by which descent is made from the integral to the differential. We have—

$$dy = g \times d \cdot g^{-p^x} = -g \times g^{-p^x} \times lg \times lp \times p^x dx,$$

$$= -g^{-p^x+1} \times \frac{\lambda g}{k} \times \frac{\lambda p}{k} \times p^x dx,$$

$$= -y \times \frac{k^2 a}{\lambda p} \times \frac{\lambda p}{k^2} \times p^x dx,$$

$$= -y \times a \times p^x dx,$$

which is the original differential equation.

In deducing his formula, $y = d \cdot g^{p^x}$, Mr. Gompertz begins from $dy = -y \times a p^x dx$, then states, as a consequence, that $\frac{dy}{y} = -ab p^x dx$ [a new quantity, (b) being introduced as arising in the process of integration], and afterwards concludes that $y = d \cdot g^{p^x}$, (g) being stated to be the number whose common logarithm is (c), and (c) being stated to be equal to the common logarithm of $\frac{1}{p}$ multiplied by the square of the hyperbolic logarithm of 10. Since (d), a particular value of (y), does not express or exhibit (a) or (b), the result arrived at is the extraordinary one, that the formula gives the number living or surviving at the age (x) in terms entirely independent of the mortality at that or any other age. In the last Number of the *Assurance Magazine* (July, 1860), Mr. De Morgan, in his office of self-constituted judge between Mr. Gompertz and me, overlooks this important error, although he corrects other errors of inferior importance. A very simple process of integration has been rendered obscure and ambiguous by the aid of two superfluous and useless indeterminate constants, (b) and (d). The imperfect corrections of Mr. De Morgan have not diminished the original obscurity and ambiguity.

Mr. De Morgan makes the following remark respecting the introduction by Mr. Gompertz of the quantity (b):—"The preceding (b) is a superfluous quantity, which is useless. It seems to have originated in the idea that the diminution of the number living is proportional to the intensity of mortality, and therefore represented by that intensity multiplied by a constant. But the constant (a) already introduced is sufficient." On this opinion of Mr. De Morgan I would observe that a factor (b), which is greater or less than unity, cannot be superfluous and useless without being erroneous also. It is true that the constant (a), if correctly assumed, would have been sufficient; but if this quantity (a) is erroneous, and differs from the quantity (a) which I have adopted, there must be introduced, for the purpose of correcting the original error, a factor (b), such that $ab = a$. This quantity (ab) has unaccountably disappeared in the result of the investigation. If this quantity had been exhibited in the result, inquiries would have been made as to the nature of the quantity (b), which inquiries, probably, could not have been satisfactorily answered.

The formula $y = 10^{\frac{k^2 a}{\lambda p}(1 - p^x)}$, by which the number living or surviving at any age (x) is expressed in terms of the mortality at that age, may be obtained without the aid of the differential calculus, by direct methods, in which no unknown quantities are introduced. The formula may be obtained by a course of investigation similar to that pursued in calculating the amount of £1 with compound interest for one year, when the periods of conversion of interest into capital are moments or equal intervals of time infinitely small. When the rate of interest (or the rate of mortality) is constant, the problem is one of which the solution is familiar to most actuaries. In the case of the rate of interest (or the rate of mortality) continually increasing at a given annual rate (p), the sum in one year of the momentary increments may be calculated on principles similar to those on which the sum in one year of momentary increments, derived from constant ratios, is commonly calculated. It may be useful here to observe that the series of survivors in tables of mortality resemble, in a great degree, tables of discount representing the values of £1 receivable (x) years hence; that the annual rate of discount of money is $\dfrac{1}{1+a}$ when the amount of £1, with compound interest for one year, is $(1+a)$; and that, by a change of sign in the exponent, annual ratios of increasing money or population become annual ratios of decreasing money or population, applicable to the same years or intervals of age, in the same table, according to the direction in which progress is measured.

In the first place, let it be supposed that the rate of interest is constant, being such that £1 increases to £$(1+a)$ in every year. Then, let it be required to find another quantity (a), such that $\left(1+\dfrac{a}{n}\right)^{n}=1+a$; the year being divided into an infinite number (n) of equal parts, $\dfrac{a}{n}$ being a quantity indefinitely small, and $\left(1+\dfrac{a}{n}\right)$ being a constant ratio repeated or multiplied (n) times in order to produce a result equal to $(1+a)$. In treatises on algebra, it is shown that when $\left(\dfrac{a}{n}\right)$ is indefinitely small, the quantity $\left(1+\dfrac{a}{n}\right)=e^{\frac{a}{n}}$, ($e$) being the base of the hyperbolic system of logarithms and equal to 2·7182818. By substituting this value of $\left(1+\dfrac{a}{n}\right)$ in the first equation, we get $\left(e^{\frac{a}{n}}\right)^{n}=1+a$, or $e^{a}=1+a$; that is to say, $a=$ hyperbolic logarithm of $(1+a)=\dfrac{1}{k}\times$ common logarithm of $(1+a)$; (k) being the modulus of the common system and equal to ·4342945. Similarly, if (p), the annual ratio at which the interest (u) increases, be taken equal to $(1+b)$, and if (β) have the same relation to (b) that (a) has to (a), we shall have

$$\beta=\text{hyp. log. }(1+b)=\text{hyp. log. }p=\frac{1}{k}\times\text{com. log. }p.$$

Let it now be assumed that the rate of interest of money, represented by (a) at the beginning of the year, becomes, by uniform increase throughout the year, (ap) at the end of the year; and let it be required to ascertain the amount of capital and interest at the end of the year in respect of £1 capital possessed at the beginning of the year. It will be seen that the rate of interest of money, which is $\dfrac{a}{n}$ at the beginning of the first moment of the year,

becomes $\dfrac{a}{n}\left(1+\dfrac{\beta}{n}\right)$ at the end of the first of the (n) equal intervals

into which the year has been divided, and $\dfrac{a}{n}\left(1+\dfrac{\beta}{n}\right)^2$ at the end of

the second of these equal intervals; for $p=\left(1+\dfrac{\beta}{n}\right)^n$, and $\left(1+\dfrac{\beta}{n}\right)$

is the constant ratio borne by the rate of interest at the end of one moment to the rate of interest which existed at the end of the preceding moment. We shall, consequently, have, for the rates of interest existing at the end of the 1st, 2nd, 3rd, to nth moments—

$$\frac{a}{n}\left(1+\frac{\beta}{n}\right), \ \frac{a}{n}\left(1+\frac{\beta}{n}\right)^2, \ \frac{a}{n}\left(1+\frac{\beta}{n}\right)^3, \ \ldots \ \frac{a}{n}\left(1+\frac{\beta}{n}\right)^n.$$

And, since the ratio of capital and interest together, at the end of any moment, to the capital at the beginning of such moment, is that of $(1+\text{rate of interest})$ to 1, we shall have for such ratios, in the 1st, 2nd, 3rd, to nth of the (n) infinitely small equal intervals of the year—

$$\left[1+\frac{a}{n}\left(1+\frac{\beta}{n}\right)\right], \ \left[1+\frac{a}{n}\left(1+\frac{\beta}{n}\right)^2\right], \ \left[1+\frac{a}{n}\left(1+\frac{\beta}{n}\right)^3\right] \ \ldots \ \left[1+\frac{a}{n}\left(1+\frac{\beta}{n}\right)^n\right],$$

which may be written thus, on putting $\dfrac{a}{n}=a'$, and $\left(1+\dfrac{\beta}{n}\right)=v$:—

$$(1+a'v), \ (1+a'v^2), \ (1+a'v^3) \ \ldots \ (1+a'v^n);$$

which last quantities may be written as follows, since $a'v$, $a'v^2$, $a'v^3$, &c., are indefinitely small—

$$e^{a'v}, \ e^{a'v^2}, \ e^{a'v^3} \ \ldots \ e^{a'v}.$$

The product of all the above factors, (n) in number, or the value of (y) at the end of the year, will be—

$$y=e^{a'v}\times e^{a'v^2}\times e^{a'v^3}\times \text{ &c. &c. } \times e^{a'v^n},$$

$$=e^{a'(v+v^2+v^3+\text{ &c. }+v^n)}=e^{a'\frac{v^n-1}{v-1}\times v}.$$

On taking the hyperbolic logarithms of both sides, we get—

$$\text{hyp. log. } y = a' \frac{v^n - 1}{v - 1} \times v = \frac{a}{n} \times \frac{\left(1 + \dfrac{\beta}{n}\right)^n - 1}{1 + \dfrac{\beta}{n} - 1} \times \left(1 + \frac{\beta}{n}\right)$$

$$= \frac{a}{\beta}\left(e^\beta - 1\right), \quad \text{omitting } \frac{\beta}{n} \text{ as indefinitely small in } \left(1 + \frac{\beta}{n}\right),$$

$$= \frac{a}{\beta}\left(p - 1\right),$$

which is of the same form as the value previously obtained (p. 180) of the constant used in multiplying (p^x) in order to yield the hyperbolic logarithm of the probability of living one year from the age (x) years.

The value just obtained of the logarithm of the amount of £1 at the end of the first year, when the rate of interest increases by equal proportions in equal infinitely small intervals of time, will, when multiplied successively by $p^1, p^2, p^3, p^4 \ldots p^{x-1}$, give the values of the logarithms of (y) for the 2nd, 3rd, 4th . . . and xth year. The numbers corresponding to these logarithms will severally represent, for successive years, the capital and interest at the end of each year in respect of £1 capital at the beginning of such year. The multiplication together of all the annual ratios, beginning with the first and ending with the (x)th year, will give the amount of £1 in (x) years when the rate of interest is (a) at the beginning and (ap^x) at the end of the term. If (a) had been taken to represent decrement by mortality, instead of increment by interest of money, a formula in the same terms as that just given would have been obtained, with the difference only of a change of sign from positive to negative in the exponent of $(1 + a'v)$.

The formula now obtained by the most direct method of investigation is in exact agreement with the formula which I have elsewhere obtained by the aid of the differential calculus, neither of them containing any unknown constant like (b), introduced by Mr. Gompertz in the process by which his formula has been obtained. Hence it may be concluded that such a quantity does not belong to the correct formula, and that Mr. De Morgan is not wrong in designating it as superfluous and useless. If Mr. Gompertz has used this unknown quantity in order to obtain a correct result, he must have previously committed an error in the process of integration, of which (b) represents the correction.

THE

ASSURANCE MAGAZINE,

AND

JOURNAL

OF THE

INSTITUTE OF ACTUARIES.

Newton's Table of Leases. By PROFESSOR DE MORGAN.

BY this I do not mean what is usually called Newton's Table, but something which has a better right. First, however, I will speak of the table which, having Newton's certificate of accuracy, is (or was) usually called by his name.

Mr. Edleston, Fellow of Trinity College, in making researches for his biography* of Newton, found out the author of the old table. His name was Mabbot, and he was manciple *(mancipium)* or caterer of King's College. Mabbot published his table in 1686, with the well-known attestation of Newton opposite to the title-page. There seems to have ·been a special reason for procuring this certificate. The church lessors were beginning to open their eyes to the leniency of their terms, and Mabbot's tables were certainly intended both to urge them to raise their fines, &c., and to point out the proper way of doing it. Subsequent editions of these tables had attached to them the well-known letter which pointed out the advantage under which the tenant was living. As it was natural that the lessees should oppose the new scheme, and they would probably question its accuracy, it seems that Newton was applied to for the testimonial which, as is well known, he gave.

The dispute about the church leases continued till about 1731, when it was at the fiercest, and then burnt out. See a list of

---

* Attached to his *Correspondence of Sir Isaac Newton and Professor Côtes;* London, 1850; 8vo. This biography is in the form of *annals,* and, besides containing a large quantity of new and curious matter, is of exceedingly convenient reference.

some of the pamphlets which it produced in *Notes and Queries,* 2nd series, vol. iv., p. 361.

Mr. Edleston added to his mention of Mabbot the following note:—

"In the treasury of Trinity College, in a book labelled ' Notitia E,' which belonged to Humphrey Babington, as Bursar (1674-78), containing ' a true particular of the rents and leases belonging to Trin. Coll. 1674-5,' there is a table and an explanation of it in Newton's handwriting, of the fines to be paid for renewing any number of years lapsed in a lease for twenty years. It is entitled *Tabula redemptionalis ad reditus Collegij Ss. Trinitatis accommodata.* It is constructed on the hypothesis that a lease for twenty years is worth seven years' purchase, and [*i.e.,* whence it follows] that for the renewal of seven years lapsed, one year's purchase must be paid. (This is equivalent to allowing the lessee between 12 and 13 per cent. for his money.) This table, which was apparently drawn up by Newton for Babington's official use, continued to be employed by the College till 1700, when Bentley, on his appointment to the Mastership, introduced the 10 per cent. tables. The innovation, however, according to Vice-Master Walter, was unpalatable to the seniors and officers, whose ' greediness for present sealing money,' superadded to ' quarrels in the College,' compelled a return to the old system, and occasionally the granting of terms still more favourable to the tenant. On Dr. Robert Smith's succeeding to the Mastership in 1742, the 10 per cent. tables were introduced, and these were replaced in 1750 by 9 per cent. tables."

Mr. Edleston had the kindness to make a copy of this table, which I subjoin:—

" *Tabula redemptionalis ad reditus Collegij Ss. Trinitatis accommodata.*

	l.	*s.*	*d.*	*l.*	*s.*	*d.*	*l.*	*s.*	*d.*	*s.*	*d.*
	100	0	0	10	0	0	1	0	0	2	0
1	9	19	8	1	0	0	0	2	0	0	2
2	21	2	10	2	2	3	0	4	3	0	5
3	33	12	2	3	7	3	0	6	9	0	8
4	47	11	0	4	15	1	0	9	6	0	11
5	63	2	7	6	6	3	0	12	8	1	3
6	80	10	9	8	1	1	0	16	1	1	7
7	100	0	0	10	0	0	1	0	0	2	0
8	121	15	1	12	3	6	1	4	4	2	5
9	146	1	1	14	12	1	1	9	2	2	11
10	173	4	5	17	6	5	1	14	8	3	6
11	203	11	8	20	7	2	2	0	9	4	1
12	237	10	2	23	15	0	2	7	6	4	9
13	275	8	9	27	10	10	2	15	1	5	6
14	317	16	7	31	15	8	3	3	7	6	4
15	365	4	1	36	10	5	3	13	0	7	4
16	418	3	0	41	16	4	4	3	7	8	4
17	477	6	6	47	14	8	4	15	6	9	6
18	543	9	4	54	6	11	5	8	8	10	10
19	617	7	8	61	14	9	6	3	6	12	4
20	700	0	0	70	0	0	7	0	0	14	0

"Construitur hæc tabula ex hypothesi quod viginti anni septuplo reditus anni unius redimendi sint; et quod illorum viginti annorum septem annis elapsis, ut alij septem addantur et sic restituatur vicennium, reditus unius anni denuò solvi debeat. Columna sinistra continet annos redimendos ad restituendum vicennium, et reliquæ columnæ e regione continent pretia redimendi sive fines ad annuos reditus qui in fronte columnarum habentur.

"Ut si reditus annuatim sit 100*l.*, et anni 14 redimendi sint, in columna sub 100*l.* 0*s.* 0*d.* e regione 14 invenietur finis 317*l.* 16*s.* 7*d.*

"Quod si reditus sit 110*l.* 10*s.* 0*d.* et anni 12 redimendi sint, colligetur finis ex numeris qui sunt e regione 12 ad hunc modum. Nempe sub 100*l.* 0*s.* 0*d.* habetur 237*l.* 10*s.* 2*d.* finis pro reditu 100*l.* Sub 10*l.* 0*s.* 0*d.* habetur 23*l.* 15*s.* 0*d.* finis pro reditu 10*l.* Sub 1*l.* 0*s.* 0*d.* habetur 2*l.* 7*s.* 6*d.* cujus dimidium 1*l.* 3*s.* 9*d.* est finis pro reditu 10*s.* Et summa omnium 262*l.* 8*s.* 11*d.* est finis pro reditu toto.

"Eodem modo si reditus sit 237*l.* 12*s.* 6*d.*, et anni redimendi 15, consulendi erunt numeri e regione 15, et sumendum duplum numeri 365*l.* 4*s.* 1*d.*, hoc est 730*l.* 8*s.* 2*d.* pro reditu 200*l.*; triplum 36*l.* 10*s.* 5*d.*, hoc est 109*l.* 11*s.* 3*d.* pro reditu 30*l.*; septuplum 3*l.* 13*s.* 0*d.*, id est 25*l.* 11*s.* 0*d.* pro reditu 7*l.*; sextuplum 7*s.* 4*d.*, id est 2*l.* 4*s.* 0*d.* pro reditu 12*s.*, et ejusdem 7*s.* 4*d.* quarta pars hoc est 1*s.* 10*d.* pro reditu 6*d.* Atque summa omnium 867*l.* 16*s.* 3*d.* erit finis pro reditu toto.

"*Computus exempli secundi præcedentis ita se habet.*

237	10	2
23	15	0
1	3	9
262	8	11

"*Computus tertij sic.*

730	8	2
109	11	3
25	11	0
2	4	0
	1	10
867	16	3 "

The little ambiguity of expression, which I have corrected in [ ] in my quotation of Mr. Edleston's note, is, it will be observed, Newton's own doing. Seven years' purchase for twenty years gives 13·08 per cent., and one year's purchase for seven years lapsed gives 13·44 per cent.

On the Construction of Life Tables, illustrated by a new Life Table of the Healthy Districts of England. By W. FARR, ESQ., M.D., F.R.S.

(Concluded from page 141.)

IV. CONSTRUCTION OF THE COLUMNS d_x, l_x, L_x, P_x, Q_x, Y_x, AND NOTICES OF SOME OF THEIR PRACTICAL APPLICATIONS.

THE series l_x has been constructed, and from that series others are deduced to complete the Life Table, consisting now of six columns.

(1.) $d_x = l_x - l_{x+1} =$ number of deaths in the year of age following, out of l_x alive at the age x. By taking x successively at 0, 1, 2, 3 . . . . to the last age in the Table, the numbers *dying* in every year of age are obtained. The numbers dying of the age x and under the age l_{x+n} are immediately derived from the column l_x, as (2) $l_x - l_{x+n} = d_x + d_{x+1} \cdots d_{x+n-1}$. When $x + n > \omega =$ the oldest age in the Table, $l_x = d_x + d_{x+1} \cdots + d_\omega$.

(3.) $L_x = l_x + l_{x+1} \cdots \cdots + l_\omega$. The series is formed by the successive addition of the series l_x, from l_ω upwards.

(3a.) $L_x - L_{x+n} = L_{x|n} = l_x + l_{x+1} \cdots + l_{x+n-1}$.

(4.) $\left. \begin{matrix} P_x = l_{x+1} + \frac{1}{2}d_x \\ P_x = l_x - \frac{1}{2}d_x \end{matrix} \right\}$ and (5) $P_x = \dfrac{l_x + l_{x+1}}{2}$.

$P_{x+1} = l_{x+1} - \frac{1}{2}d_{x+1} = l_{x+2} + \frac{1}{2}d_{x+1}$.

The series in column P_x is constructed from the two columns l_x and d_x, or from the single column l_x, as $2P_x = l_x + l_{x+1}$; and $\therefore P_x + \dfrac{l_x + l_{x+1}}{2}$, $\therefore l_x = 2P_x - l_{x+1}$; so, conversely, the series l_x can be constructed from the series P_x. The P_x is assumed to represent the population, as expressed by the Life Table, living at the age x and under the age $x+1$; thus $P_{20} =$ the population of the age 20 and under 21 years.

By substituting the successive values of P_x in the equation (5a), $P_x + P_{x+1} \cdots P_{x+n}$, we have $\frac{1}{2}l_x + l_{x+1} \cdots + l_{x+n} + \frac{1}{2}l_{x+n+1}$.

(6.) $Q_x = P_x + P_{x+1} + P_{x+2} \cdots P_{x+n-1} + P_{x+n} \cdots + P_\omega \cdots \cdots$

$Q_{x+n} = P_{x+n} + P_{x+n+1} + P_{x+n+2} \cdots + P_\omega$.

(7.) $\therefore Q_x - Q_{x+n} = Q_{x|n} = P_x + P_{x+1} + P_{x+2} \cdots P_{x+n-1}$. The column Q_x is constructed by adding up the column P_x, and transferring the successive sums to the column Q_x.

By substituting, for the series P_x, its values in l_x, we have—

(8). $Q_x = \frac{1}{2}l_x + l_{x+1} + l_{x+2} \cdots \cdots + l_\omega$.

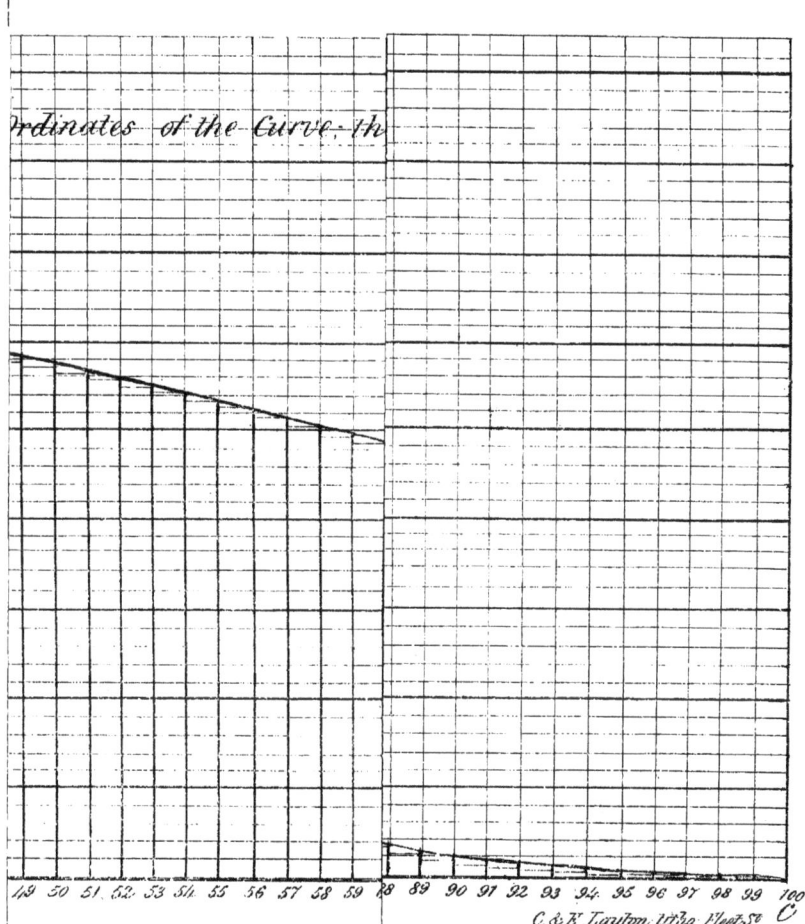

Ordinates of the Curve, th

49 50 51 52 53 54 55 56 57 58 59 88 89 90 91 92 93 94 95 96 97 98 99 100

C. & E. Layton, litho. Fleet S.^t   C

Plate XLII

DISTRICTS.
E DIAGRAMS.

And by again substituting for the series l_x its corresponding values in d_x, we have—

(9.) $Q_x = \frac{1}{2}d_x + 1\frac{1}{2}d_{x+1} + 2\frac{1}{2}d_{x+2} \ldots + (\omega + \frac{1}{2})d_\omega.$

(10.) Thus Q_x is equal to the numbers dying in each year of age after the age x, multiplied by the time (expressed in years and fractions of a year) that they have respectively lived over that age; and if $x=0$, then $Q_0 = \frac{1}{2}d_0 + 1\frac{1}{2}d_1 + 2\frac{1}{2}d_2 \ldots (n+\frac{1}{2})d_{x+n}$, when $(x+n)$ becomes $> \omega$.

(11.) This column Q_x represents, therefore, two distinct orders of facts; it represents the sum of the number of years that will be lived after the age x by the l_x persons then living, and $\therefore \dfrac{Q_x}{l_x} = $ the mean after-lifetime, of which $\dfrac{Q_{x|n}}{l_x}$ will be enjoyed before the age $x+n$ is attained, and $\dfrac{Q_{x+n}}{l_x}$ after the age $x+n$ is attained.   At birth the mean after-lifetime is $\dfrac{Q_0}{l_0}$, the unit here being one year of individual life.

(12.) Q_x also represents the sum of the numbers of men or women living at all ages over the age x, out of Q_0 living at all ages, as Q_x is in all cases the sum of the numbers living in each year of age, represented by the series P_x.   The unit is here an individual man.

(13.) Thus, on referring to the Plate appended to this paper, the lifetime of 100,000 children born simultaneously may be represented by 100,000 parallel lines, drawn from AB horizontally in the direction of CD until they cut the curved line BC.   And Q_0 is the sum of these lines expressed in the linear units of the scale on the line AC; so $\dfrac{Q_0}{l_0} = \dfrac{Q_0}{100,000} = \dfrac{4,899,665}{100,000} = 48 \cdot 99665$; the mean length of those lines = the number of years of mean lifetime.

It will be observed that, in this Table, instead of 100,000 lines, these lines are thrown into 106 groups, each comprising the variable number of lines terminating in each of 106 intervals numbered on the line AC, and representing years of age; and in these short intervals it is assumed that the mean length of the lines terminating in the eleventh interval (10 to 11) is represented by $10\frac{1}{2}$, and so on.

The relative numbers of persons living simultaneously at each interval of age will also be represented in the same Plate, fig. 1, by 106 successive vertical lines, raised from nearly the centre of each interval between the ordinates on the line AC, and measured in units of which the line AB contains 100,000. The same lines bound the figure representing the two orders of facts, and the numerical units expressing the aggregate length of the vertical lines equal in amount the units expressing the aggregate length of the horizontal lines expressed in the horizontal units.

(14.) I will now explain briefly the nature of the column Y_x, which I have added to the Life Table.* The Life Table (column P_x) exhibits a representative population, such as would be constituted by separating every year 100,000 births as they occurred, and keeping them apart in a separate community, subject to a definite law of mortality. Any population living in the tabular proportions at each year of age may, for the sake of distinction, be called a normally-constituted population.

The ages of the population represented by the Life Table amount, in the aggregate, to Y_0 years; it is the aggregate *number of years which they have already lived*, and, singularly enough, it is also, if the law of mortality remain constant, *the number of years which they will live.* Thus, Q_0 persons in such a population have lived on an average $\dfrac{Y_0}{Q_0}$ years; *that is their* MEAN AGE, and it is also their mean *after-lifetime.* Y_x is the number of years that Q_x persons have lived *over the age x,* and the mean age of such persons is $x + \dfrac{Y_x}{Q_x}$; their after-lifetime is $\dfrac{Y_x}{Q_x}$.

The series Y_x is formed by successively adding up a series of the form $\frac{1}{2}(Q_x + Q_{x+1})$, commencing at $x + 1 = \omega =$ the oldest age in the Table.

$$(15.) \; \therefore \; Y_0 = \tfrac{1}{2}Q_0 + Q_1 + Q_2 \ldots + Q_\omega,$$
$$Y_x = \tfrac{1}{2}Q_x + Q_{x+1} + Q_{x+2} \ldots + Q_\omega.$$

* *See* paper in Appendix to *Registrar-General's Sixth Annual Report*, pp. 544-552.

Extract from the Registrar-General's Sixth Annual Report (1845), p. 528.

" *Note.*—Halley's Table (1693) contained the column P. John Smart made 1,000 'born' the basis of his Table (1738), and introduced the columns *d* and *l*. Simpson adopted Smart's form of Table, which was followed by Kersseboom (1738), Deparcieux (1746), Price (1773), and Milne (1815). The columns S.*y*, *y*, and Δ*y*, in Duvillard's *Loi de Mortalité (en France) dans l'état naturel*,† correspond with the columns L, *l*, *d*, in the new Table. The S.*y* added by Duvillard is our L, and Barrett's column B. Duvillard's short Table (p. 123) has the four columns *d*, *l*, P, Q, for quinquennial or decennial ages, and the 'expectation of life.' Mathieu's Table II. is an expansion of the column Q of Duvillard's short Table, and is that column for each year of age. In a recent report on the Bengal Military Fund, Mr. Davies has a Table (1) containing columns corresponding with the *d*, *l*, L, P, Q, of the English Table, the 'Mortality per cent.', and the 'Expectation of Life' at each age."‡

I have in this paper employed *d*, *l*, L, instead of C, D, N, which have been formerly used by me and others, and should still be used where the factor v^x is introduced.

† *Influence de la Petite Vérole*, p. 161.   ‡ *See* the Note (A), p. 558.

By substituting for Q_0, for Q_1, for Q_2, and so on, their values in P_x, it will be found that—

(16.) $Y_0 = \frac{1}{2}P_0 + 1\frac{1}{2}P_1 + 2\frac{1}{2}P_2 + 3\frac{1}{2}P_3 \ldots \ldots + (n+\frac{1}{2})P_n \ldots \ldots + (\omega+\frac{1}{2})P_\omega$.

(17.) But the mean age of the persons (P_0) of the age of 0 and under 1 is nearly $\frac{1}{2}$; and so the series $\frac{1}{2}$, $1\frac{1}{2}$, $2\frac{1}{2}$, $3\frac{1}{2}$, $4\frac{1}{2}$, $5\frac{1}{2}$, $6\frac{1}{2} \ldots (n+\frac{1}{2})$, expresses nearly the mean age of all the persons in the first (P_0), second (P_1), third (P_2), and $(n+1)$th (P_n) years of age, and so for all other ages; consequently the sum of the series (16) Y_0 is the sum of the ages of all the persons living contemporaneously, as they are represented in the Life Table.

In like manner it is shown that—

(18.) $Y_x = \frac{1}{2}P_x + (1+\frac{1}{2})P_{x+1} + (2+\frac{1}{2})P_{x+2} \ldots + (\omega+\frac{1}{2}-x)P_\omega$

is the sum of the number of years that the Q_x persons in the Table have lived over the age x. They have all lived x years; and, consequently, $x + \dfrac{Y_x}{Q_x}$ gives their average age precisely as $\dfrac{Y_0}{Q_0}$ gives the average age of the whole community.

(19.) It has been shown that Q_x expresses the number of years that l_x persons will live; in the same manner it may be shown that Q_{x+1} expresses the number of years that l_{x+1} persons will live; $\therefore (l_x + l_{x+1})$ persons will live $(Q_x + Q_{x+1})$ years, $\therefore \frac{1}{2}(l_x + l_{x+1}) = P_x$ persons will live $\frac{1}{2}(Q_x + Q_{x+1})$ years. And the same may be demonstrated for each successive value of x.

But the sum of the series P_x is Q_x = the number of persons living of all ages; and the sum of the series $\frac{1}{2}(Q_x + P_{x+1})$ is Y_x = the number of years that Q_x persons will live; $\therefore \dfrac{Y_x}{Q_x}$ = the *mean after-lifetime* of all the persons living simultaneously of the age x and upwards. Thus, by the Table D, 4,899,665 persons are living contemporaneously; their mean age is $\dfrac{Y_0}{Q_0} = \dfrac{166209701}{4899665} = 33\cdot92$ years, and they will life on an average $33\cdot92$ years.

(20.) The Life Table serves to determine the value of life annuities, the value of policies, and the premiums of insurance.

This is effected by introducing a new unit, such as £1, 1 franc, 1 dollar, or any other monetary unit. Thus, if £1 is payable at each death, the series d_x will show the number of pounds falling due in each year of age; so if £1 is payable by each person on attaining the age x, and each subsequent year of age, the series l_x shows the number of *pounds* payable every year by the l_x persons; and N_x will be the number of pounds payable in the whole course of life after the age x: thus $\dfrac{N_x \pounds 1}{l_x} =$ the AVERAGE AMOUNT of an annuity of $\pounds 1$ payable on each life at and after the age x. The money-unit may be introduced into the other columns, and $\dfrac{Y_x}{Q_x}.\pounds 1$ would show the AVERAGE AMOUNT payable under an annuity of $\pounds 1$ on each of Q_x lives. The *present value* of these future payments can always be determined by assuming a given rate of interest. The estimates thus obtained are also always read subject to the qualification that, by hypothesis, the *Life Table* is based on a law of mortality actually to rule for a definite time in the population to which it is applied. The probability of the hypothesis is not here in question.

Under the same circumstances, masses of mankind appear to experience, at the same ages, the same rates of mortality; consequently, if, for several years, d_x persons have died annually on an average out of l_x persons living at the beginning of the year, other things being equal, the probability that the same number will die out of l_x persons in a year to come is greater than any other that can be named, and the fraction expressing that probability is $\dfrac{d_x}{l_x}$. We know that d_x expressing the numbers dying in a year, l_{x+1} must express the numbers surviving as $l_{x+1} + d_x = l_x$. The chances may be represented by l_x balls: l_{x+1} *white* balls in an urn will represent the chances of living, d_x *black* balls in the same urn will represent the chances of dying. Now, let each of l_x persons pay the sum z for a ticket, and each person that draws a *white* ball be entitled to $\pounds 1$. Before the drawing commences the value of each ticket is $\dfrac{l_{x+1}}{l_x}$; for l_x (the total chances) : l_{x+1} (the chances in favour of winning on one ticket) $:: 1 : \dfrac{l_{x+1}}{l_x} = z$.

Put $l_x = 30{,}007$, and $l_{x+1} = 29{,}647$; then $\dfrac{l_{x+1}.\pounds 1}{l_x} = \dfrac{29{,}647.\pounds 1}{30{,}007}$ $= \pounds{\cdot}98802$. The amount of money to be paid on l_{x+1} white balls is $\pounds 29{,}647$, and $\pounds{\cdot}9802 \times 30{,}007 = z.l_x = \pounds 29{,}647$.

In like manner it may be shown that if $\pounds 1$ is paid to each person who draws a *black* ball, the value of each ticket is $\dfrac{d\pounds 1}{l_x} = y\pounds 1$, for $y.l_x.\pounds 1 = d_x\pounds 1$, and $\pounds 1$ is to be paid on each of d_x tickets.

Should $\pounds 1$ be paid alike to those who draw white balls and to those who draw black balls, the value of a ticket will be equal to the sum of the two fractions expressing the several probabilities, namely—

$$\frac{l_{x+1}.\pounds 1}{l_x} + \frac{d_x\pounds 1}{l_x} = z + y = \frac{l_{x+1} + d_x}{l_x}\pounds 1 = \frac{l_x}{l_x}\pounds 1 = \pounds 1.$$

As one or other of the two kinds of balls *must by hypothesis be drawn*, and £1 is paid for each ball, the receipt of the £1 is certain : certainty is thus in all cases expressed by *unity*.

If every ball as it was drawn were replaced in the urn, although in 30,007 trials *white balls* were not actually drawn 29,647 times, black balls 360 times, still $\frac{29,647}{30,007}$ would express the probability of drawing a white ball, and the value of £1 contingent on that event, more accurately than any other fraction that could be named.

Again, if an urn contained, by hypothesis, an indefinite number of balls, out of which 29,647 white balls and 360 black balls were drawn and then replaced, the probability of again drawing a white ball on trial, and the value of £1 contingent on that event, would be expressed more accurately by $\frac{29,648}{30,009}$* than by any other fraction that could be named—past experience being, by hypothesis, the only means we have here of judging of the future.

Thus a Life Table applicable to the case furnishes the fractions to determine the value of any sums of money dependent on the life or death of a given person, or a certain number of given persons in a given time.

The probability of living two years, expressed by the fraction $\frac{l_{x+2}}{l_x} = \frac{l_x - (d_x + d_{x+1})}{l_x}$, is less than the probability of living one year.

Making n any number of years and fractional parts of years, the fraction $\frac{l_{x+n}}{l_x}$ will invariably express the probability of living n years after the age x. As n approaches zero, the fraction will approximate to 1, the symbol of certainty—thus a person is more likely to live a day than a year, a minute than a day. As n increases, l_{x+n} diminishes in value ; and when $x + n$ expresses a year after the age ω in the Life Table, $l_{\omega+1}$ is, by hypothesis, zero, $\therefore \frac{l_{\omega+1}}{l_x} = \frac{0}{l_x} = 0$. The chance of living so long is expressed in this

---

* The addition of 1 to the numerator, and of 2 to the denominator, may be neglected, when, as in this case, the numbers are large.

case by zero; the chance of dying in the time by 1, the symbol of certainty.

(21.) l_{x+n} expresses the number of chances in favour of surviving n years, and $l_x - l_{x+n}$ the number of chances of dying in the same time—the sum of the two together (l_x) expressing the total number of chances. Thus the fraction $\left(\dfrac{l_{x+n}}{l_x}\right)$ expressing the probability of living a given time ranges from 1 to 0, and $\dfrac{l_x - l_{x+n}}{l_x} = 1 - \dfrac{l_{x+n}}{l_x}$, or the chance of dying in a given time, also ranges from 1 to 0 as n varies. When the two fractions are equal $\dfrac{l_{x+n}}{l_x} = \dfrac{l_x - l_{x+n}}{l_x}$, then $l_{x+n} = l_x - l_{x+n}$ and $2l_{x+n} = l_x$, $\therefore l_{x+n} = \dfrac{l_x}{2}$.

To verify the equations, an age $x+n$ must be chosen at which l_{x+n} is exactly equal to $\frac{1}{2}l_x$. Thus, by the Life Table of healthy districts, 100,000 children born alive are reduced to 50,851 in 58 years, and to 49,895 in 59 years; so the chances are rather in favour of their living 58 years, as they are 50,851 to 49,149; upon the other hand, the chances of their living 59 years (49,895) are less than the chances 50,105 of their dying before attaining that age. Upon trial it will be found that the chances of living to and of dying before $58\frac{851}{956}$ years $= 58 + \dfrac{50,851 - 50,000}{d_{58}} = 58 + \dfrac{851}{956}$ years, or about $58\frac{8}{9}$ years, are nearly equal; hence this is called the *probable lifetime,* or *vie probable* by French writers, for $\dfrac{l_{58\frac{8}{9}}}{l_0} = \dfrac{1}{2}$.

At the age 20 the probable lifetime is $47\frac{1588}{1633}$, nearly 48 years. The probable lifetime at every age is immediately seen by inspection.

(22.)   V. THE THREEFOLD LIFE TABLE—PERSONS, MALES,
FEMALES.

The Life Table is threefold: a table having the six columns is made for males, another table is separately made for females; the several columns of the two tables incorporated together form the Table of Persons, which has 100,000, and may have any other number, for its basis. The basis of the Male Table in the illustration is 51,125, while the basis of the Female Table is 48,875. In that proportion males and females were born in the districts. Under this arrangement the number of contemporaneous males and females living at each age in column l_x is shown—thus, 38,388 males and 37,212 females attain the age of 20; 17,145 males attain the age of 70, and 17,133 females attain the same age. At all ages under 71 the number of males exceeds the females; at the age of 71 and upwards the females exceed the males in number; and upon referring to the columns d_x, it will be seen that the males die off in greater number than females after the age of 42. The age after the second year at which the greatest number of deaths occur is 75 in males, 76 in females.

These numbers all refer to the Life Table for Healthy Districts.

Some of the other properties of the Life Tables, admitting of innumerable applications in the solution of social phenomena, will appear in the following formulæ, which will be found useful in practice.

VI. USEFUL FORMULÆ.

The following formulæ will facilitate the use of the Life Table. The figures must be taken from the Tables of Persons, of Males or Females, applicable to the case. The formulæ are general, and are applicable to any other Life Table.

(23.) $\dfrac{d_x}{P_x} = m_x =$ the rate of mortality in the year of age following the precise age x.

(24.) $\dfrac{d_x}{l_x} = \dfrac{l_x - l_{x+1}}{l_x} = 1 - \dfrac{l_{x+1}}{l_x} =$ the probability that a person A, of the age x, in average health, will die in the following year.

(25.) $\dfrac{l_{x+1}}{l_x} = p_x = \dfrac{l_x - d_x}{l_x} = 1 - \dfrac{d_x}{l_x} =$ the probability that A, a person of the age x, will live a year; $\therefore 1 - p_x =$ the probability that A, age x, *will die in the year following*, as certainty of life $= 1$.

(26.) $\dfrac{l_x - l_{x+n}}{l_x} =$ the probability that A, age x, will die in the next n years.

(27.) $\dfrac{l_{x+n}}{l_x} =$ the probability that A, of age x, will live n years.

(28.) Put $\dfrac{l_x}{2} = l_{x+n}$, and when l_{x+n} is taken at such an age as to fulfil the conditions of the equation, then n is the *probable lifetime = vie probable =* the time that it is an even chance a person of the age x will live.

(29.) $\dfrac{Q_x}{l_x} = A_x =$ the mean *after-lifetime*, or, as it is often called, the *expectation of life*—an incorrect expression, which is rather applicable to the probable lifetime.

Note.—Upon Demoivre's hypothesis, the *probable lifetime*— that is, the time that a person may fairly expect to live, his *expectation*—was the same as the mean after-lifetime.

(30.) $G_x = x + A_x =$ the mean age at death of persons who have already lived exactly x years.

(31.) $S = c \dfrac{Q_{x \mid n}}{l_x} =$ the number of members of any Society between the ages x and $x+n$, which will be permanently sustained by $c \ldots$ annual admissions at the age x.

(32.) $c = \dfrac{S l_x}{Q_{x \mid n}} =$ annual recruits of the Society (S).

(33.) $\dfrac{S l_{x+n}}{Q_{x \mid n}} =$ annual members leaving the Society (S) on attaining the age $x+n$.

(34.) $\dfrac{S l_{x \mid n}}{Q_{x \mid n}} =$ annual deaths in such a Society (S).

(35.) $S \dfrac{Q_{x+n}}{Q_{x \mid n}} =$ the aggregate number of persons living who have left such a Society as pensioners or otherwise.

In the following formulæ it is assumed that the population is normally constituted.

(36.) $\dfrac{Y_x}{Q_x} = A'_x =$ the mean after-lifetime of all persons of the age x *and upwards.*

(37.) $\dfrac{Y_x - Y_{x+n}}{Q_x - Q_{x+n}} = \dfrac{Y_{x \mid n}}{Q_{x \mid n}} =$ the mean after-lifetime of all persons of the age of x and under the age of $x+n$.

(38.) $c \cdot \dfrac{Y_{x \mid n}}{Q_{x \mid n}} =$ the number of persons of which a Society will *ultimately consist*, recruited by c annual additions of members in the tabular proportions between the age x and $x+n$.

(39.) $c \dfrac{Y_{x \mid n} - Y_{x+m \mid n}}{Q_{x \mid m}} =$ the number of persons to which a Society joined by c persons of the tabular ages x and under $x+m$ would amount in n years. When $x+n > \omega$, this formula will be reduced to the same form as equation (38); and when $x+m$, as well as $x+n > \omega$, the equation becomes the same as (36).

VII. LIFE TABLE OF THE SIXTY-THREE HEALTHIEST ENGLISH DISTRICTS.

Upon inquiry it was found that in many districts of England the mortality of the population did not exceed the rate of 17 annual deaths to 1,000 living.

For the sake of convenience these were called "healthy districts," consisting of sixty-four, or nearly a tenth part of the total registration districts of England and Wales, and inhabited by nearly a million of people.   Sixty-three of these districts have been taken as the basis of the new Life Table, constructed according to the methods previously described.

It will be seen that these districts, generally conterminous with Poor Law Unions, are distributed over the various parts of the country.   They comprise:—*Hendon* (with Harrow*) (17), *Lewisham* (17), and *Bromley* (17), in the neighbourhood of London ; *Hambledon* (16), *Dorking* (17), *Reigate* (16), and *Godstone* (17), on the southern slope of the Surrey hills ; *East Ashford* (17), in East Kent ; *Blean* (including Herne Bay) (17), between Canterbury and the sea.   Ten districts of Sussex—*Battle* (16), near Hastings ; *Eastbourne,* around Beachy Head (15) ; *Hailsham* (17), *Uckfield* (17), *East Grinstead* (17), *Cuckfield* (16), *Steyning,* near Brighton, (16) ; *Petworth* (17), *Worthing* (17), and *Midhurst* (17).   Seven districts of Hampshire—The *Isle of Wight,* separated from the mainland by the sea (17) ; *Lymington* (17), *Christchurch* (16), *Ringwood* (17), *New Forest* (17), *Catherington* (17), and *Alresford* (17). *Wokingham* (17), and *Easthampstead* (16), in Berkshire, south of the Thames ; *Ongar* (17), in Essex, east of Epping Forest ; *Mutford* (17), including Lowestoft, on the Suffolk coast ; *Henstead* (17), south of Norwich ; *Kingsbridge* (17), on the south coast of Devon ; *Okehampton* (16), *Crediton* (17), *Barnstaple* (17), *Torrington* (17), *Bideford* (17), *Holsworthy* (16), stretching from the centre over Dartmouth and Exmoor, along the coast of the Bristol Channel ; *Stratton* (17), *Camelford* (17), and *Launceston* (17), in the adjacent parts of Cornwall, and, further south, *St. Columb* (17) ; *Williton* (17), in Somerset, also on the Bristol Channel ; *Winchcomb* (17), to the east of Cheltenham and the Cotswold Hills around the sources of the Thames ; *King's Norton* (17), in Worcestershire, adjoining Birmingham ; *Melton Mowbray* (17), in Leicestershire ; *Southwell* (17), about Sherwood Forest, in the centre of Nottinghamshire ; *Garstang* (16), in Lancashire, looking northward over Lancaster Bay ; *Easingwold* (17), in the North Riding of Yorkshire ; *Guisborough* (16), on the eastern coast north of Whitby. Then follow five border districts of Northumberland on the southern face of the Cheviot Hills—*Belford* (17), *Glendale* (15), *Rothbury*

---

* The annual deaths to 1,000 living of all ages, inserted in parentheses, are deduced from returns of the living at the Censuses of 1841 and 1851, and the deaths registered in the ten years 1841 to 1850.   (*See* Registrar-General's *Sixteenth Report,* pp. 141 to 153.)

(15), *Bellingham* (17), *Haltwhistle* (16) (is omitted in the Table), *Longtown* (17) and *Brampton* (17) on the border, and *Bootle* (16) on the coast of Cumberland; the *East Ward* (17) of Westmoreland; *Haverfordwest* (17), on the western point of South Wales; *Builth* (16), *Corwen* (17), *Pwllheli* (17), on Carnarvon Bay, and *Anglesey* (17) complete the list. These districts, and others nearly equally healthy, have been thus described:—

"Such is the variety of the soil of England, that, tested by the rates of mortality, the children reared out of a given number born, the longevity of the inhabitants, the freedom from common epidemics, or the immunity from cholera, healthy districts are found in nearly every county. Large tracts of country are, however, so much healthier than the rest, that they may be justly called Salubrious Fields; and it is remarkable that here the finest races of animals are bred. The north districts of Northumberland, around the beautiful Cheviot Hills, covered with grasses, ferns, and wild thyme, extending from the region of the heaths to the rich cultivated land at their bases, touching each other or intersected by narrow valleys—the districts extending from the Tees, over the North and East Ridings of York, to Leicestershire, Herefordshire, and parts of Shropshire—some of the districts of Gloucestershire about the Cotswold Hills—parts of Wales—North Devon, including Dartmoor and Exmoor—the Surrey and Sussex hills, with the Southdowns—have given names to the best breeds of sheep, fowls, cattle, and horses in the kingdom. . . . . .

"The dry and most inland are not always the healthiest regions of the country. The salubrious fields are sometimes watered by running streams, and diversified by lakes. The dew is abundant. They are often veiled, not by infectious fogs, but by mists drawn from the sky as it breathes over them. The mountains rise above, the ocean rolls at the distance below them, as on the coast of Sussex, North Devon, the western region of Wales, extending under Snowdon and Cader Idris in a vast amphitheatre round Cardigan Bay—the lake land and moors of the North, rising between the Irish Sea and the German Ocean. The land is sometimes heathy, but may be covered by the sweetest herbage, and bees feeding on the flowers. The cereal grains, the hop, the timber, are often of the finest quality. The animals are healthy, the native breeds are vigorous, and those fine varieties are produced at intervals which men of the genius of Bakewell, Ellman, Tomkins, Colling, and O'Kelly, make the permanent stock of the country. Industry and the army receive their best recruits from the population, while they get their worst from the people of the low parts of sickly towns. Agriculture has reclaimed many unhealthy districts on the plains, so that a considerable extent of the cultivated land is now in a state of comparative salubrity; and vast systems of drainage have subdued the noxious fens, although carried out less efficiently than is desirable, and interfered with by milldams on the rivers, descending like the Nene from the inland high -lands."*

The sanitary condition of the people in these districts is, however, still in many respects defective.

* *Report to the Registrar-General on Cholera*, pp. xcv., xcvi.

CONCLUSION.

Halley first pointed out the financial applications of the Life Table, and first calculated the values of life annuities. That branch of science, in the various forms of life insurance, has since received great developments. The new Table shows that the duration of life, among large classes of the population, by no means in unexceptionable sanitary conditions, exceeds the term of the ordinary tables, and proves that life annuities cannot be sold advantageously by Offices, or by the Government, to large classes of lives, for less than the values deducible from the new Table.

A new branch of science has been developed since Halley's day: it is the science of public health—and here a new application of the Life Table is found.

It is probable, upon physiological grounds, that man goes through all the phases of his natural development in a hundred years; and that the period of active life seldom extends beyond eighty years. But this is a very indefinite measure, as the rates of mortality in all the intermediate ages are left undetermined after it has been ascertained in what proportions men attain the extreme limits.

Generations of men, under all circumstances, die at all ages; but the proportions vary indefinitely under different conditions, from a slight tribute to death each year, down to the point of extermination by pestilence. If we ascertain at what rate a generation of men dies away under the least unfavourable existing circumstances, we obtain a standard by which the loss of life, under other circumstances, is measured; and this I have endeavoured to determine in the Life Table of English Healthy Districts. And recollecting that the science of public health was almost inaugurated in England by a former president of this Society,* who encouraged and crowned the sanitary discoveries of Captain Cook, I feel assured that it will receive with favour this imperfect attempt to supply sanitary inquirers with a scientific instrument.

In a subsequent paper I hope to be able to lay before the Society the mortality by different kinds of diseases at each age, as they have been deduced from the same series of observations.

* Sir John Pringle.

HEALTHY DISTRICTS.

TABLE A.—*Population,* 1851—*Deaths in the Five Years* 1849 *to* 1853—*Average Annual Mortality per Cent., and Logarithms of the Mortality.*

AGES.	POPULATION.			DEATHS.			AVERAGE ANNUAL MORTALITY TO 100 LIVING (m).			LOGARITHMS OF THE MORTALITY (λm).		
	Persons.	Males.	Females.	Persons.	Males.	Females.	Persons.	Males.	Females.	Persons.	Males.	Females.
1.	2.	3.	4.	5.	6.	7.	8.	9.	10.	11.	12.	13.
All ages ..	996,773	493,525	503,248	87,345	43,736	43,609	1·753	1·772	1·733	$\bar{2}$·2436718	$\bar{2}$·2485599	$\bar{2}$·2388240
Under 5 ..	130,635	65,700	64,935	26,361	14,282	12,079	4·036	4·348	3·720	$\bar{2}$·6059323	$\bar{2}$·6382536	$\bar{2}$·5705821
5–	122,406	61,733	60,673	4,209	2,080	2,129	·688	·674	·702	$\bar{3}$·8374062	$\bar{3}$·8285759	$\bar{3}$·8462102
10–	110,412	56,651	53,761	2,377	1,087	1,290	·431	·384	·480	$\bar{3}$·6340429	$\bar{3}$·5840519	$\bar{3}$·6811523
15–	181,339	90,066	91,273	6,603	3,113	3,490	·728	·691	·765	$\bar{3}$·8622801	$\bar{3}$·8396482	$\bar{3}$·8835130
25–	136,892	65,422	71,470	5,869	2,675	3,194	·857	·818	·894	$\bar{3}$·9332160	$\bar{3}$·9126300	$\bar{3}$·9512411
35–	108,056	52,734	55,322	5,208	2,447	2,761	·964	·928	·998	$\bar{3}$·9840521	$\bar{3}$·9675733	$\bar{3}$·9991985
45–	85,244	42,383	42,861	5,252	2,698	2,554	1·232	1·273	1·192	$\bar{2}$·0906909	$\bar{2}$·1048802	$\bar{2}$·0761886
55–	62,857	31,105	31,752	7,001	3,568	3,433	2·228	2·594	2·162	$\bar{2}$·3478365	$\bar{2}$·3606246	$\bar{2}$·3349327
65–	39,453	18,860	20,593	10,313	5,173	5,140	5·228	5·486	4·992	$\bar{2}$·7183350	$\bar{2}$·7392308	$\bar{2}$·6982734
75–	16,737	7,718	9,019	10,297	4,946	5,351	12·304	12·817	11·866	$\bar{1}$·0900631	$\bar{1}$·1077793	$\bar{1}$·0743066
85–	2,614	1,097	1,517	3,581	1,555	2,026	27·399	28·350	26·711	$\bar{1}$·4377287	$\bar{1}$·4525536	$\bar{1}$·4266838
95 & upwds.	128	56	72	274	112	162	42·813	40·000	45·000	$\bar{1}$·6315706	$\bar{1}$·6020600	$\bar{1}$·6532125

Note.—The ages at death of 146 persons—viz., 123 males and 23 females—were not stated; in calculating the mortality they have been distributed proportionally over the several ages in the Table. The Table may be read ·thus:—136,892 persons (of whom 65,422 were males, 71,470 were females at the age of 25 and under 35) were enumerated in 1851; at the same ages, 5,869 (2,675 males and 3,194 females) died in the five years 1849 to 1853; consequently the annual rates of mortality per cent. were ·857, ·818, and ·894.

Number of Deaths, at Five Periods of Age, in the Healthy Districts, in 1848 *to* 1855.

YEARS.	AGES.														
	Persons.					Males.					Females.				
	0.	1.	2.	3.	4.	0.	1.	2.	3.	4.	0.	1.	2.	3.	4.
1848	2,935	832	458	371	312	1,678	442	244	204	162	1,257	390	214	167	150
1849	2,932	858	541	427	292	1,637	452	263	207	154	1,295	406	278	220	138
1850	2,969	859	466	331	301	1,676	453	231	164	144	1,293	406	235	167	157
1851	3,185	932	543	341	288	1,769	502	274	179	148	1,416	430	269	162	140
1852	3,405	860	567	389	297	1,913	446	273	206	140	1,492	414	294	183	157
1853	3,370	946	554	376	287	1,888	514	293	179	137	1,482	432	261	197	150
1854	3,404	1,047	601	386	311	1,903	539	317	197	165	1,501	508	284	189	146
1855	3,350	907	533	445	297	1,948	483	257	230	156	1,402	424	276	215	141

Number of Births in Sixty-three Healthy Districts of England, 1848
to 1855.

Years.	Persons.	Males.	Females.
1848	28,679	14,756	13,923
1849	29,128	14,751	14,377
1850	29,699	15,176	14,523
1851	30,163	15,465	14,698
1852	30,370	15,557	14,813
1853	29,214	15,010	14,204

Age.	Males.
	2)29,507=births in 1848 and 1849.
0	14,754=births on January 1, 1849.
1	13,117=living on January 1, 1850.
2	12,664=living on January 1, 1851.
3	12,390=living on January 1, 1852.
4	12,184=living on January 1, 1853.
5	12,047=living on January 1, 1854.

Age.	Males.
0	1,637=deaths in 1849.
1	453=deaths in 1850.
2	274=deaths in 1851.
3	206=deaths in 1852.
4	137=deaths in 1853.

TABLE B.—*The several Values of* λp_x *on which the Life Table of Healthy Districts is based; also the corresponding Values of* p_x *and* $(1-p_x)$.

Age x.	λp_x =logarithms of the probability of living one year after the age x.		p_x =probability of *living* a year.		$(1-p_x)$ =probability of *dying* in a year.	
	Males.	Females.	Males.	Females.	Males.	Females.
0	$\bar{1}\cdot9480215$	$\bar{1}\cdot9577796$	·88720	·90736	·11280	·09264
1	$\bar{1}\cdot9844929$	$\bar{1}\cdot9859276$	·96492	·96812	·03508	·03188
2	$\bar{1}\cdot9904341$	$\bar{1}\cdot9904679$	·97821	·97829	·02179	·02171
3	$\bar{1}\cdot9932422$	$\bar{1}\cdot9932928$	·98456	·98467	·01544	·01533
7	$\bar{1}\cdot9970729$	$\bar{1}\cdot9969512$	·99328	·99300	·00672	·00700
12	$\bar{1}\cdot9984539$	$\bar{1}\cdot9980197$	·99645	·99545	·00355	·00455
20	$\bar{1}\cdot9969724$	$\bar{1}\cdot9966528$	·99305	·99232	·00695	·00768
30	$\bar{1}\cdot9964260$	$\bar{1}\cdot9960967$	·99180	·99105	·00820	·00895
40	$\bar{1}\cdot9959051$	$\bar{1}\cdot9956263$	·99062	·98998	·00938	·01002
50	$\bar{1}\cdot9943048$	$\bar{1}\cdot9946669$	·98697	·98780	·01303	·01220
60	$\bar{1}\cdot9895894$	$\bar{1}\cdot9902049$	·97631	·97770	·02369	·02230
70	$\bar{1}\cdot9751357$	$\bar{1}\cdot9773538$	·94436	·94919	·05564	·05081
80	$\bar{1}\cdot9420680$	$\bar{1}\cdot9463182$	·87512	·88373	·12488	·11627
90	$\bar{1}\cdot8747315$	$\bar{1}\cdot8809176$	·74943	·76018	·25057	·23982

Note.—Age x is in this Table the precise age. Age 12 is applied frequently to all persons of the age of 12 and under the age of 13; but in this Table it applies only to persons of the precise age of 12 years, neither more nor less. The λp_7 was, in both cases, derived from the formula $\left(\dfrac{2-m}{2+m}\right)$. The λp_{12}, deduced from this formula, is for males $\bar{1}\cdot9983497$, and for females $\bar{1}\cdot9979153$, which may be regarded either as the constant or the mean values of λp_{10}, λp_{11}, λp_{12}, λp_{13}, and λp_{14}; but as these are the terminations of an ascending and a descending series, it is probable, and quite in conformity with other observations, that one, two, or more of these values will exceed the mean value. The logarithms of p_{12} adopted are given above; and the two arithmetical means of the five logarithms, λp_{10}, λp_{11}, λp_{12}, λp_{13}, and λp_{14}, resulting from the interpolation, are $\overline{1}\cdot9983688$ for males, and $\overline{1}\cdot9979435$ for females.

The values of λp_{20}, λp_{30} . . . . . are derived from the formula

$$y_z = 10^{\frac{k^2 m}{\lambda r}(1 - r^z)}.$$

Note on the two Hypotheses.

Let b be the decrement of the ordinate y in a unit of time, then the decrement Δy of the ordinate in the time x, represented by the abscissa, will be $\Delta y = -bx$, on Demoivre's hypothesis; and as it is always proportional to the time, it will be, in an infinitely short time, $dy = -bdx$.

Passing to the integral $y = c - bx$; and if $y = a$ at the origin, when $x = 0$, $c = a$, $\therefore y = a - bx$; and if $b = 1$, then $y = a - x$. This evidently represents very closely short portions of the Life Table curve; and the smaller x is taken, the nearer is the approximation to the corresponding value of y.

Again: let Δy be the decrement of the ordinate y in the indefinite time Δx, represented by the abscissa; and let the mortality (m), represented by the ratio of the area $abfg$ to the area dfg, be $\dfrac{d_0}{P_0} = m_0$; let also m_0 increase at the rate r in a unit of time, so that $\dfrac{geh}{bcgh} = \dfrac{d_1}{P_1}$ $= m_1 = m_0 r$, and generally, within given limits, $m_0 r^x = m_x$; then $\Delta y = -y m_x \Delta x$ nearly, Δx being any small portion of time.

The error increases as the time Δx is extended, from the circumstance that on the one hand m_x varies by hypothesis momentarily, and that y, from which the varying proportional part is taken, constantly grows shorter. But by passing to the

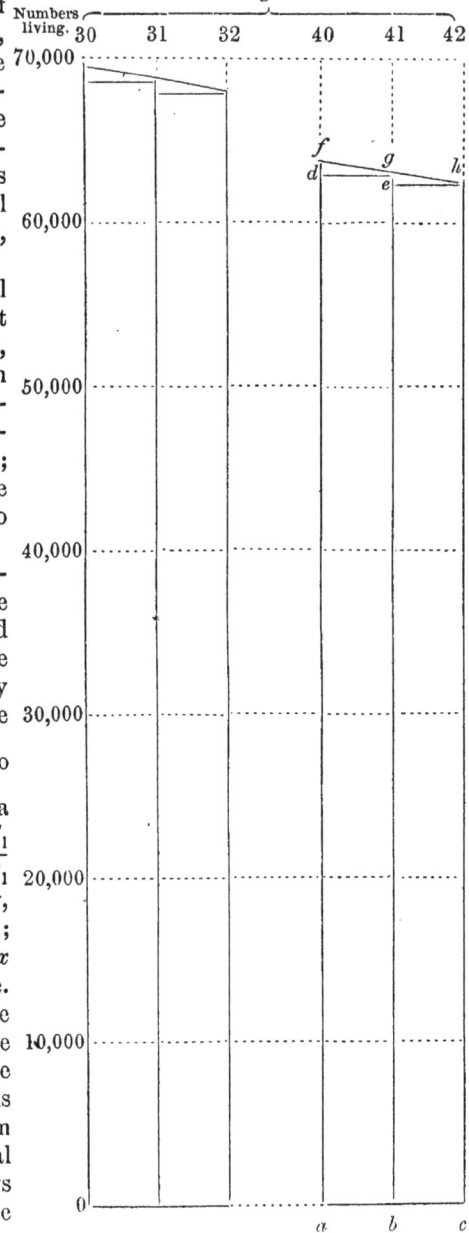

limit, and making the time dx infinitely short, m_x and y during that infinitely short time may be considered constant, and $dy = -ym_x dx$ will be the true decrement. Substituting $m_0 r^x$ for m_x, the equation becomes $dy = -ym_0 r^x dx$, from which the value of y can be derived, as before shown; for $\dfrac{dy}{y} = -m_0 r^x dx$, and, integrating both sides, $\lambda_\epsilon y = \lambda_\epsilon c - \dfrac{m_0 r^x}{\lambda_\epsilon r}$; here λ_ϵ stands for the logarithm having ϵ for its base.

At the origin of the curve, when $x=0$, let $y=1$, and then $\lambda_\epsilon c = \dfrac{m_0}{\lambda_\epsilon r}$.

Now, substituting this value for $\lambda_\epsilon c$, we have $\lambda_\epsilon y = \dfrac{m_0}{\lambda_\epsilon r} - \dfrac{m_0 r^x}{\lambda_\epsilon r}$,

$\therefore \lambda_\epsilon y = \dfrac{m_0}{\lambda_\epsilon r}(1 - r^x)$; and, passing to the number, $y = \epsilon^{\frac{m_0}{\lambda_\epsilon r}(1-r^x)}$. Putting k for the modulus of the common logarithm (λ) having 10 for its base, we have $\lambda_\epsilon y = \dfrac{\lambda y}{k}$, and $\lambda_\epsilon r = \dfrac{\lambda r}{k}$, $\therefore \dfrac{\lambda y}{k} = \dfrac{km}{\lambda r}(1 - r^x)$; or, passing to the number, $y = 10^{\frac{k^2 m}{\lambda^2 r}(1-r^x)}$

Upon the one hypothesis, out of a generation of men an *equal quantity of life** is destroyed in equal times, out of diminishing quantities in existence, the *proportion* that perishes of the residual life constantly *increasing*.

Upon the other hypothesis, a *decreasing proportion* of the residual life is destroyed from birth down to the age of puberty; in the after ages, a *proportion increasing* at different rates is destroyed in equal times. The *quantity* of life *destroyed* in equal times may be the same, or different upon this hypothesis; and in very short intervals of age the differences between *the quantities of life destroyed* may be so inconsiderable that they may be neglected.

The two hypotheses may be illustrated. Assume that at every beat of the heart an equal quantity of vital force on an average is consumed in excess of that produced; or if this does not happen at distant ages, assume that it happens during two consecutive years, two consecutive days, two consecutive pulses of a generation of men, and is represented by the deaths in the two intervals. This will give an idea of the first hypothesis.

The second hypothesis will be represented by assuming that, in addition to the existing force, a certain amount of vital force is produced, while a certain amount is also destroyed at every beat of the heart—the quantity destroyed exceeding the quantity produced in a diminishing ratio, and then in an increasing ratio—the proportional part destroyed being for this purpose always represented by the proportional number of hearts beating to the number of hearts ceasing to beat at every instant of age among a generation of men. The respirations, the sensations, the secretions, nutrition, and all the vital acts, may be conceived, like the heart, to influence the continuance of the vital force, implying here simply the force which sustains life.

---

* The quality or the intensity of life at different ages is purposely left out of consideration.

TABLE B 1.—*Life Table of Healthy English Districts.—Logarithms of the Numbers of Males and Females living at each Year of Age.*

$\lambda l_x.$				$\lambda l_x.$			
Age. x.	Males.	Age x.	Females.	Age x.	Males.	Age x.	Females.
0	4·7086364	0	4·6890835	55	4·4351998	55	4·4177773
1	4·6566579	1	4·6468631	56	4·4279544	56	4·4116015
2	4·6411508	2	4·6327907	57	4·4203212	57	4·4052190
3	4·6315849	3	4·6232586	58	4·4122719	58	4·3981522
4	4·6248271	4	4·6165514	59	4·4037768	59	4·3901691
5	4·6193109	5	4·6110606	60	4·3943905	60	4·3812819
6	4·6148376	6	4·6065737	61	4·3839799	61	4·3714868
7	4·6112225	7	4·6028950	62	4·3725154	62	4·3607637
8	4·6082954	8	4·5998462	63	4·3599518	63	4·3490765
9	4·6059001	9	4·5972658	64	4·3462281	64	4·3363727
10	4·6038946	10	4·5950094	65	4·3312678	65	4·3225837
11	4·6021511	11	4·5929497	66	4·3149786	66	4·3076249
12	4·6005560	12	4·5909763	67	4·2972528	67	4·2913951
13	4·5990100	13	4·5889960	68	4·2779668	68	4·2737774
14	4·5974279	14	4·5869326	69	4·2569814	69	4·2546384
15	4·5957387	15	4·5847269	70	4·2341418	70	4·2338287
16	4·5938855	16	4·5823368	71	4·2092775	71	4·2111825
17	4·5918259	17	4·5797373	72	4·1822024	72	4·1865180
18	4·5895314	18	4·5769202	73	4·1527146	73	4·1596372
19	4·5869878	19	4·5738947	74	4·1205968	74	4·1303259
20	4·5841951	20	4·5706868	75	4·0856157	75	4·0983537
21	4·5811675	21	4·5673396	76	4·0475228	76	4·0634741
22	4·5780527	22	4·5639166	77	4·0060534	77	4·0254242
23	4·5748607	23	4·5604237	78	3·9609277	78	3·9839252
24	4·5716008	24	4·5568665	79	3·9118498	79	3·9386819
25	4·5682808	25	4·5532498	80	3·8585083	80	3·8893831
26	4·5649078	26	4·5495779	81	3·8005763	81	3·8357013
27	4·5614874	27	4·5458546	82	3·7377111	82	3·7772929
28	4·5580244	28	4·5420830	83	3·6695542	83	3·7137979
29	4·5545223	29	4·5382656	84	3·5957318	84	3·6448405
30	4·5509835	30	4·5344046	85	3·5158541	85	3·5700284
31	4·5474095	31	4·5305013	86	3·4295159	86	3·4889532
32	4·5438005	32	4·5265566	87	3·3362962	87	3·4011904
33	4·5401557	33	4·5225708	88	3·2357583	88	3·3062992
34	4·5364730	34	4·5185435	89	3·1274500	89	3·2038228
35	4·5327494	35	4·5144739	90	3·0109034	90	3·0932880
36	4·5289808	36	4·5103606	91	2·8856349	91	2·9742056
37	4·5251620	37	4·5062016	92	2·7511453	92	2·8460701
38	4·5212864	38	4·5019942	93	2·6069196	93	2·7083599
39	4·5173467	39	4·4977353	94	2·4524273	94	2·5605372
40	4·5133342	40	4·4934212	95	2·2871223	95	2·4020479
41	4·5092393	41	4·4890475	96	2·1104426	96	2·2323219
42	4·5050512	42	4·4846093	97	1·9218108	97	2·0507729
43	4·5007579	43	4·4801012	98	1·7206337	98	1·8567982
44	4·4963465	44	4·4755172	99	1·5063024	99	1·6497793
45	4·4918029	45	4·4708506	100	1·2781926	100	1·4290811
46	4·4871119	46	4·4660943	101	1·0356640	101	1·1940526
47	4·4822570	47	4·4612404	102	0·7780608	102	0·9440265
48	4·4772210	48	4·4562807	103	0·5047118	103	0·6783194
49	4·4719852	49	4·4512061	104	0·2149296	104	0·3962318
50	4·4665301	50	4·4460074	105	9·9080117	105	0·0970476
51	4·4608349	51	4·4406743	106	9·5832396	106	9·7800351
52	4·4548778	52	4·4351962	107	9·2398792	107	9·4444460
53	4·4486358	53	4·4295620	108	8·8771808	108	9·0895160
54	4·4420848	54	4·4237598	109	8·4943792	109	8·7144646

TABLE C.—*Healthy Districts.*

Age.	LIVING AT EACH AGE (l_x).			DYING IN EACH YEAR OF AGE (d_x).			Age.
x.	Persons.	Males.	Females.	Persons.	Males.	Females.	x.
0	100,000	51,125	48,875	10,295	5,767	4,528	0
1	89,705	45,358	44,347	3,005	1,591	1,414	1
2	86,700	43,767	42,933	1,885	953	932	2
3	84,815	42,814	42,001	1,305	661	644	3
4	83,510	42,153	41,357	1,051	532	519	4
5	82,459	41,621	40,838	847	427	420	5
6	81,612	41,194	40,418	682	341	341	6
7	80,930	40,853	40,077	555	275	280	7
8	80,375	40,578	39,797	459	223	236	8
9	79,916	40,355	39,561	391	186	205	9
10	79,525	40,169	39,356	347	161	186	10
11	79,178	40,008	39,170	324	146	178	11
12	78,854	39,862	38,992	319	142	177	12
13	78,535	39,720	38,815	328	144	184	13
14	78,207	39,576	38,631	350	154	196	14
15	77,857	39,422	38,435	379	168	211	15
16	77,478	39,254	38,224	414	186	228	16
17	77,064	36,068	37,996	451	205	246	17
18	76,613	38,863	37,750	489	227	262	18
19	76,124	38,636	37,488	524	248	276	19
20	75,600	38,388	37,212	552	267	285	20
21	75,048	38,121	36,927	562	272	290	21
22	74,486	37,849	36,637	571	277	294	22
23	73,915	37,572	36,343	577	281	296	23
24	73,338	37,291	36,047	583	284	299	24
25	72,755	37,007	35,748	588	287	301	25
26	72,167	36,720	35,447	591	288	303	26
27	71,576	36,432	35,144	593	289	304	27
28	70,983	36,143	34,840	595	290	305	28
29	70,388	35,853	34,535	596	291	305	29
30	69,792	35,562	34,230	598	292	306	30
31	69,194	35,270	33,924	599	292	307	31
32	68,595	34,978	33,617	599	292	307	32
33	67,996	34,686	33,310	601	293	308	33
34	67,395	34,393	33,002	601	293	308	34
35	66,794	34,100	32,694	603	295	308	35
36	66,191	33,805	32,386	604	296	308	36
37	65,587	33,509	32,078	608	298	310	37
38	64,979	33,211	31,768	610	300	310	38
39	64,369	32,911	31,458	613	302	311	39
40	63,756	32,609	31,147	618	306	312	40
41	63,138	32,303	30,835	623	310	313	41
42	62,515	31,993	30,522	630	315	315	42
43	61,885	31,678	30,207	633	320	318	43
44	61,247	31,358	29,889	645	326	319	44
45	60,602	31,032	29,570	656	334	322	45
46	59,946	30,698	29,248	666	341	325	46
47	59,280	30,357	28,923	679	350	329	47
48	58,601	30,007	28,594	692	360	332	48
49	57,909	29,647	28,262	706	370	336	49
50	57,203	29,277	27,926	722	381	341	50
51	56,481	28,896	27,585	740	394	346	51
52	55,741	28,502	27,239	758	407	351	52
53	54,983	28,095	26,888	777	420	357	53

TABLE C (*continued*).

Age.	LIVING AT EACH AGE (l_x).			DYING IN EACH YEAR OF AGE (d_x).			Age.
x.	Persons.	Males.	Females.	Persons.	Males.	Females.	*x.*
54	54,206	27,675	26,531	798	435	363	54
55	53,408	27,240	26,168	820	451	369	55
56	52,588	26,789	25,799	843	467	376	56
57	51,745	26,322	25,423	894	483	411	57
58	50,851	25,839	25,012	956	501	455	58
59	49,895	25,338	24,557	1,040	542	498	59
60	48,855	24,796	24,059	1,123	587	536	60
61	47,732	24,209	23,523	1,205	631	574	61
62	46,527	23,578	22,949	1,281	672	609	62
63	45,246	22,906	22,340	1,356	712	644	63
64	43,890	22,194	21,696	1,430	752	678	64
65	42,460	21,442	21,018	1,501	789	712	65
66	40,959	20,653	20,306	1,571	826	745	66
67	39,388	19,827	19,561	1,638	861	777	67
68	37,750	18,966	18,784	1,705	895	810	68
69	36,045	18,071	17,974	1,767	926	841	69
70	34,278	17,145	17,133	1,825	954	871	70
71	32,453	16,191	16,262	1,876	978	898	71
72	30,577	15,213	15,364	1,921	999	922	72
73	28,656	14,214	14,442	1,955	1,013	942	73
74	26,701	13,201	13,500	1,980	1,022	958	74
75	24,721	12,179	12,542	1,991	1,023	968	75
76	22,730	11,156	11,574	1,987	1,016	971	76
77	20,743	10,140	10,603	1,966	1,000	966	77
78	18,777	9,140	9,637	1,931	977	954	78
79	16,846	8,163	8,683	1,875	943	932	79
80	14,971	7,220	7,751	1,803	902	901	80
81	13,168	6,318	6,850	1,713	851	862	81
82	11,455	5,467	5,988	1,608	794	814	82
83	9,847	4,673	5,174	1,491	731	760	83
84	8,356	3,942	4,414	1,360	662	698	84
85	6,996	3,280	3,716	1,224	591	633	85
86	5,772	2,689	3,083	1,084	520	564	86
87	4,688	2,169	2,519	943	448	495	87
88	3,745	1,721	2,024	805	380	425	88
89	2,940	1,341	1,599	675	316	359	89
90	2,265	1,025	1,240	555	257	298	90
91	1,710	768	942	444	204	240	91
92	1,266	564	702	350	159	191	92
93	916	405	511	269	122	147	93
94	647	283	364	201	89	112	94
95	446	194	252	146	65	81	95
96	300	129	171	104	45	59	96
97	196	84	112	71	31	40	97
98	125	53	72	48	21	27	98
99	77	32	45	31	13	18	99
100	46	19	27	19	8	11	100
101	27	11	16	12	5	7	101
102	15	6	9	7	3	4	102
103	8	3	5	4	1	3	103
104	4	2	2	2	1	1	104
105	2	1	1	1	1	..	105
106	1	..	1	1	..	1	106

TABLE D.—*Healthy Districts.*—*Persons.*

Age.	Dying in each Year of Age, 0-1, 1-2, to 105-106.	Born and Living at each Age.	Sum of the Numbers Born and Living at each Age (x) from x to the last Age in the Table.	Population, or the living in each Year of Age 0 to 1, 1 to 2, &c.	(1) Sum of the Living, and of the Living of every Age (x) and upwards to the last Age in the Table; also (2) the Years which the Persons (l_x) will live.	(1) The Years which the Persons at the Age (x) and upwards will live; also (2) the Years which they have lived over x.	Age.
	$\Sigma d_x.$	$\Sigma l_x.$	$\mathbf{L}_x.$	$\frac{1}{2}(l_x+l_{x+1})$ $=l_{x+1}+\frac{1}{2}d_x.$	$\Sigma \mathbf{P}_x.$	$\Sigma\frac{1}{2}(\mathbf{Q}_x+\mathbf{Q}_{x+1})$ $=\mathbf{Y}_{x+1}+(\mathbf{Q}_{x+1}+\frac{1}{2}\mathbf{P}_x).$	
$x.$	$d_x.$	$l_x.$	$\mathbf{L}_x.$	$\mathbf{P}_x.$	$\mathbf{Q}_x.$	$\mathbf{Y}_x.$	$x.$
0	10,295	100,000	4,951,908	92,611	4,899,665	166,209,701	0
1	3,005	89,705	4,851,908	88,202	4,807,054	161,356,341	1
2	1,885	86,700	4,762,203	85,758	4,718,852	156,593,388	2
3	1,305	84,815	4,675,503	84,162	4,633,094	151,917,415	3
4	1,051	83,510	4,590,688	82,985	4,548,932	147,326,402	4
5	847	82,459	4,507,178	82,036	4,465,947	142,818,963	5
6	682	81,612	4,424,719	81,270	4,383,911	138,394,034	6
7	555	80,930	4,343,107	80,653	4,302,641	134,050,757	7
8	459	80,375	4,262,177	80,145	4,221,988	129,788,443	8
9	391	79,916	4,181,802	79,721	4,141,843	125,606,527	9
10	347	79,525	4,101,886	79,352	4,062,122	121,504,545	10
11	324	79,178	4,022,361	79,016	3,982,770	117,482,099	11
12	319	78,854	3,943,183	78,694	3,903,754	113,538,837	12
13	328	78,535	3,864,329	78,371	3,825,060	109,674,430	13
14	350	78,207	3,785,794	78,032	3,746,689	105,888,556	14
15	379	77,857	3,707,587	77,668	3,668,657	102,180,882	15
16	414	77,478	3,629,730	77,271	3,590,989	98,551,059	16
17	451	77,064	3,552,252	76,838	3,513,718	94,998,706	17
18	489	76,613	3,475,188	76,369	3,436,880	91,523,407	18
19	524	76,124	3,398,575	75,862	3,360,511	88,124,711	19
20	552	75,600	3,322,451	75,323	3,284,649	84,802,131	20
21	562	75,048	3,246,851	74,767	3,209,326	81,555,144	21
22	571	74,486	3,171,803	74,201	3,134,559	78,383,202	22
23	577	73,915	3,097,317	73,626	3,060,358	75,285,743	23
24	583	73,338	3,023,402	73,047	2,986,732	72,262,198	24
25	588	72,755	2,950,064	72,461	2,913,685	69,311,989	25
26	591	72,167	2,877,309	71,872	2,841,224	66,434,535	26
27	593	71,576	2,805,142	71,279	2,769,352	63,629,247	27
28	595	70,983	2,733,566	70,685	2,698,073	60,895,535	28
29	596	70,388	2,662,583	70,091	2,627,388	58,232,804	29
30	598	69,792	2,592,195	69,493	2,557,297	55,640,462	30
31	599	69,194	2,522,403	68,894	2,487,804	53,117,911	31
32	599	68,595	2,453,209	68,296	2,418,910	50,664,554	32
33	601	67,996	2,384,614	67,695	2,350,614	48,279,792	33
34	601	67,395	2,316,618	67,095	2,282,919	45,963,025	34
35	603	66,794	2,249,223	66,492	2,215,824	43,713,654	35
36	604	66,191	2,182,429	65,889	2,149,332	41,531,076	36
37	608	65,587	2,116,238	65,283	2,083,443	39,414,688	37
38	610	64,979	2,050,651	64,674	2,018,160	37,363,887	38
39	613	64,369	1,985,672	64,062	1,953,486	35,378,064	39
40	618	63,756	1,921,303	63,447	1,889,424	33,456,609	40
41	623	63,138	1,857,547	62,827	1,825,977	31,598,909	41
42	630	62,515	1,794,409	62,200	1,763,150	29,804,345	42
43	638	61,885	1,731,894	61,566	1,700,950	28,072,295	43
44	645	61,247	1,670,009	60,925	1,639,384	26,402,128	44
45	656	60,602	1,608,762	60,274	1,578,459	24,793,206	45
46	666	59,946	1,548,160	59,612	1,518,185	23,244,885	46
47	679	59,280	1,488,214	58,941	1,458,573	21,756,505	47
48	692	58,601	1,428,934	58,255	1,399,632	20,327,403	48
49	706	57,909	1,370,333	57,556	1,341,377	18,956,899	49
50	722	57,203	1,312,424	56,842	1,283,821	17,644,300	50
51	740	56,481	1,255,221	56,111	1,226,979	16,338,099	51
52	758	55,741	1,198,740	55,362	1,170,868	15,189,976	52
53	777	54,983	1,142,999	54,594	1,115,506	14,046,789	53

Table D (continued).

Age.	Dying in each Year of Age, 0-1, 1-2, to 105-106. Σd_x.	Born and Living at each Age. Σl_x.	Sum of the Numbers Born and Living at each Age (x) from x to the last Age in the Table. Σl_x.	Population, or the Living in each Year of Age 0 to 1, 1 to 2, &c. $\frac{1}{2}(l_x + l_{x+1})$ $= l_{x+1} + \frac{1}{2}d_x$.	(1) Sum of the Living, and of the Living of every Age (x) and upwards to the last Age in the Table; also (2) the Years which the Persons (l_x) will live. ΣP_x.	(1) The Years which the Persons at the Age (x) and upwards will live; also (2) the Years which they have lived over x. $\Sigma \frac{1}{2}(Q_x + Q_{x+1})$ $= Y_{x+1} + (Q_{x+1} + \frac{1}{2}P_x)$.
x.	d_x.	l_x.	L_x.	P_x.	Q_x.	Y_x.
54	798	54,206	1,088,016	53,808	1,060,912	12,958,580
55	820	53,408	1,033,810	52,997	1,007,104	11,924,572
56	843	52,588	980,402	52,167	954,107	10,943,966
57	894	51,745	927,814	51,298	901,940	10,015,943
58	956	50,851	876,069	50,373	850,642	9,139,652
59	1,040	49,895	825,218	49,375	800,269	8,314,197
60	1,123	48,855	775,323	48,293	750,894	7,538,615
61	1,205	47,732	726,468	47,130	702,601	6,811,867
62	1,281	46,527	678,736	45,887	655,471	6,132,831
63	1,356	45,246	632,209	44,568	609,584	5,500,304
64	1,430	43,890	586,963	43,175	565,016	4,913,004
65	1,501	42,460	543,073	41,709	521,841	4,369,575
66	1,571	40,959	500,613	40,173	480,132	3,868,589
67	1,638	39,388	459,654	38,570	439,959	3,408,544
68	1,705	37,750	420,266	36,897	401,389	2,987,869
69	1,767	36,045	382,516	35,161	364,492	2,604,929
70	1,825	34,278	346,471	33,366	329,331	2,258,017
71	1,876	32,453	312,193	31,515	295,965	1,945,369
72	1,921	30,577	279,740	29,617	264,450	1,665,162
73	1,955	28,656	249,163	27,678	234,833	1,415,520
74	1,980	26,701	220,507	25,711	207,155	1,194,527
75	1,991	24,721	193,806	23,726	181,444	1,000,227
76	1,987	22,730	169,085	21,736	157,718	830,646
77	1,966	20,743	146,355	19,760	135,982	683,796
78	1,931	18,777	125,612	17,811	116,222	557,694
79	1,875	16,846	106,835	15,909	98,411	450,377
80	1,803	14,971	89,989	14,070	82,502	359,921
81	1,713	13,168	75,018	12,311	68,432	284,454
82	1,608	11,455	61,850	10,651	56,121	222,178
83	1,491	9,847	50,395	9,102	45,470	171,382
84	1,360	8,356	40,548	7,676	36,368	130,463
85	1,224	6,996	32,192	6,383	28,692	97,933
86	1,084	5,772	25,196	5,230	22,309	72,432
87	943	4,688	19,424	4,217	17,079	52,739
88	805	3,745	14,736	3,342	12,862	37,768
89	675	2,940	10,991	2,603	9,520	26,577
90	555	2,265	8,051	1,988	6,917	18,358
91	444	1,710	5,786	1,488	4,929	12,436
92	350	1,266	4,076	1,090	3,441	8,251
93	269	916	2,810	782	2,351	5,355
94	201	647	1,894	547	1,569	3,395
95	146	446	1,247	372	1,022	2,099
96	104	300	801	249	650	1,263
97	71	196	501	160	401	737
98	48	125	305	101	241	416
99	31	77	180	61	140	226
100	19	46	103	37	79	116
101	12	27	57	21	42	56
102	7	15	30	11	21	24
103	4	8	15	7	10	9
104	2	4	7	2	3	3
105	1	2	3	1	1	..
106	1	1	1	..	..	..

TABLE E —*Healthy Districts.—Males.*

Age.	Dying in each Year of Age 0–1, 1–2, to 104–105.	Born and Living at each Age.	Sum of the Numbers Born and Living at each Age (x) from x to the last Age in the Table.	Population, or the Living in each Year of Age 0 to 1, 1 to 2, &c.	(1) Sum of the Living, and of the Living of every Age (x) and upwards to the last Age in the Table; also (2) the Years which the Males (l_x) will live.	(1) The Years which the Males at the Age (x) and upwards will live; also (2) the Years which they have lived over x.	Age.
x.	Σd_x.	l_x.	Σl_x.	$\frac{1}{2}(l_x+l_{x+1})$ $=l_{x+1}+\frac{1}{2}d_x$.	ΣP_x.	$\Sigma\frac{1}{2}(Q_x+Q_{x+1})$ $=Y_{x+1}+(Q_{x+1}+\frac{1}{2}P_x)$.	x.
	d_x.	l_x.	L_x.	P_x.	Q_x.	Y_x.	
0	5,767	51,125	2,509,635	46,915*	2,482,745	84,008,921	0
1	1,591	45,358	2,458,510	44,562	2,435,830	81,549,633	1
2	953	43,767	2,413,152	43,291	2,391,268	79,136,084	2
3	661	42,814	2,369,385	42,483	2,347,977	76,766,462	3
4	532	42,153	2,326,571	41,887	2,305,494	74,439,726	4
5	427	41,621	2,284,418	41,408	2,263,607	72,155,176	5
6	341	41,194	2,242,797	41,023	2,222,199	69,912,273	6
7	275	40,853	2,201,603	40,716	2,181,176	67,710,585	7
8	223	40,578	2,160,750	40,466	2,140,460	65,549,767	8
9	186	40,355	2,120,172	40,262	2,099,994	63,429,540	9
10	161	40,169	2,079,817	40,089	2,059,732	61,349,677	10
11	146	40,008	2,039,648	39,935	2,019,643	59,309,990	11
12	142	39,862	1,999,640	39,791	1,979,708	57,310,314	12
13	144	39,720	1,959,778	39,648	1,939,917	55,350,502	13
14	154	39,576	1,920,058	39,499	1,900,269	53,430,409	14
15	168	39,422	1,880,482	39,338	1,860,770	51,549,889	15
16	186	39,254	1,841,060	39,161	1,821,432	49,708,788	16
17	205	39,068	1,801,806	38,965	1,782,271	47,906,937	17
18	227	38,863	1,762,738	38,750	1,743,306	46,144,148	18
19	248	38,636	1,723,875	38,512	1,704,556	44,420,217	19
20	267	38,388	1,685,239	38,254	1,666,044	42,734,917	20
21	272	38,121	1,646,851	37,985	1,627,790	41,088,000	21
22	277	37,849	1,608,730	37,711	1,589,805	39,479,203	22
23	281	37,572	1,570,881	37,431	1,552,094	37,908,253	23
24	284	37,291	1,533,309	37,149	1,514,663	36,374,875	24
25	287	37,007	1,496,018	36,864	1,477,514	34,878,786	25
26	288	36,720	1,459,011	36,576	1,440,650	33,419,704	26
27	289	36,432	1,422,291	36,287	1,404,074	31,997,342	27
28	290	36,143	1,385,859	35,998	1,367,787	30,611,412	28
29	291	35,853	1,349,716	35,708	1,331,789	29,261,624	29
30	292	35,562	1,313,863	35,416	1,296,081	27,947,689	30
31	292	35,270	1,278,301	35,124	1,260,665	26,669,316	31
32	292	34,978	1,243,031	34,832	1,225,541	25,426,213	32
33	293	34,686	1,208,053	34,539	1,190,709	24,218,088	33
34	293	34,393	1,173,367	34,247	1,156,170	23,044,648	34
35	295	34,100	1,138,974	33,952	1,121,923	21,905,602	35
36	296	33,805	1,104,874	33,657	1,087,971	20,800,655	36
37	298	33,509	1,071,069	33,360	1,054,314	19,729,512	37
38	300	33,211	1,037,560	33,061	1,020,954	18,691,878	38
39	302	32,911	1,004,349	32,760	987,893	17,687,455	39
40	306	32,609	971,438	32,456	955,133	16,715,942	40
41	310	32,303	938,829	32,148	922,677	15,777,037	41
42	315	31,993	906,526	31,836	890,529	14,870,434	42
43	320	31,678	874,533	31,518	858,693	13,995,823	43
44	326	31,358	842,855	31,195	827,175	13,152,889	44
45	334	31,032	811,497	30,865	795,980	12,341,311	45
46	341	30,698	780,465	30,527	765,115	11,560,764	46
47	350	30,357	749,767	30,182	734,588	10,810,912	47
48	360	30,007	719,410	29,827	704,406	10,091,415	48
49	370	29,647	689,403	29,462	674,579	9,401,923	49
50	381	29,277	659,756	29,087	645,117	8,742,075	50
51	394	28,896	630,479	28,699	616,030	8,111,501	51

* P_0 is $\frac{1}{2}(l_0+l_1) \times (\cdot9725)$. The factor $\cdot9725$ has been introduced, as the number living in the first year is less than the arithmetical mean of those born and surviving a year.

TABLE E (*continued*).

Age.	Dying in each Year of Age 0–1, 1–2, to 104–105.	Born and Living at each Age.	Sum of the Numbers Born and Living at each Age (x) from x to the last Age in the Table.	Population, or the Living in each Year of Age 0 to 1, 1 to 2, &c.	(1) Sum of the Living, and of the Living of every Age (x) and upwards to the last Age in the Table; also (2) the Years which the Males (l_x) will live.	(1) The Years which the Males at the Age x and upwards will live; also (2) the Years which they have lived over x.	Age.
		$\Sigma d_x.$	$\Sigma l_x.$	$\frac{1}{2}(l_x + l_{x+1})$ $= l_{x+1} + \frac{1}{2}d_x.$	$\Sigma P_x.$	$\Sigma\frac{1}{2}(Q_x + Q_{x+1})$ $= Y_{x+1} + (Q_{x+1} + \frac{1}{2}P_x).$	
$x.$	$d_x.$	$l_x.$	$L_x.$	$P_x.$	$Q_x.$	$Y_x.$	$x.$
52	407	28,502	601,583	28,298	587,331	7,509,821	52
53	420	28,095	573,081	27,885	559,033	6,936,639	53
54	435	27,675	544,986	27,458	531,148	6,391,548	54
55	451	27,240	517,311	27,014	503,690	5,874,129	55
56	467	26,789	490,071	26,556	476,676	5,383,946	56
57	483	26,322	463,282	26,080	450,120	4,920,548	57
58	501	25,839	436,960	25,589	424,040	4,483,468	58
59	542	25,338	411,121	25,067	398,451	4,072,223	59
60	587	24,796	385,783	24,502	373,384	3,686,305	60
61	631	24,209	360,987	23,894	348,882	3,325,172	61
62	672	23,578	336,778	23,242	324,988	2,988,237	62
63	712	22,906	313,200	22,550	301,746	2,674,870	63
64	752	22,194	290,294	21,818	279,196	2,384,399	64
65	789	21,442	268,100	21,047	257,378	2,116,112	65
66	826	20,653	246,658	20,240	236,331	1,869,258	66
67	861	19,827	226,005	19,397	216,091	1,643,047	67
68	895	18,966	206,178	18,518	196,694	· 1,436,654	68
69	926	18,071	187,212	17,608	178,176	1,249,219	69
70	954	17,145	169,141	16,668	160,568	1,079,847	70
71	978	16,191	151,996	15,702	143,900	927,613	71
72	999	15,213	135,805	14,714	128,198	791,564	72
73	1,013	14,214	120,592	13,707	113,484	670,723	73
74	1,022	13,201	106,378	12,690	99,777	564,093	74
75	1,023	12,179	93,177	11,668	87,087	470,661	75
76	1,016	11,156	80,998	10,648	75,419	389,408	76
77	1,000	10,140	69,842	9,640	64,771	319,313	77
78	977	9,140	59,702	8,651	55,131	259,362	78
79	943	8,163	50,562	7,692	46,480	208,556	79
80	902	7,220	42,399	6,769	38,788	165,922	80
81	851	6,318	35,179	5,892	32,019	130,519	81
82	794	5,467	28,861	5,070	26,127	101,446	82
83	731	4,673	23,394	4,308	21,057	77,854	83
84	662	3,942	18,721	3,611	16,749	58,951	84
85	591	3,280	14,779	2,984	13,138	44,007	85
86	520	2,689	11,499	2,429	10,154	32,361	86
87	448	2,169	8,810	1,945	7,725	23,422	87
88	380	1,721	6,641	1,531	5,780	16,669	88
89	316	1,341	4,920	1,183	4,249	11,655	89
90	257	1,025	3,579	897	3,066	7,997	90
91	204	768	2,554	666	2,169	5,380	91
92	159	564	1,786	484	1,503	3,544	92
93	122	405	1,222	344	1,019	2,283	93
94	89	283	817	239	675	1,436	94
95	65	194	534	161	436	880	95
96	45	129	340	107	275	525	96
97	31	84	211	68	168	303	97
98	21	53	127	43	100	169	98
99	13	32	74	25	57	91	99
100	8	19	42	15	32	46	100
101	5	11	23	9	17	22	101
102	3	6	12	4	8	9	102
103	1	3	6	3	4	3	103
104	1	2	3	1	1	1	104
105	1	1	1	..	..	..	105
106	..		..	..	..	..	106

TABLE F.—*Healthy Districts.*—*Females.*

Age.	Dying in each Year of Age 0–1, 1–2, to 105–106.	Born and Living at each Age.	Sum of the Numbers Born and Living at each Age (x) from x to the last Age in the Table.	Population, or the Living in each Year of Age 0 to 1, 1 to 2, &c.	(1) Sum of the Living, and of the Living of every Age (x) and upwards to the last Age in the Table; also (2) the Years which the Females (l_x) will live.	(1) The Years which the Females at the Age (x) and upwards will live; also (2) the Years which they have lived over x.	Age.
	$\Sigma d_x.$	$\Sigma l_x.$	$\frac{1}{2}(l_x + l_{x+1})$ $= l_{x+1} + \frac{1}{2}d_x.$	$\Sigma P_x.$	$\Sigma\frac{1}{2}(Q_x + Q_{x+1})$ $= Y_{x+1} + (Q_{x+1} + \frac{1}{2}P_x).$		
$x.$	$d_x.$	$l_x.$	$L_x.$	$P_x.$	$Q_x.$	$Y_x.$	$x.$
0	4,528	48,875	2,442,273	45,696*	2,416,920	82,200,780	0
1	1,414	44,347	2,393,398	43,640	2,371,224	79,806,708	1
2	932	42,933	2,349,051	42,467	2,327,584	77,457,304	2
3	644	42,001	2,306,118	41,679	2,285,117	75,150,953	3
4	519	41,357	2,264,117	41,098	2,243,438	72,886,676	4
5	420	40,838	2,222,760	40,628	2,202,340	70,663,787	5
6	341	40,418	2,181,922	40,247	2,161,712	68,481,761	6
7	280	40,077	2,141,504	39,937	2,121,465	66,340,172	7
8	236	39,797	2,101,427	39,679	2,081,528	64,238,676	8
9	205	39,561	2,061,630	39,459	2,041,849	62,176,987	9
10	186	39,356	2,022,069	39,263	2,002,390	60,154,868	10
11	178	39,170	1,982,713	39,081	1,963,127	58,172,109	11
12	177	38,992	1,943,543	38,903	1,924,046	56,228,523	12
13	184	38,815	1,904,551	38,723	1,885,143	54,323,928	13
14	196	38,631	1,865,736	38,533	1,846,420	52,458,147	14
15	211	38,435	1,827,105	38,330	1,807,887	50,630,993	15
16	228	38,224	1,788,670	38,110	1,769,557	48,842,271	16
17	246	37,996	1,750,446	37,873	1,731,447	47,091,769	17
18	262	37,750	1,712,450	37,619	1,693,574	45,379,259	18
19	276	37,488	1,674,700	37,350	1,655,955	43,704,494	19
20	285	37,212	1,637,212	37,069	1,618,605	42,067,214	20
21	290	36,927	1,600,000	36,782	1,581,536	40,467,144	21
22	294	36,637	1,563,073	36,490	1,544,754	38,903,999	22
23	296	36,343	1,526,436	36,195	1,508,264	37,377,490	23
24	299	36,047	1,490,093	35,898	1,472,069	35,887,323	24
25	301	35,748	1,454,046	35,597	1,436,171	34,433,203	25
26	303	35,447	1,418,298	35,296	1,400,574	33,014,831	26
27	304	35,144	1,382,851	34,992	1,365,278	31,631,905	27
28	305	34,840	1,347,707	34,687	1,330,286	30,284,123	28
29	305	34,535	1,312,867	34,383	1,295,599	28,971,180	29
30	306	34,230	1,278,332	34,077	1,261,216	27,692,773	30
31	307	33,924	1,244,102	33,770	1,227,139	26,448,595	31
32	307	33,617	1,210,178	33,464	1,193,369	25,233,341	32
33	308	33,310	1,176,561	33,156	1,159,905	24,061,704	33
34	308	33,002	1,143,251	32,848	1,126,749	22,918,377	34
35	308	32,694	1,110,249	32,540	1,093,901	21,808,052	35
36	308	32,386	1,077,555	32,232	1,061,361	20,730,421	36
37	310	32,078	1,045,169	31,923	1,029,129	19,685,176	37
38	310	31,768	1,013,091	31,613	997,206	18,672,009	38
39	311	31,458	981,323	31,302	965,593	17,690,609	39
40	312	31,147	949,865	30,991	934,291	16,740,667	40
41	313	30,835	918,718	30,679	903,300	15,821,872	41
42	315	30,522	887,883	30,364	872,621	14,933,911	42
43	318	30,207	857,361	30,048	842,257	14,076,472	43
44	319	29,889	827,154	29,730	812,209	13,249,239	44
45	322	29,570	797,265	29,409	782,479	12,451,895	45
46	325	29,248	767,695	29,085	753,070	11,684,121	46
47	329	28,923	738,447	28,759	723,985	10,945,593	47
48	332	28,594	709,524	28,428	695,226	10,235,988	48
49	336	28,262	680,930	28,094	666,798	9,554,976	49
50	341•	27,926	652,668	27,755	638,704	8,902,225	50
51	346	27,585	624,742	27,412	610,949	8,277,398	51
52	351	27,239	597,157	27,064	583,537	7,680,155	52
53	357	26,888	569,918	26,709	556,473	7,110,150	53

* P_0 is $\frac{1}{2}(l_0 + l_1) \times (\cdot98037)$. The factor ·98037 has been introduced, as the number living in the first year is less than the arithmetical mean of those born and surviving a year.

TABLE F (continued).

Age.	Dying in each Year of Age, 0–1, 1–2, to 105–106.	Born and Living at each Age.	Sum of the Numbers Born and Living at each Age (x) from x to the last Age in the Table.	Population, or the Living in each Year of Age 0 to 1, 1 to 2, &c.	(1) Sum of the Living, and of the Living of every Age (x) and upwards to the last Age in the Table; also (2) the Years which the Females (l_x) will live.	(1) The Years which the Females at the Age (x) and upwards will live; also (2) the Years which they have lived over x.	Age.
	$\Sigma d_x.$	$\Sigma l_x.$	$L_x.$	$\frac{1}{2}(l_x + l_{x+1})$ $= l_{x+1} + \frac{1}{2}d_x.$	$\Sigma P_x.$	$\Sigma\frac{1}{2}(Q_x + Q_{x+1})$ $= Y_{x+1} + (Q_{x+1} + \frac{1}{2}P_x).$	
$x.$	$d_x.$	$l_x.$	$L_x.$	$P_x.$	$Q_x.$	$Y_x.$	$x.$
54	363	26,531	543,030	26,350	529,764	6,567,032	54
55	369	26,168	516,499	25,983	503,414	6,050,443	55
56	376	25,799	490,331	25,611	477,431	5,560,020	56
57	411	25,423	464,532	25,218	451,820	5,095,395	57
58	455	25,012	439,109	24,784	426,602	4,656,184	58
59	498	24,557	414,097	24,308	401,818	4,241,974	59
60	536	24,059	389,540	23,791	377,510	3,852,310	60
61	574	23,523	365,481	23,236	353,719	3,486,695	61
62	609	22,949	341,958	22,645	330,483	3,144,594	62
63	644	22,340	319,009	22,018	307,838	2,825,434	63
64	678	21,696	296,669	21,357	285,820	2,528,605	64
65	712	21,018	274,973	20,662	264,463	2,253,463	65
66	745	20,306	253,955	19,933	243,801	1,999,331	66
67	777	19,561	233,649	19,173	223,868	1,765,497	67
68	810	18,784	214,088	18,379	204,695	1,551,215	68
69	841	17,974	195,304	17,553	186,316	1,355,710	69
70	871	17,133	177,330	16,698	168,763	1,178,170	70
71	898	16,262	160,197	15,813	152,065	1,017,756	71
72	922	15,364	143,935	14,903	136,252	873,598	72
73	942	14,442	128,571	13,971	121,349	744,797	73
74	958	13,500	114,129	13,021	107,378	630,434	74
75	968	12,542	100,629	12,058	94,357	529,566	75
76	971	11,574	88,087	11,088	82,299	441,238	76
77	966	10,603	76,513	10,120	71,211	364,483	77
78	954	9,637	65,910	9,160	61,091	298,332	78
79	932	8,683	56,273	8,217	51,931	241,821	79
80	901	7,751	47,590	7,301	43,714	193,999	80
81	862	6,850	39,839	6,419	36,413	153,935	81
82	814	5,988	32,989	5,581	29,994	120,732	82
83	760	5,174	27,001	4,794	24,413	93,528	83
84	698	4,414	21,827	4,065	19,619	71,512	84
85	633	3,716	17,413	3,399	15,554	53,926	85
86	564	3,083	13,697	2,801	12,155	40,071	86
87	495	2,519	10,614	2,272	9,354	29,317	87
88	425	2,024	8,095	1,811	7,082	21,099	88
89	359	1,599	6,071	1,420	5,271	14,922	89
90	293	1,240	4,472	1,091	3,851	10,361	90
91	240	942	3,232	822	2,760	7,056	91
92	191	702	2,290	606	1,938	4,707	92
93	147	511	1,588	438	1,332	3,072	93
94	112	364	1,077	308	894	1,959	94
95	81	252	713	211	586	1,219	95
96	59	171	461	142	375	738	96
97	40	112	290	92	233	434	97
98	27	72	178	58	141	247	98
99	18	45	106	36	83	135	99
100	11	27	61	22	47	70	100
101	7	16	34	12	25	34	101
102	4	9	18	7	13	15	102
103	3	5	9	4	6	6	103
104	1	2	4	1	2	2	104
105	1*	1*	2	1	1	..	105
106	1*	1*	1	..	..	..	106

* The values of l_{104}, l_{105}, and l_{106}, decimally carried out, are 2·490, 1·250, and 0·603; and their differences are 1·240, 0·647, and 0·325. The apparent anomaly, that no death happens between the ages 105 and 106, arises from the omission of decimals.

TABLE G.—*Healthy Districts Life Table.*—*The Mean After-Lifetime (or the Expectation of Life) at the Age x, and at the Age x and upwards; also the Mean Ages of the Living and the Mean Ages at Death (constructed from Tables D, E, F).*

				Mean Age at Death.	
Age (or past Lifetime).	Mean After-Lifetime of Persons of the Age x.	Mean After-Lifetime of Persons of the Age x and upwards.	Mean Age of Persons living of the Age x and upwards.	Of Persons actually living at the Age x.	Of Persons actually living at the Age x and upwards.
$x.$	$A_x = \dfrac{Q_x}{D_x}.$	$A'_x = \dfrac{Y_x}{Q_x}.$	$x + A'_x.$	$x + A_x.$	$x + 2A'_x.$
0	49·00	33·92	33·92	49·00	67·84
5	54·16	31·98	36·98	59·16	68·96
10	51·08	29·91	39·91	61·08	69·82
15	47·12	27·85	42·85	62·12	70·70
20	43·45	25·82	45·82	63·45	71·64
25	40·05	23·79	48·79	65·05	72·58
30	36·64	21·76	51·76	66·64	73·52
35	33·17	19·73	54·73	68·17	74·46
40	29·64	17·71	57·71	69·64	75·42
45	26·05	15·71	60·71	71·05	76·42
50	22·44	13·74	63·74	72·44	77·48
55	18·86	11·84	66·84	73·86	78·68
60	15·37	10·04	70·04	75·37	80·08
65	12·29	8·37	73·37	77·29	81·74
70	9·61	6·86	76·86	79·61	83·72
75	7·34	5·51	80·51	82·34	86·02
80	5·51	4·36	84·36	85·51	88·72
85	4·10	3·41	88·41	89·10	91·82
90	3·05	2·65	92·65	93·05	95·30
95	2·29	2·05	97·05	97·29	99·10
100	1·72	1·47	101·47	101·72	102·94

	MALES.		FEMALES.	
Age (or past Lifetime).	Mean After-Lifetime of Males of the Age x.	Mean Age at Death of Males actually living at the Age x.	Mean After-Lifetime of Females of the Age x.	Mean Age at Death of Females actually living at the Age x.
$x.$	$A_x = \dfrac{Q_x}{D_x}.$	$x + A_x.$	$A_x = \dfrac{Q_x}{D_x}.$	$x + A_x.$
0	48·56	48·56	49·45	49·45
5	54·39	59·39	53·93	58·93
10	51·28	61·28	50·88	60·88
15	47·20	62·20	47·04	62·04
20	43·40	63·40	43·50	63·50
25	39·93	64·93	40·18	65·18
30	36·45	66·45	36·85	66·85
35	32·90	67·90	33·46	68·46
40	29·29	69·29	30·00	70·00
45	25·65	70·65	26·46	71·46
50	22·03	72·03	22·87	72·87
55	18·49	73·49	19·24	74·24
60	15·06	75·06	15·69	75·69
65	12·00	77·00	12·58	77·58
70	9·37	79·37	9·85	79·85
75	7·15	82·15	7·52	82·52
80	5·37	85·37	5·64	85·64
85	4·01	89·01	4·19	89·19
90	2·99	92·99	3·11	93·11
95	2·25	97·25	2·32	97·32
100	1·69	101·69	1·75	101·75

The Table may be read thus:—Persons in the Healthy Districts of England of the precise age 20 will live, on an average, 43·45 years; while persons of the age 20 and *upwards*, living in a normally-constituted population of the same character, will live on an average 25·82 years. The mean age of persons of the age 20 and *upwards* is 45·82 years; the mean age at death of persons living at the precise age 20 will be 63·45, while the mean age at death of persons actually living at the age x and *upwards* will be 71·64 years.

On Gompertz's Law of Mortality. By Professor De Morgan.

I HAVE a suspicion that Mr. Edmonds intends the fifteen pages on the " Law of Human Mortality," which appear in the last Number of this *Journal,* to stand in place of an answer to my remarks " On an unfair suppression of due acknowledgment to the writings of Mr. Benjamin Gompertz," printed in the July Number; for though no allusion is made to the charge, still less statement of it or answer to the evidence produced, there is one mention of me which looks so like a distortion of my paper, that I think the suspicion is justified.   Mr. Edmonds says (p. 181), " Mr. De Morgan, in his office of self-constituted judge between Mr. Gompertz and me, overlooks this important error . . . "   Now, though my paper does not deal with the truths or errors of either, but only with the question whether Mr. Edmonds's mention of Mr. Gompertz was suppressive; and though I never said, and certainly never thought, that there was or could be any question pending between Mr. Gompertz and Mr. Edmonds; and though I was not the judge, either self-constituted or otherwise, but only the promoter of an accusation of unfair suppression for others to judge of;—there is in the quotation that remote likeness to an account of my proceeding which often exists between that which cannot be answered and that which it is convenient to substitute for it. Some acute adviser seems to have whispered, " When you cannot answer what needs answer, answer something else, and keep what you ought to answer out of sight; this will do for all who are to see only one side, and nothing will do for those who are to see both."   I shall make two short remarks, and then leave the whole to those who have read the whole.

It will be asked, in turning over page after page, " What has all this talk about the better and worse of this and that method to do with the year 1832 and the question whether the account then given by Mr. Edmonds was, or was not, an unfair suppression of what had been done by Mr. Gompertz?"   What did Mr. Edmonds *then* bring forward?   What *had* Mr. Gompertz brought forward? What is there in what Mr. Edmonds then brought forward which will redeem his description of Mr. Gompertz's method from the imputation of unfair suppression?   If Mr. Edmonds had given all the description he has now given, weak as it is, there would have been foolish and unfounded self-assertion, but at least there would not have been suppression.   With the contents of Mr. Edmonds's paper I have nothing to do ; and this because no account he may *now* give, be it true or be it false, can affect the question whether what he *then* gave was or was not unfair suppression. This want of allusion to my accusation relieves me from all necessity of further rejoinder.

Secondly : the process of bringing Dr. Price into the paper is one which has been repeated many and many a time. When A is charged with dealing unfairly with the writings of B, he tries to prove—sometimes he does prove—that he has dealt just as unfairly with C. Dr. Price calls attention to the manner in which the Holy-Cross data and others exhibit the periods of infancy, manhood, and old age. Mr. Edmonds calls this a *discovery*, and then adds, that, " to perfect his discovery," he should have remarked " that the rate was *constant* throughout each of the periods." There was no discovery in what Dr. Price *did* remark, and as to what he *should* have remarked, all we can say is, not that it *does* prove a discovery, but that it *would have* done so if the remark *had been* made. O! these auxiliary verbs! What queer auxiliaries they sometimes are! The introduction of Dr. Price and Mr. Gompertz, as a pair of imperfect predecessors to Mr. Edmonds, can only be explained thus :—Dr. Price was no predecessor at all ; Mr. Gompertz was a full and entire predecessor. Take the mean, and each of them was half a predecessor. The policy of such an introduction is of a very questionable stamp, though, no doubt, there is assurance enough in it for a whole life.

[We have considered this question with a good deal of care, and we cannot but think that there is great truth and justness in the remarks which Professor De Morgan has thought fit to make upon it. When Mr. Edmonds admits " that the honour belongs to Mr. Gompertz of first discovering that some connexion existed between tables of mortality and the algebraic expression a^{b^x}," it seems to us that there is an end of the matter. What else of any moment in connexion with it has been discovered? The giving another form to an expression or another phase to an hypothesis already suggested, cannot be looked upon as a discovery. Mr. Edmonds has applied or made use of Mr. Gompertz's suggestion or " discovery" with great ingenuity, neatness, and effect. We do not see how it can be said with truth that he has done more than this. Of the less commendable features of his work we say nothing.—ED. *A. M.*]

On the Stability of Results based upon Average Calculations, considered with reference to the Number of Transactions embraced. By ROBERT CAMPBELL, M.A., *Advocate, Edinburgh, and Fellow of Trinity Hall, Cambridge.*

[Read before the Institute, 31st December, 1860, and ordered by the Council to be printed.]

AN article published in this *Journal* last January (vol. viii., p. 316), proposed itself the problem of furnishing a test whether the degree of uniformity, or the reverse, observed in a table of statistics, required to be assigned to any other cause than the relative magnitude of the numbers of combinations which each figure in the table represented. The formulæ obtained in that article gave a general solution to the problem proposed, by giving an expression for the ratios of the numbers of such combinations which would be embraced under any figure in the table; and, consequently, expressing the relative frequency with which each number should be expected to occur from the mere consideration of counting the combinations represented by it.

These formulæ will aid in the solution of another and a much more useful problem— namely, to give us the power of prediction, in some very important practical cases, as to the *stability* to be expected in future results calculated from the observation of an average. The simplest case is that where the data on which the calculation is founded are taken from so large sources as to be considered complete.

The problem proposed in the following pages is this : *To show the degree of stability to be expected in the numbers representing the deaths in a given period to be experienced in the course of a life assurance business of given amount*—the problem, for the present, being treated on the supposition that the data on which the calculations of the Office are based are sufficiently large to furnish the true average.

The relation, it may be remarked, between the extensiveness of the data and the accuracy of the average deduced from them, is implicitly given by the formulæ in page 321 of the above article, which may be the subject of future expansion. Enough, for the present, to observe, that when *a* (the number representing the number of years for which observations are taken) is very large, the formulæ approximate to expressions independent of *a* altogether —which, looking at the considerations on which these formulæ are obtained, means to say that the true average has been approximately arrived at. Again: the true average being arrived at, the formulæ on page 324 represent the relative frequency with which numbers differing from the average by any given amount may be expected to occur; but as these formulæ may be obtained more simply and from less complicated considerations than those involved in the above article, I shall proceed to obtain them by a more direct method.

Suppose that we have a sufficient number of observations for obtaining with certainty the *average* percentage of persons of a given class to whom a certain event happens in a given time.

Let a society consist of n persons, who are supposed so far unconnected with each other that the event contemplated happening to one does not make it more or less likely to happen to another, and that of these n persons we know of no reason why it should happen to one rather than another, but that they are all of the given class on which the above observations are made. And suppose that, according to the average above found, the number out of n persons to which the event should happen is b.

Let us designate the above persons by $A_1, A_2 \ldots A_n$; then, by the above hypothesis, the probability of the event happening to A_1 is

$$\frac{b}{n},$$

and the probability of its *not* happening to A_2 is

$$1 - \frac{b}{n}.$$

The probability, therefore, of its happening to A_1 and not to A_2 is

$$\frac{b}{n}\left(1 - \frac{b}{n}\right);$$

and that of its happening to A_1 and neither to A_2, A_3, nor A_n, is

$$\frac{b}{n}\left(1 - \frac{b}{n}\right)^{n-1}.$$

But, of course, the same will be the probability of its happening to A_2 and not to A_1, A_3, nor A_n; and, therefore, the probability of its happening to one person only out of the whole society is

$$n \cdot \frac{b}{n}\left(1 - \frac{b}{n}\right)^{n-1} = b \cdot \left(1 - \frac{b}{n}\right)^{n-1}.$$

The probability of its happening to both A_1 and A_2 is

$$\left(\frac{b}{n}\right)^2,$$

and that of its happening to both A_1 and A_2, and not to any of the rest,

$$\left(\frac{b}{n}\right)^2\left(1-\frac{b}{n}\right)^{n-2}.$$

The probability, therefore, of its happening to *two persons only* out of the society is

$$\frac{n(n-1)}{2}\cdot\left(\frac{b}{n}\right)^2\left(1-\frac{b}{n}\right)^{n-2};$$

that of its happening to *three persons only*,

$$\frac{n(n-1)(n-2)}{1.2.3}\cdot\left(\frac{b}{n}\right)^3\left(1-\frac{b}{n}\right)^{n-3}.$$

The law is evident, and we shall only further write down the probability of its happening to exactly b persons out of the Society, and no more nor less, which is

$$\frac{n(n-1)(n-2)\ldots(n-b+1)}{1.2.3\ldots\ldots\ b}\left(\frac{b}{n}\right)^b\left(1-\frac{b}{n}\right)^{n-b}.*$$

This will be the *most probable number*, for it has been obtained from the last by multiplying by

$$\frac{n-b+1}{b}\cdot\frac{b}{n}\cdot\left(1-\frac{b}{n}\right)^{-1}=\frac{n-b+1}{b}\cdot\frac{b}{n-b}=\frac{n-b+1}{n-b},\ \text{which is}>1;$$

and the next would be obtained from it by multiplying by

$$\frac{n-b}{b+1}\cdot\frac{b}{n}\cdot\left(1-\frac{b}{n}\right)^{-1}=\frac{n-b}{b+1}\cdot\frac{b}{n-b}=\frac{b}{b+1},\ \text{which is}<1.$$

If we now compare the probability of the number b being the exact number out of the society to whom the event happens, with that of $b-1$, $b-2$, and so on being the number, and then with that of $b+1$, $b+2$, and so on being the number, the ratios will be—

---

* So far, this investigation is substantially identical with one which is well known— the language of the above being, for the sake of clearness, adapted to the case here contemplated. The simplifications which follow, admitting the definite working out of the problem in the most useful practical form, are believed to be new.

The mathematical demonstration of the stability of average results, a principle on the practical truth of which the whole fabric of assurance business rests, has, I believe, hitherto rested on very general grounds borrowed from investigations of La Place, which were chiefly worked out in a totally different direction, and apply only to the problem now under consideration in the case where the numbers are very large. In other words, it has been proved that the results of extensive transactions of this kind approach to stability, but no definite relation has hitherto been given between the extent of transactions and the degree of stability to be expected.

Probability of b being the number to that of $(b-1)$

$$=\frac{n-b+1}{b}\cdot\frac{b}{n-b} \quad \cdot \cdot \cdot \cdot \cdot \cdot \cdot \cdot \cdot \quad a_1$$

Probability of b being the number to that of $(b-2)$

$$=\frac{(n-b+1)(n-b+2)}{b\quad(b-1)}\cdot\left(\frac{b}{n-b}\right)^2 \quad \cdot \cdot \cdot \cdot \quad a_2$$

Probability of b being the number to that of $(b-3)$

$$=\frac{(n-b+1)(n-b+2)(n-b+3)}{b\quad(b-1)\quad(b-2)}\cdot\left(\frac{b}{n-b}\right)^3 \quad \cdot \quad a_3$$

and so on; and

Probability of b being the number to that of $(b+1)$

$$=\frac{b+1}{n-b}\cdot\frac{n-b}{b} \quad \cdot \cdot \cdot \cdot \cdot \cdot \cdot \cdot \cdot \quad a_1$$

Probability of b being the number to that of $(b+2)$

$$=\frac{(b+1)(b+2)}{(n-b)(n-b-1)}\cdot\left(\frac{n-b}{b}\right)^2 \quad \cdot \cdot \cdot \cdot \cdot \quad a_2$$

Probability of b being the number to that of $(b+3)$

$$=\frac{(b+1)\ (b+2)\ (b+3)}{(n-b)\ (n-b-1)(n-b-2)}\cdot\left(\frac{n-b}{b}\right)^3 \quad \cdot \cdot \quad a_3$$

and so on.

It is evident that, after getting a certain length, these numbers will increase with enormous rapidity. That rapidity of increase is the test of the stability of calculations proceeding upon average results, and we shall best show how it may be applied to practical cases, by dealing with actual numbers and taking advantage of all the approximate simplifications which they will supply to us.

It is plain that if n is very large compared to b, we may neglect such factors as $\dfrac{n-b+1}{n-b}$, $\dfrac{n-b+2}{n-b}$, $\dfrac{n-b}{n-b-1}$; in which case the above ratios will be very much simplified, and will *depend on the value of b alone.* They will become

$$a_1=\frac{b}{b}=1, \qquad\qquad a_1=\frac{b+1}{b},$$

$$a_2=\frac{b}{b-1}, \qquad\qquad a_2=\frac{(b+1)(b+2)}{b^2},$$

$$a_3=\frac{b^2}{(b-1)\,(b-2)}, \qquad a_3=\frac{(b+1)(b+2)(b+3)}{b^3},$$

and so on.

The practical extent to which we may extend this simplification will soon appear from a numerical example. Suppose $n=1{,}000$, and let $b=7$, which will very nearly represent the mortality among persons of 20 years of age according to the Carlisle Tables; we have then $n-b=993$, and, substituting these values in the expression first given for a_1, a_a, &c., we have—

$$a_1 = \frac{994}{7} \cdot \frac{7}{993} = \frac{994}{993} \times 1,$$

$$a_2 = \frac{994.995}{7 \cdot 6} \times \left(\frac{7}{993}\right)^2 = \frac{994}{993} \cdot \frac{995}{993} \times \frac{7}{6},$$

$$a_3 = \frac{994.995.996}{7 \cdot 6 \cdot 5} \cdot \left(\frac{7}{993}\right)^3 = \frac{994}{993} \cdot \frac{995}{993} \cdot \frac{996}{993} \times \frac{7}{6} \cdot \frac{7}{5},$$

and so on; and

$$a_1 = \frac{8}{993} \cdot \frac{993}{7} = \frac{8}{7},$$

$$a_2 = \frac{8 \cdot 9}{993.992} \cdot \left(\frac{993}{7}\right)^2 = \frac{8}{7} \cdot \frac{9}{7} \cdot \frac{993}{992},$$

$$a_3 = \frac{8 \cdot 9 \cdot 10}{993.992.991} \cdot \left(\frac{993}{7}\right)^3 = \frac{8}{7} \cdot \frac{9}{9} \cdot \frac{10}{7} \times \frac{993}{992} \cdot \frac{993}{991},$$

and so on.

Now, it will not make much difference in the rapidity with which these ratios, a_1, a_2, a_3, &c., and a_1, a_2, a_3, increase, *especially when we get beyond the first two or three*, if we neglect such a fraction as $\frac{996}{993}$ in comparison with $\frac{7}{5}$, or $\frac{993}{991}$ in comparison with $\frac{10}{7}$. We need only remember, that if we had not neglected these fractions the stability of results would have been found to be rather greater.

Again: let $n=1,000$, as before, and let $b=96$, which will nearly represent the mortality at the age of 75 from the same tables; then $n-b=904$, and we have—

$$a_1 = \frac{905}{96} \cdot \frac{96}{904} = \frac{905}{904} \times 1,$$

$$a_2 = \frac{905.906}{96.95} \left(\frac{96}{904}\right)^2 = \frac{905}{904} \cdot \frac{906}{904} \times \frac{96}{95},$$

$$a_3 = \frac{905.906.907}{96.95.94} \left(\frac{96}{904}\right)^3 = \frac{905}{904} \cdot \frac{906}{904} \cdot \frac{907}{904} \times \frac{96}{95} \cdot \frac{96}{94},$$

and

$$a_1 = \frac{97}{904} \cdot \frac{904}{96} = \frac{97}{97},$$

$$a_2 = \frac{97.98}{904.903} \cdot \left(\frac{904}{96}\right)^2 = \frac{97}{96} \cdot \frac{98}{96} \times \frac{904}{903},$$

$$a_3 = \frac{97.98.99}{904.903.902} \cdot \left(\frac{904}{96}\right)^3 = \frac{97}{96} \cdot \frac{98}{96} \cdot \frac{99}{96} \times \frac{904}{903} \cdot \frac{904}{902}.$$

Now, as before, and having an eye to the use we are about to make of these figures, it will not materially affect our results if we neglect the fraction involving the large figures in comparison of the others corresponding to them. We might, perhaps, retain all through a_1, a_2, a_3, the factor $\dfrac{905}{904}$; but even this may, for the present, be neglected—only remembering that in what follows we shall slightly underestimate the stability of results to be expected from considerations based on an average.

Now, suppose the lives assured in an Office comprise various ages between 20 and 75 (if there are a few over the latter age, which we fixed on merely for illustration, it will not materially affect our results); and suppose that, according to the tables, taking the due proportion of lives to each age, the whole number of lives that should fall next year is 96. It scarcely requires proof that what I call the stability of this expectation—that is, the rapidity of increase of the numbers representing the ratio of the probability of this result to the respective results of 95, 94, 93, &c., 97, 98, 99, &c., lives falling—will be intermediate to the stability of expected results in two hypothetical cases—namely, these:—

Take first the case of an Office which should have nothing but lives aged 75 assured, and whose expected number of deaths next year was 96.

Then take the case of an Office which should have nothing but lives aged 20 assured, and whose expected number of deaths next year was also 96.

The first hypothetical Office must have just 1,000 lives assured, and the values of a_1, a_2, &c., in its case, are just those found above on p. 220.

The second hypothetical Office must have a number of lives assured equal to $96 \times \dfrac{1,000}{7} = 13,914$. We should then have to make $n = 13,914$, $b = 96$, and $n - b = 13,818$.

But from what has been shown above, in estimating the stability of either, we might have, for all practical purposes, neglected the value of n. The stability in either case, therefore, is nearly the same—depends on $b = 96$ alone, and not on n—and may be estimated from the values of a_1, a_2, &c., on p. 219; and the stability thus estimated, being a sufficient approximation to the stability of results to be expected in either of the hypothetical cases, is a sufficient approximation to that in the real case which lies between them.

The formulæ on p. 219, then, easily supply the means of constructing tables by which it may be shown at a glance, in the case of an Office expecting (according to average calculations) that a certain number (say 50) of lives will fall in the course of the year, what are the relative probabilities of the number of lives actually falling being 50, 49, 48, &c., 51, 52, 53, &c.

This being done, it will easily appear how, finally, we can form a table showing, if the number on an average calculation be given, what is the probability of the true number being found to differ from the calculated one by more than a given percentage of its amount, which I conceive to be the best form in which the relation between the stability of an assurance business and the number of its transactions can be exhibited.

After what has been said, the tables which follow will sufficiently explain themselves. In what follows we shall adopt the following notation :—

$$\phi_1(b) = \frac{b}{b} = 1,$$

$$\phi_2(b) = \frac{b-1}{b},$$

$$\phi_3(b) = \frac{(b-1)(b-2)}{b^2},$$

$$\cdots \cdots \cdots$$

$$\phi_n(b) = \frac{(b-1)(b-2)\ldots(b-n+1)}{b^{n-1}},$$

and

$$\psi_1(b) = \frac{b}{b+1},$$

$$\psi_2(b) = \frac{b^2}{(b+1)(b+2)},$$

$$\psi_3(b) = \frac{b^3}{(b+1)(b+2)(b+3)},$$

$$\cdots \cdots \cdots$$

$$\psi_n(b) = \frac{b^n}{(b+1)(b+2)\ldots(b+n)}.$$

the reciprocals of the expressions on p. 219.

TABLE I.—*Preparatory Division Table, by Logarithms, giving the value of* log. $\dfrac{b+x}{b}$ *or* $\dfrac{b}{b-x}$, *for all values of b from 1 to 20 and x from 1 to 20.*

Dividend (b)

Divisors (x)

x	1.	2.	3.	4.	5.	6.	7.	8.	9.	10.
1	·0000000	·3010300	·4771213	·6020600	·6989700	·7781513	·8450980	·9030900	·9542425	1·0000000
2		·0000000	·1760913	·3010300	·3979400	·4771213	·5440680	·6020600	·6532125	·6989700
3			·0000000	·1249387	·2218487	·3010300	·3679767	·4259687	·4771212	·5228787
4				·0000000	·0969100	·1760913	·2430380	·3010300	·3521825	·3979400
5					·0000000	·0791813	·1461280	·2041200	·2552726	·3010300
6						·0000000	·0669467	·1249387	·1760912	·2218487
7							·0000000	·0579920	·1091445	·1549020
8								·0000000	·0511525	·0969100
9									·0000000	·0457575

x	b=11.	b=12.	b=13.	b=14.	b=15.	b=16.	b=17.	b=18.	b=19.	b=20.
1	1·0413927	1·0791812	1·1139434	1·1461280	1·1760913	1·2041200	1·2304489	1·2552725	1·2787536	1·3010300
2	·7403627	·7781512	·8129134	·8450980	·8750613	·9030900	·9294189	·9542425	·9777236	1·0000000
3	·5642714	·6020599	·6368221	·6690067	·6989700	·7269987	·7533276	·7781512	·8016323	·8239087
4	·4393327	·4771212	·5118834	·5440680	·5740313	·6020600	·6283889	·6532125	·6766936	·6989700
5	·3424227	·3802112	·4149734	·4471580	·4771213	·5051500	·5314789	·5565025	·5797836	·6020600
6	·2632414	·3010300	·3357921	·3679767	·3979400	·4259687	·4522976	·4771212	·5006023	·5228787
7	·1962947	·2340832	·2688454	·3010300	·3309933	·3590220	·3853509	·4101745	·4336556	·4559320
8	·1383027	·1760912	·2108534	·2430380	·2730013	·3010300	·3273589	·3521825	·3756636	·3979400
9	·0871502	·1249387	·1597009	·1908855	·2218488	·2498775	·2762064	·3010300	·3245111	·3467875
10	·0413927	·0791812	·1139434	·1461280	·1760913	·2041200	·2304489	·2552725	·2787536	·3010300
11	·0000000	·0377885	·0725507	·1047353	·1346986	·1627273	·1890562	·2138798	·2373609	·2596373
12		·0000000	·0347622	·0669463	·0969101	·1249388	·1512677	·1760913	·1995724	·2218488
13			·0000000	·0321846	·0621479	·0901766	·1165055	·1413291	·1648102	·1870866
14				·0000000	·0299633	·0579920	·0843209	·1091445	·1326256	·1549020
15					·0000000	·0280287	·0543576	·0791812	·1026623	·1249387
16						·0000000	·0263289	·0511525	·0746336	·0969100
17							·0000000	·0248236	·0483047	·0705811
18								·0000000	·0234811	·0457575
19									·0000000	·0222764
20										·0000000

TABLE I. (*continued*).

$x=$	$b=21$	$b=22$	$b=23$	$b=24$	$b=25$	$b=26$	$b=27$	$b=28$	$b=29$	$b=30$	$b=31$
1	1·3222193	1·3424227	1·3617278	1·3802112	1·3979400	1·4149733	1·4313638	1·4471580	1·4623980	1·4771213	1·4913617
2	1·0211893	1·0413927									
3	·8450980	·8650140	·8846065								
4	·7201593	·7403627	·7596678	·7781513							
5	·6324937	·6434527	·6627573	·6812412	·6989700						
6	·5440680	·5642714	·5835765	·6020600	·6197887	·6368221					
7	·4771213	·4973247	·5166298	·5351132	·5528420	·5698754	·5862658				
8	·4191293	·4393327	·4586378	·4771213	·4948500	·5118834	·5282738	·5440680			
9	·3679768	·3881802	·4074853	·4259687	·4436975	·4607309	·4771213	·4929155	·5081555		
10	·3222193	·3424227	·3617278	·3802112	·3979400	·4149734	·4313638	·4471580	·4623980	·4771213	
11	·2808266	·3010300	·3203351	·3388185	·3565473	·3735806	·3899711	·4057653	·4210053	·4357286	·4499690
12	·2430381	·2632414	·2825466	·3010300	·3187588	·3357921	·3521826	·3679767	·3832168	·3979410	·4121805
13	·2082759	·2284793	·2477844	·2662678	·2839966	·3010300	·3174204	·3332146	·3484546	·3631779	·3774183
14	·1759687	·1962947	·2155998	·2340832	·2518120	·2688454	·2852358	·3010300	·3162700	·3309933	·3452337
15	·1461280	·1663314	·1856365	·2041199	·2218487	·2388820	·2552725	·2710667	·2863067	·3010300	·3152704
16	·1180993	·1386065	·1576078	·1760912	·1938200	·2108534	·2272438	·2430380	·2582780	·2730013	·2872417
17	·0917704	·1129738	·1312789	·1497623	·1674911	·1845244	·2009149	·2167091	·2319491	·2466724	·2609128
18	·0669468	·0871502	·1064553	·1249387	·1426675	·1597009	·1760913	·1908855	·2071255	·2218488	·2360892
19	·0434657	·0636691	·0829742	·1014576	·1191864	·1362197	·1526102	·1684044	·1836444	·1983677	·2126081
20	·0211893	·0413927	·0606978	·0791812	·0969100	·1139434	·1303338	·1461280	·1613680	·1760913	·1903317

$x=$	$b=32$	$b=33$	$b=34$	$b=35$	$b=36$	$b=37$	$b=38$	$b=39$	$b=40$	$b=41$	$b=42$
1	1·5051500	1·5185139	1·5314789	1·5440680	1·5563025	1·5682017	1·5797836	1·5910646	1·6020600		
12	·4259687										
13	·3912066	·4045705									
14	·3590220	·3723859	·3853509								
15	·3290587	·3424426	·3553876	·3679767							
16	·3010300	·3143939	·3276589	·3339480	·3521825						
17	·2747011	·2880650	·3010300	·3136191	·3258536	·3377528	·3493347				
18	·2498775	·2632414	·2762064	·2887955	·3010300	·3129292	·3245111	·3357921			
19	·2263964	·2397593	·2527253	·2653144	·2775489	·2894481	·3010300	·3123110	·3233064	·3340303	
20	·2041200	·2174839	·2304489	·2430380	·2552725	·2671717	·2787536	·2900346	·3010300	·3117539	·3222192

TABLE II.—*Logarithms found by adding those in the last Table as indicated.*

	b=1.	b=2.	b=3.	b=4.	b=5.	b=6.	b=7.	b=8.	b=9.	b=10.	b=11.	b=12.	b=13.	b=14.
1 − log. φ14														
1 − log. φ13													4·6869838	3·9582556
1 − log. φ12												4·2698375	3·5730404	3·1131576
1 − log. φ11											3·8441639	3·1906563	2·7601270	2·4441509
1 − log. φ10										3·4402369	2·8027712	2·4125051	2·1233049	1·9000829
1 − log. φ9									3·0284194	2·4402369	2·0624035	1·8104452	1·6114215	1·4529249
1 − log. φ8								2·6191994	2·046(?)	1·7412669	1·4981371	1·3333240	1·1964481	1·0849482
1 − log. φ7							2·2132554	1·7161094	1·4209664	1·2183882	1·0588044	·9531128	·8606560	·7839182
1 − log. φ6						1·8155752	1·3681574	1·1140494	·9438432	·8204482	·7163817	·6520828	·5918106	·5408802
1 − log. φ5					1·4156687	1·0334239	·8240894	·6880807	·5916607	·5194182	·4531403	·4179996	·3809572	·3499947
1 − log. φ4				1·0280287	·7166987	·5563026	·4561127	·3870507	·3363882	·2975695	·2568456	·2419084	·2212563	·2038667
1 − log. φ3			·6532126	·4259687	·3187587	·2552726	·2130747	·1829307	·1602970	·1426675	·1285429	·1169697	·1073129	·0991314
1 − log. φ2		·3010300	·1760913	·1249387	·0969100	·0791813	·0669467	·0579920	·0511525	·0457575	·0413927	·0377885	·0347622	·0321846
1 − log. φ1	·0000000	·0000000	·0000000	·0000000	·0000000	·0000000	·0000000	·0000000	0	·0000000	·0000000	·0000000	·0000000	·0000000
1 − log. ψ1	·3010300	·1760913	·1249387	·0909(?)	·0791813	·0669467	·0579920	·0511525	·0457575	·0413927	·0377885	·0347622	·0321846	·0299633
1 − log. ψ2	·7781513	·4771213	·3467874	·2730013	·2253093	·1918854	·1671365	·1480625	·1329077	·1205739	·1103492	·1017000	·0943325	·0879553
1 − log. ψ3	1·3802113	·8750613	·6478174	·5160393	·4294293	·3679766	·3220385	·2763652	·2578464	·2345173	·2150845	·1985(?)	·1845091	·1722762
1 − log. ψ4	2·0791813	1·3826506(?)	1·0(?)	·8093(?)	·6847018	·5898253	·5183332	·4524564	·4175473	·3806153	·3497831	·3235579	·3010146	·2814207
1 − log. ψ5	2·8573326	1·8962506	1·4417628	1·1692518	·9857318	·8530667	·7524164	·6633098	·6084328	·5567366	·5125104	·4748256	·4423437	·4140463
1 − log. ψ6	3·7024360	2·4983106	1·9188840	1·6518(?)	1·3281545	1·1540967	1·0212618	·9063478	·8302816	·7608566	·7015666	·6509169	·6071539	·5589483
1 − log. ψ7	4·6055206	3·1515231	2·4417627	2·0065245	1·7083657	1·4898888	1·3222618	1·2793491	1·1091(?)	·9913055	·920(?)	·8504893	·7942405	·7450396
1 − log. ψ8		3·8504931	3·0060341	2·4836457	2·1233391	1·8578655	1·6532851	1·4803791	1·3563655	1·2457580	1·1528073	1·0723381	1·0025164	·9413343
1 − log. ψ9		4·5908558	3·6080940	2·9955291	2·5704971	2·2558055	2·0123071	1·8077380	1·6573955	1·5253316	1·41(?)	1·3153762	1·2309957	1·1693410(?)
1 − log. ψ10			4·2449161	3·5395971	3·0476184	2·6817742	2·3976580	2·1599205	1·9819066	1·8263616	1·6932712	1·5786176	1·4787801	1·3910173
1 − log. ψ11				4·1136384	3·5527684	3·1340718	2·8088325	2·5335241	2·3286941	2·1485809	1·9943012(?)	1·86(?)	1·7450479	1·6123293
1 − log. ψ12					4·0842473	3·6111930	3·2424881	2·9335241	2·6967709	2·4910036	2·3146363	2·1621942	2·0290445	1·9116747
1 − log. ψ13						4·1117953	3·6984201	3·3526534	3·0848511	2·8527314	2·6534548	2·4809530	2·3300745	2·1969105
1 − log. ψ14							4·1755414	3·7919861	3·4923364	3·2329426	3·0100021	2·8167451	2·6474919	2·4979405
1 − log. ψ15								4·2406239	3·9183051	3·6308826	3·3835827	3·1689277	2·9807095	2·8142105
1 − log. ψ16									4·3620026	4·0458560	3·7735538	3·5369044	3·3291641	3·1452038
1 − log. ψ17											4·1793191	3·9201212	3·6923420	3·4943755(?)
1 − log. ψ18												4·3180622	4·0697603	3·8494595
1 − log. ψ19														4·2218454

In any column corresponding to a particular value of b placed at the top, the numbers from the centre upwards are successively the values of 1 − log. φ(b), and those successively from the centre downwards are the values of 1 − log. ψ(b).

TABLE II. (*continued*).

$1-\log \phi_n$	$b=15$	$b=16$	$b=17$	$b=18$	$b=19$	$b=20$	$b=25$	$b=30$	$b=40$	$b=50$	$b=70$	$b=100$	$b=200$	$b=500$
$1-\log \phi_{100}$														4·616996
$1-\log \phi_{90}$														3·707105
$1-\log \phi_{80}$														2·903012
$1-\log \phi_{70}$														2·202209
$1-\log \phi_{60}$													4·294021	1·602296
$1-\log \phi_{50}$												3·9501612	2·911505	1·100982
$1-\log \phi_{40}$											6·149176	2·1083913	1·817768	·696080
$1-\log \phi_{30}$										4·872150	3·1861803	1·4245980	·994793	·385494
$1-\log \phi_{25}$								4·1556512	4·256354	3·181821	2·1268572	·8847140	·679801	·264940
$1-\log \phi_{20}$							4·2275419	3·6785299	2·5156797	1·919986	1·3066302	·7931990	·4266829	·1672201
$1-\log \phi_{19}$						4·5095366	3·6746999	3·2428013	2·2358390	1·7123767	1·1691024	·7070129	·3838315	·1503952
$1-\log \phi_{18}$						3·8105666	3·1798499	2·8448603	1·9762017	1·5185567	1·0400077	·6260910	·3423729	·1344722
$1-\log \phi_{17}$						4·5095366	3·1798499	2·4816824	1·7358695	1·3381006	·9191856	·5503703	·3037940	·1194493
$1-\log \phi_{16}$				4·5790489	4·1531142	3·8105666	2·7361524	2·1506891	1·5140207	1·1706095	·8064814	·4797992	·2675818	·1053247
$1-\log \phi_{15}$		4·7411803	4·2066949	3·8008977	3·4764206	3·2085066	2·3382124	1·8496591	1·3099007	1·0157075	·7017461	·4142977	·2337235	·0920964
$1-\log \phi_{14}$	4·3487785	3·8380903	3·4533673	3·1476852	2·8966370	2·6856279	1·9816651	1·5766578	1·1228140	·8730402	·6048361	·3538170	·2022064	·0797627
$1-\log \phi_{13}$	3·3737172	3·1110916	2·8249784	2·5913827	2·3960347	2·2296959	1·6629063	1·3300854	·9521178	·7422719	·5155130	·2982997	·1730180	·0683217
$1-\log \phi_{12}$	2·7774472	2·5090316	2·2934995	2·1142615	1·9623791	1·8317559	1·3789097	1·1082366	·7972158	·6230855	·4339430	·2476897	·1461458	·0577715
$1-\log \phi_{11}$	2·2007159	2·0038816	1·8412019	1·7040870	1·5867155	1·4849684	1·1270977	·9098689	·6575538	·5151801	·3596970	·2019322	·1215776	·0481104
$1-\log \phi_{10}$	1·7235946	1·5779129	1·4558510	1·3519045	1·2622044	1·1839384	·9052490	·7337776	·5326151	·4182701	·2927503	·1609736	·0993012	·0393365
$1-\log \phi_{9}$	1·3256546	1·2188909	1·1284921	1·0508745	·9834508	·9243011	·7114290	·5788756	·4219168	·3320840	·2329821	·1247614	·0793046	·0314480
$1-\log \phi_{8}$	·9946613	·9178609	·8522857	·7956020	·7460899	·7024523	·5439379	·4441770	·3250068	·2563633	·1807258	·0932443	·0615758	·0244431
$1-\log \phi_{7}$	·7216600	·6679834	·6218368	·5817222	·5465175	·5153657	·4012704	·3287835	·2414607	·1908618	·1345183	·0663722	·0461031	·0183200
$1-\log \phi_{6}$	·4998112	·4638634	·4327806	·4056309	·3817073	·3604637	·2820840	·2317734	·1708796	·1353445	·0955997	·0440958	·0328748	·0130769
$1-\log \phi_{5}$	·3237199	·3011361	·2815129	·2643018	·2490817	·2355250	·1851740	·1525921	·1128876	·0895870	·0634151	·0263670	·0218794	·0087121
$1-\log \phi_{4}$	·1890213	·1761973	·1650074	·1551573	·1464194	·1386150	·1094533	·0904441	·0671301	·0533748	·0378610	·0131387	·0131055	·0052238
$1-\log \phi_{3}$	·0921112	·0860207	·0806865	·0759761	·0717858	·0630339	·0539410	·0446866	·0332718	·0265027	·0188380		·0065417	·0026102
$1-\log \phi_{2}$	·0299633	·0280287	·0263289	·0248236	·0234811	·0222764	·0177288	·0147233	·0109954	·0087739	·0062489	·0043648	·0021769	·0008695
$1-\log \phi_{1}$	·0000000	·0000000	·0000000	·0000000	·0000000	·0000000	·0000000	·0000000	·0000000	·0000000	·0000000	·0000000	·0000000	·0000000

	1	2	3	4	5	6	7	8	9	10	11	12	13	14
$1 - \log \psi_1$	·0008677	·0021661	·0043214	·0061603	·0086002	·0107239	·0142404	·0170333	·0211893	·0222764	·0234811	·0248236	·0263289	·0280287
$1 - \log \psi_2$	·0026014	·0064875	·0129216	·0183948	·0256335	·0319132	·0422691	·0504571	·0625820	·0657421	·0692386	·0731283	·0774814	·0823863
$1 - \log \psi_3$	·0051994	·0129535	·0257588	·0366197	·0509394	·0633217	·0836617	·0996751	·1232798	·1294112	·1361854	·1437094	·1521150	·1615675
$1 - \log \psi_4$	·0086599	·0215537	·0427921	·0607534	·0843632	·1047144	·1380193	·1641331	·2024610	·2123854	·2232356	·2354798	·2490250	·2642298
$1 - \log \psi_5$	·0129813	·0322776	·0639814	·0907167	·1257559	·1558669	·2049660	·2433444	·2993710	·3138430	·3296909	·3484536	·3671243	·3891685
$1 - \log \psi_6$	·0181618	·0451148	·0892873	·1264323	·1749759	·2165647	·2841472	·3367361	·4133144	·4330294	·4546296	·4797825	·4954270	·5352965
$1 - \log \psi_7$	·0241998	·0600551	·1186711	·1678250	·2318788	·2866026	·3752276	·4439461	·5436482	·5692491	·5972971	·6294948	·6530348	·7016279
$1 - \log \psi_8$	·0310935	·0770884	·1520949	·2148216	·2963368	·3657838	·4778899	·5645200	·6897762	·7218593	·7569980	·7969859	·8291260	·8872644
$1 - \log \psi_9$	·0388413	·0962047	·1895214	·2673507	·3682188	·4539199	·5918332	·6980589	·8511442	·8902637	·9330893	·9815103	1·0229460	1·0913843
$1 - \log \psi_{10}$	·0474415	·1173940	·2309141	·3253427	·4474001	·5508299	·7167719	·8441869	1·0272355	1·0739081	1·1239748	1·1824252	1·2337994	1·3132330
$1 - \log \psi_{11}$	·0568924	·1406465	·2762371	·3887297	·5337598	·6563401	·8524345	1·0025494	1·2175672	1·2722758	1·3311003	1·3991343	1·4610432	1·5521150
$1 - \log \psi_{12}$	·0671924	·1659524	·3254451	·4574456	·6271815	·7702834	·9985625	1·1728111	1·4216872	1·4848839	1·5529491	1·6310834	1·7040812	1·8073875
$1 - \log \psi_{13}$	·0783390	·1933020	·3785335	·5314257	·7275520	·8924993	1·1549097	1·3546547	1·6391711	1·7112803	1·7890383	1·8777558	1·9623592	2·0784542
$1 - \log \psi_{14}$	·0903329	·2226858	·4354384	·6106070	·8347620	1·0228331	1·3212411	1·5477793	1·8696200	1·9510396	2·0389158	2·1386686	2·2353605	2·3647609
$1 - \log \psi_{15}$	·1031701	·2540943	·4961362	·6949279	·9487054	1·1611358	1·4973323	1·7518993	2·1126580	2·2037649	2·3021572	2·4133697	2·5226022	2·6657909
$1 - \log \psi_{16}$	·1178498	·2875181	·5605942	·7843284	1·0692793	1·3072638	1·6829688	1·9777432	2·3679305	2·4690793	2·5783636	2·7014347	2·8236322	2·9810613
$1 - \log \psi_{17}$	·1313703	·3229478	·6287801	·8787497	1·1963841	1·4610787	1·8779454	2·1920525	2·6351022	2·7466282	2·8671591	3·0024647	3·1380261	3·3101200
$1 - \log \psi_{18}$	·1467301	·3603743	·7006621	·9781344	1·3299230	1·6224467	2·0820653	2·4275810	2·9138558	3·0360763	3·1681891	3·3160838	3·4653850	3·6525426
$1 - \log \psi_{19}$	·1629275	·3997884	·7762091	1·0824264	1·4698021	1·7912387	2·2951401	2·6730937	3·2038904	3·3371063	3·4811183	3·6419374	3·7963330	4·0079302
$1 - \log \psi_{20}$	·1799608	·441181	·8553903	1·1915709	1·6159301	1·9673300	2·5169888	2·9283662	3·5049204	3·6494173	3·8056294	3·9796902	4·1575155	4·3290249
$1 - \log \psi_{21}$									3·8166743	3·9727287	4·1414215			
$1 - \log \psi_{22}$									4·1388935	4·3067540				
$1 - \log \psi_{25}$	·276761	·677512	1·304772	1·808244	2·437287	2·953195	3·7519882	4·343929						
$1 - \log \psi_{30}$	·395923	·961994	1·840690	2·538658	3·402553	4·104960	5·1828759	5·974835						
$1 - \log \psi_{40}$	·693775	1·671280	3·159106	4·318592	5·730061									
$1 - \log \psi_{50}$	1·071943	2·561198	4·786890											
$1 - \log \psi_{60}$	1·529025	3·624532	6·703453											
$1 - \log \psi_{70}$	2·063595	4·854624												
$1 - \log \psi_{80}$	2·674295													
$1 - \log \psi_{90}$	3·359811													
$1 - \log \psi_{100}$	4·118878													

The numbers above and below ϕ_{20} and ψ_{20} are found with the help of the Table for $\lceil(x+1)$ in the *Encyclopædia Metropolitana*, on "Probabilities."

TABLE III.—Giving the Successive Values of φ, ψ, for any Value of b.

	b=1.	b=2.	b=3.	b=4.	b=5.	b=6.	b=7.	b=8.	b=9.	b=10.	b=11.	b=12.	b=13.	b=14.	b=15.	b=16.	b=17.
ϕ_{16}																	
ϕ_{15}																·00002	·00006
ϕ_{14}															·00004	·00015	·00035
ϕ_{13}														·00011	·00034	·00077	·00150
ϕ_{12}												·00005	·00027	·00077	·00163	·00310	·00509
ϕ_{11}											·00013	·00065	·00174	·00360	·00630	·00911	·01441
ϕ_{10}										·00036	·00128	·00387	·00753	·01259	·01890	·02643	·03501
ϕ_{9}									·00094	·00363	·00866	·01547	·02447	·03524	·04724	·06041	·07439
ϕ_{8}								·00240	·00843	·01814	·03176	·04642	·06361	·08223	·10124	·12082	·14051
ϕ_{7}							·00612	·01920	·03793	·06048	·08734	·11140	·13783	·16446	·18982	·21390	·23887
ϕ_{6}						·01543	·04284	·07690	·11380	·15120	·19214	·22280	·25597	·28752	·31631	·34367	·36916
ϕ_{5}					·03840	·09259	·14993	·20508	·25606	·30240	·35226	·38194	·41595	·44669	·47455	·49988	·52298
ϕ_{4}				·09375	·19200	·27145	·34985	·41016	·46091	·50400	·53355	·57292	·60082	·62536	·64711	·66650	·68390
ϕ_{3}			·22222	·37500	·48000	·55556	·61225	·65625	·69136	·72000	·74380	·76389	·78106	·79592	·80889	·82031	·83045
ϕ_{2}		·50000	·66667	·75000	·80000	·83333	·85714	·87500	·88889	·90000	·90909	·91667	·92307	·92857	·93333	·93750	·94118
ϕ_{1}	1·00000	1·00000	1·00000	1·00000	1·00000	1·00000	1·00000	1·00000	1·00000	1·00000	1·00000	1·00000	1·00000	1·00000	1·00000	1·00000	1·00000
0	1·00000	1·00000	1·00000	1·00000	1·00000	1·00000	1·00000	1·00000	1·00000	1·00000	1·00000	1·00000	1·00000	1·00000	1·00000	1·00000	1·00000

ψ_1	·50000	·66667	·75000	·80000	·83333	·85714	·87500	·88889	·90000	·90909	·91667	·92307	·92857	·93333	·93750	·94118	·94444
ψ_2	·16667	·33333	·45000	·53052	·59524	·64286	·68056	·71111	·73636	·75758	·77570	·79121	·80476	·81667	·82721	·83660	·84503
ψ_3	·04167	·13333	·22453	·30476	·37202	·42857	·47639	·52922	·55227	·58275	·60942	·63297	·65387	·67255	·68934	·70451	·71828
ψ_4	·00833	·04444	·09643	·15238	·20668	·25714	·30316	·35281	·38234	·41625	·44691	·47473	·50002	·52309	·54421	·56360	·58146
ψ_5	·00139	·01270	·03616	·06773	·10334	·14026	·17684	·21712	·24636	·27750	·30725	·33510	·36112	·38544	·40816	·42941	·44828
ψ_6	·00020	·00318	·01205	·02709	·04697	·07013	·09522	·12407	·14782	·17344	·19881	·22340	·24708	·26931	·29154	·31958	·33133
ψ_7	·00002	·00071	·00362	·00985	·01957	·03236	·04761	·06556	·08315	·10202	·12150	·14109	·16060	·17987	·19873	·22231	·23470
ψ_8		·00014	·00099	·00328	·00753	·01587	·02222	·03308	·04401	·05668	·07034	·08466	·09942	·11446	·12964	·14821	·15959
ψ_9		·00003	·00025	·00101	·00269	·00555	·00972	·01557	·02201	·02983	·03869	·04838	·05875	·06967	·08102	·09485	·10435
ψ_{10}			·00006	·00029	·00090	·00208	·00400	·00693	·01043	·01492	·02026	·02639	·03321	·04064	·04861	·05838	·06570
ψ_{11}				·00008	·00028	·00073	·00155	·00291	·00469	·00710	·01013	·01377	·01799	·02276	·02805	·03469	·03989
ψ_{12}					·00008	·00024	·00057	·00117	·00201	·00323	·00485	·00688	·00935	·01227	·01558	·01977	·02338
ψ_{13}						·00008	·00020	·00044	·00082	·00140	·00227	·00330	·00500	·00644	·00834	·01090	·01325
ψ_{14}							·00006	·00016	·00032	·00059	·00098	·00153	·00225	·00318	·00432	·00582	·00727
ψ_{15}								·00006	·00012	·00023	·00041	·00068	·00105	·00153	·00216	·00300	·00386
ψ_{16}									·00004	·00009	·00017	·00029	·00047	·00072	·00121	·00150	·00198
ψ_{17}											·00007	·00012	·00020	·00032	·00050	·00073	·00099
ψ_{18}												·00005	·00009	·00014	·00022	·00034	·00048
ψ_{19}														·00006	·00010	·00016	·00023
ψ_{20}																·00007	·00011

It is sufficiently obvious how this last Table is formed, from the one preceding it. It requires, of course, the extension of this Table to give for the particular values required all the successive values of $\phi_1(b)$, $\phi_2(b)$, &c., $\psi_1(b)$, $\psi_2(b)$, &c., till they become so small as to be safely neglected. All the ψ's below the given percentage being added for the numerator, and the whole sum of the ϕ's and ψ's, + unity, to include the chance of the average number itself being taken for the denominator, the numbers in the final Table are obtained.

Table III. (continued).

	b=18.	b=19.	b=20.	b=25.	b=30.	b=40.	b=50.	b=70.	b=100.	b=200.	b=500.
ϕ_{100}											·00002
ϕ_{90}											·00019
ϕ_{80}											·00125
ϕ_{70}											·00628
ϕ_{60}										·00005	·02499
ϕ_{50}										·00102	·07925
ϕ_{40}									·00011	·01521	·20133
ϕ_{30}							·00001	·00747	·00779	·10121	·41163
ϕ_{25}						·00006	·00066	·04936	·03762	·20902	·54332
ϕ_{20}					·00021	·00305	·01203	·06775	·13040	·37438	·68042
ϕ_{19}				·00006	·00057	·00581	·01939	·09120	·16099	·41321	·70730
ϕ_{18}				·00021	·00143	·01057	·03030	·12045	·19633	·45460	·73372
ϕ_{17}			·00003	·00066	·00330	·01837	·04590	·15614	·23654	·49683	·75954
ϕ_{16}	·00003	·00007	·00015	·00184	·00707	·03062	·06751	·19873	·28160	·54003	·78465
ϕ_{15}	·00016	·00033	·00062	·00459	·01414	·04899	·09645	·24841	·33128	·58382	·80892
ϕ_{14}	·00071	·00127	·00206	·01043	·02651	·07537	·13396	·30506	·38521	·62776	·83222
ϕ_{13}	·00256	·00402	·00589	·02173	·04676	·11166	·18102	·36801	·44277	·67140	·85443
ϕ_{12}	·00769	·01100	·01473	·04179	·07794	·15951	·23818	·43682	·50315	·71426	·87544
ϕ_{11}	·01977	·02590	·03274	·07463	·12306	·21900	·30537	·50962	·56534	·75583	·89514
ϕ_{10}	·04447	·05468	·06547	·12438	·18460	·29335	·38171	·58481	·62816	·79567	·91341
ϕ_{9}	·08895	·10388	·11904	·19434	·26371	·37852	·46550	·66027	·69028	·83310	·93015
ϕ_{8}	·16010	·17943	·19840	·28580	·35960	·47314	·55416	·73364	·75031	·86781	·94527
ϕ_{7}	·26199	·28411	·30523	·39694	·46905	·57351	·64437	·80242	·80678	·89928	·95869
ϕ_{6}	·39298	·41523	·43605	·52229	·58644	·67472	·73224	·86414	·85828	·92710	·97034
ϕ_{5}	·54412	·56353	·58140	·65287	·70373	·77110	·81360	·91651	·90345	·95087	·98014
ϕ_{4}	·69846	·71381	·72675	·77722	·81200	·85678	·88435	·95755	·94109	·97027	·98804
ϕ_{3}	·83951	·84765	·85500	·88320	·90222	·92625	·94080	·98571	·97020	·98505	·99401
ϕ_{2}	·94444	·94737	·95000	·96000	·96667	·97500	·98000	—	·99000	·99500	·99800
ϕ_{1}	1·00000	1·00000	1·00000	1·00000	1·00000	1·00000	1·00000	1·00000	1·00000	1·00000	1·00000

	C1	C2	C3	C4	C5	C6	C7	C8	C9	C10	C11
0	1·00000	1·00000	1·00000	1·00000	1·00000	1·00000	1·00000	1·00000	1·00000	1·00000	1·00000
1	·99800	·99502	·99010	·98592	·98039	·97561	·96774	·96154	·95238	·95000	·94737
2	·99403	·98517	·97068	·95853	·94268	·92915	·90735	·89031	·86580	·85952	·85263
3	·98810	·97061	·94241	·91914	·88933	·86433	·82478	·79492	·75287	·74232	·73083
4	·98026	·95158	·90617	·86945	·82345	·78575	·72775	·68528	·62739	·61322	·59809
5	·97055	·92837	·86302	·81149	·74859	·69845	·62378	·57102	·50191	·48546	·46807
6	·95904	·90133	·81417	·74742	·66838	·60734	·51983	·46054	·38609	·36895	·35105
7	·94580	·87085	·76090	·67948	·58630	·51689	·42148	·35979	·28599	·26962	·25276
8	·93091	·83736	·70454	·60979	·50543	·43074	·33275	·27257	·20428	·17707	·17499
9	·91445	·80130	·64637	·54032	·42833	·35163	·25596	·20042	·14088	·12875	·11166
10	·89652	·76314	·58761	·47278	·35694	·28130	·19196	·14316	·09392	·08435	·07517
11	·87722	·72336	·52937	·40857	·29258	·22063	·14046	·09942	·06059	·05342	·04665
12	·85666	·68241	·47266	·34878	·23595	·16971	·10033	·06717	·03787	·03274	·02799
13	·83495	·64076	·41828	·29415	·18726	·12809	·06000	·04419	·02295	·01944	·01625
14	·81221	·59884	·36691	·24513	·14630	·09488	·04773	·02833	·01350	·01129	·00914
15	·78855	·55706	·31905	·20187	·11254	·06900	·03182	·01771	·00772	·00626	·00499
16	·76234	·51580	·27505	·16431	·08526	·04929	·02075	·01053	·00431	·00339	·00264
17	·73897	·47539	·23508	·13221	·06362	·03459	·01325	·00643	·00232	·00179	·00136
18	·71330	·43614	·19922	·10516	·04678	·02385	·00828	·00374	·00122	·00092	·00068
19	·68718	·39830	·16741	·08271	·03390	·01617	·00507	·00212	·00063	·00046	·00033
20	·66075	·36209	·13951	·06433	·02421	·01078	·00304	·00118	·00031	·00022	·00016
25	·52770	·21013	·04957	·01555	·00367	·00111	·00018	·00005			
30	·40186	·10914	·01443	·00289	·00040	·00008					
40	·20242	·02132	·00069	·00005							
50	·08473	·00275	·00002								
60	·02958	·00024									
70	·00864	·00001									
80	·00212										
90	·00044										
100	·00008										

Finally, TABLE IV., *showing the probability on the hypothesis of the preceding investigation of the occurrences in a given period exceeding the average by more than a named percentage.*

Per Cent.	Average Numbers.									
	20.	25.	30.	40.	50.	70.	100.	200.	500.	1,000.
5	·35544	·37062	·38137	·33820	·35417	·33189	·28711	·23407	·19821	·05604
10	·27939	·29981	·25568	·23413	·21553	·18387	·14711	·07755	·02002	·00087
15	·21250	·23659	·20270	·11705	·14156	·10658	·06314	·01766	·00074	·00001 –
20	·15677	·13668	·11963	·09247	·07216	·04483	·02263	·00275	·00001 –	
25	·11218	·09998	·08897	·03551	·04239	·02114	·00675	·00029		
30	·07788	·07144	·04625	·02806	·01726	·00672	·00166	·00002		
40	·03433	·02244	·01482	·00658	·00297	·00063	·00006			
50	·01347	·00919	·00396	·00120	·00037	·00004				

The numbers above the columns denote the numbers that should occur according to the observed average; the numbers on the left, the percentage of deviation in excess of the average. The numbers in a column (say the one headed 50) opposite a certain percentage (say 25), give the probability of the occurrences in a particular case, in which, according to the average observations, they should be 50, exceeding this number by 25 per cent.—that is, in this case, of their being 63 or more. A table such as the above appears to be the best form of representing, at a glance, the relation between the stability of the probable results and the magnitude of the field on which they are looked for.

It is easy to see how the table (IV.) thus finally arrived at, gives the solution of the problem with which we commenced.

Table III. gives a more detailed view of the probabilities of each particular number occurring, and will also itself furnish a very good general idea of what I call the stability of these results.

The case chiefly contemplated in the foregoing pages is that of life assurance; but the columns corresponding to small values of *b* will give a good idea of the stability to be expected in such a business as marine assurance (for instance).

The somewhat large values of *b* will express the kind of risks incurred in fire and life businesses, which have not secured a large connexion. The columns for higher values of *b* explain how the risk arising from fluctuations of luck becomes diminished as the number of transactions increase—becoming, indeed, in the case of very large Offices, an element of much subordinate importance to the fluctuations in the money market, which appear less reducible to law.

Take, for example, the case of an Office which calculates upon losing 70 lives in the course of the year. The odds against the losses in the course of the year *exceeding* this number by 20 per cent.—that is, being greater than 84, are about 24 to 1—that is, a result which we may expect to occur only about once in 25 years.

But, looking at the possible losses in a septennial valuation of assets, we shall have to look at the result of taking 490 (say 500) as the calculated average. I must remark, by the way, that this case does not precisely fulfil the conditions of the problem, because the losses in each year are not entirely independent of one another; but the result of the distinction will simply be, practically to increase slightly the stability to be expected.

In the case now contemplated, the chances are about 5 to 1 against the losses exceeding the average by 5 per cent.—and about 50 to 1 against their exceeding the average by 10 per cent. Again: in the case now contemplated, it is considerably more than 10,000 to 1 against the number exceeding the average by 20 per cent.; and the enormous ratio in which these numbers afterwards diminish, is just the mathematical expression for the certainty of which practice has assured us that the results of such contingent transactions on a sufficiently large scale, will, within certain limits, compensate each other with unerring certainty.

I must remark, that the method by which the general part of the problem is treated in the present article, might have been offered with some hesitation, but for the identity of its result with the limiting case of the problem as more completely analysed in the former article referred to. The former method seems unexceptionable, as it affords a complete analysis of the combinations of events represented by the different numbers, and shows completely what an enormous proportion of the possible combinations of occurrences is represented by the arrangements, which nearly equalize the number of occurences in each year.

I believe the foregoing pages will be of some interest in a practical as well as a theoretical point of view. No doubt, those who are deeply conversant with the practical working of Offices, may acquire a kind of empirical knowledge of the relation between the stability and extent of such business; but, even to the experienced, it is thought that greater precision of ideas may be attained by the expression of this relation in figures; and, moreover, it is something to be assured that the stability thus experienced requires for its explanation no recondite principle of moral or physical laws, but is simply demonstrated by an arithmetical computation of the number of possible events, between which there are no data for assigning preference of expectation.

For the sake of readers less familiar with the principles on which the reasoning with which we started is founded, it may be well to state the result of the present problem in its most elementary form.

Take the case above supposed, where 500 is the average. The proposition is this :—Out of various parts of the country, and from various occupations, &c., a number of people has been taken, of whom, according to the laws of mortality, which (for the present) we suppose known with sufficient accuracy, 500 would die in the course of the year. For every combination of circumstances which would involve the death of a number greater than 10 per cent. of the above, there are 50 which would involve the number being below this limit. For every combination of circumstances involving the death of a number exceeding the 500 by 20 per cent., there are more than 10,000 equally probable combinations of circumstances which will keep the number within that limit. It must be kept in view that the problem assumes—what, of course, is not quite realized—the true ascertainment of the average law. This, no doubt, in the first form of the problem, is implicitly involved along with the oscillations about the average here investigated; but, I think, the assumption already made will be scarcely disputed—namely, that it may be sufficiently ascertained, in order to give the problem of the oscillations here treated a sufficient standing ground of its own. That being admitted, the stability of the results, practically important as it is, is simply the arithmetical consequence, and the law of the oscillations about the average is that expressed by the above tables.

The American Life Underwriters' Convention.

[We drew attention to this important movement in a previous Number,* and now present our readers with the Report of a committee of the Convention on vital statistics. It will be observed from this Report that data have been already collected for the required purpose more extensive than those from which our "Combined Experience" Table has been deduced, and that the records of many Companies are still to be supplied. This we need hardly say is only as it should be, since it seems probable that divisions of the experience must be made on account of differences of

* *See* vol. viii., p. 268.

climate in so wide a range of country, and since also so many of the Companies are but of recent origination and can afford but slender information as to the lives of persons of advanced age.   Great skill and judgment will evidently be required in the general arrangement of the data, to say nothing of that demanded in the ordinary processes of the work.   The carrying out of the whole, it appears, has been confided to the committee, and we have reason to believe that no better or wiser course could have been adopted. In a future Number we hope to give the Reports sent in by other committees of the Convention.—ED. *A. M.*]

Report of the Committee on Vital Statistics.—At a Convention of American Life Underwriters, held in this city one year ago, the following preamble and resolutions, offered by Mr. Chickering, of Pittsfield, Mass., were unanimously adopted:—

" Whereas it is the opinion of this Convention (in accordance with the Report already adopted) that it is very desirable that each of the Companies in the United States should unite in contributing towards the determination of the general experience of the rate of mortality among assured lives in the United States—

" Resolved—That Messrs. Homans and Eadie, of New York, and Russell, of Hartford, be appointed a committee to carry into effect the Report on Vital Statistics, adopted by the Convention.

" Resolved—That in conferring with the different Life Companies, and soliciting statements from those not represented in this Convention, the committee assure them that such statements shall be regarded as strictly confidential.

" Resolved—That this committee be instructed to treat all statements furnished by the several Companies as strictly confidential, and to make public only the combined experience ascertained as the result of such statements.   That as soon as the necessary blanks can be procured, said committee furnish them to the different Companies, requesting as early replies, and up to as recent dates, as the committee shall see fit.

" Resolved—That the expenses of collecting and publishing said statistics be paid by the Companies jointly, and each Company contributing their experience shall be entitled to one or more copies of the same."

The committee thus appointed have diligently considered the important subjects committed to their care, and have spared no pains to induce all American Life Companies to unite in a project of such practical utility and interest.

In order to carry out the objects thus contemplated, the committee addressed a circular letter to each of the following Companies:—Union Mutual, Portland, Maine; National, Montpelier, Vermont; Massachusetts Hospital, Boston, Mass.; New England Mutual, Boston, Mass.; State Mutual, Worcester, Mass.; Berkshire Mutual, Pittsfield, Mass.; Massachusetts Mutual, Springfield, Mass.; American Temperance, Hartford, Connecticut; Charter Oak, Hartford, Connecticut; Ætna, Hartford, Connecticut; Connecticut Mutual, Hartford, Connecticut; American Mutual, New Haven, Connecticut; Mutual Life, New York; New York, New York; United States, New York; Manhattan, New York; Knickerbocker, New York; New York Life and Trust, New York; Guardian, New York; Washington, New York; Equitable, New York; Home, Brooklyn, N. Y.; Royal, of London and Liverpool, England; International, London, England; Liverpool and London, England; Eagle and Albion, London, England; British Commercial, London, England; Mutual Benefit, Newark, N. J.; Penn Mutual, Philadelphia, Penn.; Pennsylvania, Philadelphia, Penn.; Girard, Philadelphia, Penn.; American, Philadelphia, Penn.; United States, Philadelphia, Penn.; Philadelphia, Philadelphia, Penn.; Baltimore, Baltimore, Md.; Southern Mutual, Columbia, S. C.; Wisconsin Mutual, Milwaukie, Wisconsin; Canada, Hamilton, Canada West; in all, thirty-nine Companies.

To this circular favourable replies were received from the following Companies, who have agreed to contribute from the records of their policies the data requested to a general fund:—Union Mutual, Portland, Maine; National, Montpelier, Vermont; N. E. Mutual, Boston, Mass.; Berkshire, Pittsfield, Mass.; Am. Temperance, Hartford, Connecticut; Ætna, Hartford, Connecticut; Charter Oak, Hartford, Connecticut; Connecticut Mutual, Hartford, Connecticut; New York Life, New York, N. Y.; Manhattan, New York, N. Y.; Knickerbocker, New York, N. Y.; Mutual Life, New York, N. Y.; Washington, New York, N. Y.; Equitable, New York, N.Y.; Guardian, New York, N. Y.; Mutual Benefit, Newark, N. J.; Penn Mutual, Philadelphia, Penn.; Girard Life and Trust, Philadelphia, Penn.; Am. Life and Trust, Philadelphia, Penn.; Southern Mutual, Columbia, S.C.; Canada Life, Canada; Royal, Liverpool and London, England; in all, twenty-two Companies.

The United States, of New York, contemplate the publication of their experience in a separate form.

Completed returns from the following Companies have been received by the committee:—National, Etna, Connecticut Mutual, Royal (of Liverpool and London), Equitable, Girard Life and Trust, American Temperance, Charter Oak, New York Life, Mutual Life of New York, Washington, American Life and Trust, and Southern Mutual; in all, thirteen Companies.

No replies have been received from other Companies, with the exception of one letter from the Secretary of the American Mutual, of New Haven, declining to furnish the *data*, but offering to furnish *results* deduced by themselves, which proposition the committee would respectfully refer to the Convention for action.

The committee have thought that the object contemplated in their appointment would be the better attained if the contributions from the various Companies were to remain unopened until further action by the Convention, when it was hoped that all contributing Companies, including those *not* represented in the last Convention, would be present, and have a voice in the final disposition of the data. The committee have then confined themselves to the duty of inducing as many Companies as possible to contribute their quotas to a general fund, and now place the various contributions at the disposal of the Convention in the same order as when first received.

The committee have thus collected in a simple, concise, yet comprehensive form, the elementary data according to the official record of thirteen Life Companies, and have, in addition, data promised to them from nine other Companies. It is not yet known how many lives and policies, or rather *years of life*, are thus embraced, but the number is undoubtedly larger than that from which was deduced the celebrated " Actuaries'," or " Combined Experience" Table of Mortality.

It now remains for the Convention to decide upon what shall be done with the valuable materials thus gathered, at the cost of so much time and labour.

There is at the present moment a fund of no less than 22,000,000 dollars held in trust and invested by American Life Companies, with annual premiums amounting to more than 7,000,000 dollars, covering policies of about 180,000,000 dollars on the lives of nearly 60,000 American citizens. Is it not time, in a mere pecuniary point of view, that some efforts were made to ascertain the laws governing the duration of life in the localities thus covered? It has been well said that a knowledge of the laws of mortality is the very essence and foundation of the system of life insurance; in fact, a table of mortality may be called the keystone of the arch upon which the vast superstructure of life insurance is based, and upon the accuracy of our tables depend, in a great measure, the stability and safety of the great institution (amounting, in this country, to so many millions of a most sacred fund), in which we are all so much interested.

We have now, Mr. President, from the best of all sources—viz., abstracts from the actual records of the Companies themselves, the preliminary data for constructing tables and deducing sound and reliable information of the greatest practical value. In short, Sir, we have now the opportunity of substituting certainty for uncertainty, facts for mere conjecture, and of placing the whole system of insurance, heretofore greatly dependent upon individual judgment, or at least upon foreign observations, upon the broad foundation of scientific investigation. If the contributing Companies would so arrange their books as that, at stated times—once in five or ten years, for instance—a *census* might be taken, so that we then could correct or corroborate our tables from actual experience, little more could be desired.

In placing the contributions in the hands of the Convention, the committee would recommend that the task of deducing practical tables having special reference to the comparative mortality in the different sections into which our country has been divided, at different ages or epochs of life, both among native and foreign assured, the determination of the extra risk on voyages, &c., and the value of selection, &c., be confided to a proper committee.

SHEPPARD HOMANS.
THOMAS W. RUSSELL.
New York, May, 1860.   JOHN EADIE.

A long discussion seems to have followed the reading of the Report. We have space only for some observations by Mr. Homans, the chairman of the committee, with reference to their *modus operandi.* That gentleman says, " The committee have not consulted together very much on the best plan to be adopted in relation to the matter before the Convention; but it seems to me that the best plan would be, perhaps, to appoint a committee to deduce the results, and give them authority to employ a person to arrange these data in a convenient form. I take it for granted that no person employed in a Life Company could give the time necessary for such a work. I know it would be impossible for me to do it, but it could be done by a clerk at a salary of, say 50 dollars a month, who would arrange the matters in such a way that the committee could at once deduce the information they require, and the different tables that they think requisite, or that the Convention thinks requisite. That, I think, would be the cheapest and the most certain way of arriving at the information sought for. Of course, the tables of mortality could only be deduced by persons who have given some little attention to that subject. There are many persons in this country who would be willing to give their assistance freely—and, indeed, this is not a very small labour—for the sake of helping on the cause of Life Insurance. I have no doubt that any committee which the Convention will appoint will freely give their information and labour."

CORRESPONDENCE.

AUTHORSHIP OF THE TREATISE ON PROBABILITY PUBLISHED BY THE SOCIETY FOR THE DIFFUSION OF USEFUL KNOWLEDGE.

To the Editor of the Assurance Magazine.

DEAR SIR,—In your last Number (p. 143) my friend Sir John Lubbock says, in a note relative to the Useful Knowledge treatise on probability by himself and the late Mr. Drinkwater Bethune, that " if anyone shall pretend that this work was written by De Morgan, I can produce the letter of my lamented friend with which he furnished our manuscript to Mr. Coates." This is a sly joke about the state of things which I shall describe; but to those who are not in the secret there will be something mysterious about it. I am glad of the opportunity of once more setting right a mistake which has now lasted at least fifteen years.

When anything anonymous on mathematics turns up in the Useful Knowledge publications, it was written by me: this is one of the fundamental laws of thought. Bacon says that the mind delights in springing up to general maxims, that it may find rest; and the mind of the catalogue-maker finds rest in the maxim that I am the author of all the anonymous mathematics published by the Useful Knowledge Society. About 1845, a binder, in obedience to the general law, stamped an issue of Lubbock and Drinkwater (Bethune) on *Probability* with the title "De Morgan on Probabilities," in gold letters. In the *Arabian Nights*, when a traveller tells the Sultan some particularly veracious story, it is ordered to be written in letters of gold and deposited in the archives of the kingdom. The story above is as well entitled to the bright alphabet as any which the Princess Schehezerade (or Scheherezade, I forget which) ever told for one day more of life; but our manners, though not wholly averse from lying, properly gilt, do not absolutely require us to perpetuate false titles; I therefore printed one correction of this mistake in my "Arithmetical Books" (1847), and others in other places. But what can a sober black-and-white statement, shut up inside a book, do against a brilliant misstatement on the outside? The only chance is to make the statement long enough to attract attention, and I have spun this letter out accordingly.

Yours truly,

November 8, 1860.   A. DE MORGAN.

THE LATE MR. HILLMAN'S TABLES OF THE VALUES OF LIFE ASSURANCE POLICIES.

To the Editor of the Assurance Magazine.

SIR,—Having recently had occasion to use the values of policies at 4 per cent. "Experience" rate of mortality, I very naturally sought to avail myself of the labours of the late Mr. Hillman, who, in his *Value of Life Policies* published the values at the rate above mentioned—that being the only published table, according to the "combined experience" at 4 per cent., with which I am acquainted. For the object I had in hand this promised to be a valuable assistance, inasmuch as it would save the necessity of my forming the values which I had there hoped to find were at my service. Instead, however, of obtaining the expected aid, I discovered that a considerable number of the values I had based upon Mr. Hillman's figures were untrue, owing to the extreme inaccuracy of the table referred to, not a page of which is without numerous errors.

It is painful to criticise the work of a deceased member of the profession, but I am impelled to do so as a duty to others who, like myself, may be tempted to avail themselves of these tables, which are stated to be the result of many years' labour, and to have been printed "for circulation among those engaged in the business of life assurance," in consequence of several of Mr. Hillman's more immediate friends being desirous of possessing a copy. The work is commended on the ground that, "if it will not much abridge the labour of the computations, it will form results by which they may be verified." The compiler also expresses his obligation to four gentlemen named, and others, for assistance rendered him during the compilation of the work. It therefore naturally follows that a certain amount of confidence will be reposed in the values thus accredited, which are there so extensively collected, and which, if trustworthy, would be of great value. Unfortunately, such confidence would be sadly misplaced, inasmuch as in the table at 4 per cent. Experience (to which my examination has been limited), at pages 69 to the end, the number of errors is, on an average, 10 to each page; besides nearly nine hundred differences which I have noted to the extent of $+$ or -1 and 2 in the final figures, and which appear to have arisen from incorrect interpolation of the logarithmic results.

That this communication may not be considered undeserving the notice of the profession of actuaries, and to show that I am not influenced by any feeling towards the deceased gentleman by whose name the tables are sanctioned, and with whom I was in no way acquainted, I append a list of the incorrect values above referred to, in order that others may possibly be spared the inconvenience to which I have been subjected in correcting the erroneous results at which I had arrived by confiding in so untrustworthy a pioneer. Allow me, in conclusion, to observe that some of the errors are so obvious as to be readily detected by inspection of the differences, by which means I first alighted upon a portion of them—sufficient, indeed, to throw distrust upon the whole. A recomputation of the entire table by myself has shown how far that distrust was well founded.

As further evidence of the want of attention in the compilation, I may instance Tables IV., V., and VI., being the preparatory tables for computing the present values of policies at the rates of 3, $3\frac{1}{2}$, and 4 per cent. Experience, the logarithms of the annuities payable immediately ($\overline{\lambda 1 + a}$) from age 93 to the final age, in each of the three tables, are given with a negative index—an error manifest to the merest tyro, and which, had those logarithms been used in the computation of the Tables of Values Nos. X., XI., and XII., founded on them, would have produced values of one-tenth only of the correct results.

It is very probable that the errors to which I thus draw attention may have been observed by others, who may have felt, as I do, a delicacy in making the fact public. I trust, however, that I shall be acquitted of having transgressed the requirements of good feeling, for I may truly say that I have no other object in view than to save others from falling into a difficulty similar to that which I have encountered.

I am, Sir,

Your obedient servant,

Eagle Life Office,   SAMUEL L. LAUNDY.
    10*th December*, 1860.

A Schedule of Ninety-nine Errors referred to in the foregoing Letter.

Page 69,	Age 14,	column 3	years,	·01517	should be		·01523
,,	,, 15	,, 5	,,	·02770	,,		·02735
,,	,, 17	,, 2	,,	·01122	,,		·01127
,,	,, 18	,, 2	,,	·01172	,,		·01169
	,, 23	,, 3	,,	·02185	,,		·02176
	,, 23	,, 5	,,	·01748	,,		·03748
,,	,, 27	,, 5	,,	·04400	,,		·04397
,,	,, 29	,, 1	,,	·00894	,,		·00890
,,	,, 31	,, 3	,,	·02994	,,		·03005
,,	,, 33	,, 2	,,	·02141	,,		·02146
,,	,, 39	,, 5	,,	·07321	,,		·07342
,,	,, 40	,, 2	,,	·02927	,,		·02933
,,	,, 45	,, 1	,,	·01800	,,		·01797
,,	,, 49	,, 5	,,	·10705	,,		·10715
,,	,, 50	,, 3	,,	·06567	,,		·06570
,,	,, 51	,, 2	,,	·04501	,,		·04507
,,	,, 52	,, 3	,,	·07133	,,		·07032
,,	,, 55	,, 2	,,	·05067	,,		·05168
,,	,, 56	,, 3	,,	·08146	,,		·08046
Page 70,	,, 16	,, 9	,,	·05596	,,		·05494
,,	,, 28	,, 9	,,	·08790	,,		·08802
,,	,, 42	,, 9	,,	·15644	,,		·15633
,,	,, 45	,, 8	,,	·15295	,,		·15292
,,	,, 46	,, 6	,,	·11792	,,		·11693
,,	,, 46	,, 9	,,	·17912	,,		·17903
,,	,, 49	,, 7	,,	·15160	,,		·15170
,,	,, 57	,, 10	,,	·27641	,,		·27661
Page 71,	,, 18	,, 11	,,	·07473	,,		·07515
,,	,, 27	,, 11	,,	·10688	,,		·10685
,,	,, 33	,, 15	,,	·19861	,,		·19869

Page 71,	Age	34,	column	15	years,	·20823	should be	·20640
,,	,,	40	,,	15	,,	·25587	,,	·25570
,,	,,	47	,,	15	,,	·31533	,,	·31690
Page 72,	,,	18	,,	18	,,	·13989	,,	·13886
,,	,,	38	,,	17	,,	·27355	,,	·27555
,,	,,	43	,,	16	,,	·30101	,,	·30200
,,	,,	45	,,	18	,,	·36400	,,	·36299
,,	,,	47	,,	20	,,	·42633	,,	·42607
,,	,,	60	,,	20	,,	·55258	,,	·55247
Page 73,	,,	16	,,	24	,,	·19126	,,	·19114
,,	,,	20	,,	22	,,	·19790	,,	·19691
,,	,,	30	,,	25	,,	·32679	,,	·33603
,,	,,	33	,,	21	,,	·29812	,,	·29892
,,	,,	42	,,	24	,,	·45435	,,	·45426
,,	,,	56	,,	23	,,	·58011	,,	·58021
Page 74,	,,	22	,,	26	,,	·26914	,,	·26895
,,	,,	25	,,	28	,,	·33969	,,	·33069
,,	,,	38	,,	28	,,	·48452	,,	·48440
,,	,,	40	,,	26	,,	·47039	,,	·47027
,,	,,	41	,,	26	,,	·48294	,,	·48194
,,	,,	44	,,	30	,,	·59198	,,	·59098
,,	,,	52	,,	26	,,	·60163	,,	·60082
,,	,,	53	,,	27	,,	·62943	,,	·62964
,,	,,	58	,,	28	,,	·60871	,,	·69871
Page 75,	,,	14	,,	32	,,	·27231	,,	·27398
,,	,,	22	,,	33	,,	·37638	,,	·37628
,,	,,	27	,,	35	,,	·47242	,,	·47169
,,	,,	35	,,	35	,,	·57323	,,	·57320
,,	,,	43	,,	31	,,	·59765	,,	·59776
,,	,,	49	,,	32	,,	·67836	,,	·67805
,,	,,	50	,,	34	,,	·72049	,,	·72079
,,	,,	52	,,	34	,,	·74176	,,	·74170
,,	,,	54	,,	34	,,	·76182	,,	·76382
,,	,,	57	,,	31	,,	·74441	,,	·74461
Page 76,	,,	28	,,	38	,,	·53658	,,	·53568
,,	,,	31	,,	37	,,	·55124	,,	·55724
,,	,,	31	,,	38	,,	·57043	,,	·57402
,,	,,	32	,,	39	,,	·60311	,,	·60301
,,	,,	34	,,	37	,,	·59460	,,	·59469
,,	,,	34	,,	40	,,	·64426	,,	·64326
,,	,,	39	,,	36	,,	·63740	,,	·63749
,,	,,	39	,,	38	,,	·67019	,,	·66942
,,	,,	41	,,	38	,,	·69098	,,	·69113
,,	,,	46	,,	38	,,	·74212	,,	·74222
,,	,,	47	,,	39	,,	·76285	,,	·76755
Page 77,	,,	22	,,	43	,,	·53005	,,	·53995
,,	,,	26	,,	42	,,	·57504	,,	·57594
,,	,,	27	,,	41	,,	·57245	,,	·57255
,,	,,	33	,,	44	,,	·69312	,,	·69212
,,	,,	33	,,	45	,,	·69646	,,	·70646
,,	,,	34	,,	41	,,	·65980	,,	·65878
,,	,,	34	,,	45	,,	·71733	,,	·71739

Page 77,	Age	35	column	44	years,	·71524	should	be	·71424
,,	,,	41	,,	44	,,	·77667	,,		·77656
	,,	42	,,	45	,,	·80076	,,		·80065
,,	,,	45	,,	42	,,	·78552	,,		·79040
,,	,,	45	,,	43	,,	·80480	,,		.80474
Page 78,	,,	22	,,	47	,,	·60449	,,		·60357
,,	,,	22	,,	49	,,	·63498	,,		·63414
	,,	23	,,	46	,,	·60080	,,		·60089
	,,	23	,,	50	,,	·66130	,,		·66139
,,	,,	26	,,	46	,,	·64703	,,		·63880
	,,	30	,,	47	,,	·70160	,,		·70100
,,	,,	32	,,	47	,,	·71674	,,		·72319
,,	,,	34	,,	46	,,	·73150	,,		·73112
..	,,	34	,,	49	,,	·77046	,,		·77041
	,,	35	,,	46	,,	·74267	,,		·74166
	,,	37	,,	50	,,	·81826	,,		·81402
	,,	40	,,	46	,,	·79303	,,		·79321

MISS FLORENCE NIGHTINGALE'S "NOTES ON NURSING."[*]

To the Editor of the Assurance Magazine.

SIR,—The important bearing which all matters connected with social and sanitary science have upon the health and longevity of the people, will be my apology for offering a few remarks on the above work, which, though entitled by the estimable Author *Notes on Nursing*, is devoted, to a considerable extent, to household *hygiène*—a subject entirely cognate to the questions usually discussed in the pages of the *Assurance Magazine*, as affecting, in the first place, the question of sanitary reform, in which we are all alike interested, and, in the second, the deepest interests of the assurance community.

With respect to that portion of the *Notes* which refers to nursing *proper*, I have only to say that the experience of the Author most fully qualifies her to lay down rules on the subject, and that every word she utters teems with valuable advice to those who have the personal charge of the health of others entrusted to them.

In connexion with *hygiène*, Miss Nightingale lays great stress upon the importance of procuring in our houses a due supply of fresh external air, by means of proper ventilation, a matter far too little attended to in general in this country, and she shows how it is possible to secure efficient ventilation without necessarily producing draughts of cold air—by means of proper arrangements as to the opening of the windows of houses, which, even at night, she considers to be unobjectionable.

Referring to the popular idea that exposure to the night air is generally undesirable, she says, that "in great cities, night air is often the best and purest air to be had in the twenty-four hours," and quotes the opinion of one of the highest medical authorities on consumption and climate in confirmation of the fact, that "the air in London is never so good as after ten o'clock at night."

* *Notes on Nursing: What it is, and What it is not.* By Florence Nightingale. London, Harrison, 59, Pall Mall.

While giving Miss Nightingale credit for having fully considered any subject on which she expresses an opinion, I must admit that her views militate against all my preconceived notions on this subject. The fact is patent to most of us, that a change from a warm room to the cold air has a most determined effect in producing irritation of the air passages of the lungs, if there should chance to exist the least tendency to weakness in this respect; and though I do not pretend to deny that the air at night may, in large cities, be purer than in the day-time, owing to the fact that a less amount of carbon is likely to be floating in the atmosphere after the extinction of the greater proportion of the fires used for domestic purposes—still the air being at night unquestionably colder than it is in the day-time, and the irritation produced by it on the delicate passages of the respiratory apparatus being clearly in proportion to the temperature, we have to consider how far the greater purity of the night air may compensate for evils likely to be produced by its lower temperature, and we must then strike a balance between the advantages and disadvantages of exposure to it.

There are five essential points, we are told, to which due attention must be paid, if we desire to secure the elements of health in our houses. These are—pure air, pure water, efficient drainage, cleanliness, and light.

In connexion with this subject, Miss Nightingale refers to the interests of Life Offices in sanitary matters, and says—" The object in building a house is to obtain the largest interest for the money, not to save doctors' bills to the tenants. But, if tenants should ever become so wise as to refuse to occupy unhealthily constructed houses, and if Insurance Companies should ever come to understand their interest so thoroughly as to pay a Sanitary Surveyor to look after the houses where their clients live, speculative architects would speedily be brought to their senses."

And again : " In Life Insurance and such like societies, were they instead of having the persons examined by a medical man, to have the houses, conditions, and ways of life, of these persons examined, at how much truer results would they arrive!"

" Undoubtedly," she adds, " a person of no scientific knowledge whatever but of observation and experience in these kinds of conditions, will be able to arrive at a much truer guess as to the probable duration of life of members of a family or inmates of a house, than the most scientific physician to whom the same persons are brought to have their pulse felt; no enquiry being made into their conditions."

There is some truth in this; though, I fear, it would be impracticable to carry into effect our author's suggestions.

" Minute enquiries into conditions," she goes on to say, " enable us to know that in such a district, nay, in such a street—or even on one side of that street, in such a particular house, or even on one floor of that particular house, will be the excess of mortality, that is, the person will die who ought not to have died before old age."

" Now, would it not very materially alter the opinion of whoever were endeavouring to form one, if he knew that from that floor, of that house, of that street the man came."

There can, certainly, be no doubt that with reference to one large class of diseases—zymotic diseases—as has been clearly shown in communications in these pages, we hold in our own hands the key to the mode of conduct by which alone we are likely to be enabled to grapple with them—and this key is, sanitary reform.

The principal diseases of this class are, fever, small-pox, scarlatina, whooping-cough, measles, croup, and diarrhœa.

That these diseases are, to no inconsiderable extent, of a preventable nature, there is, I apprehend, no reason to doubt. I cannot, however, entirely coincide with Miss Nightingale in her views on this subject.

"It is commonly thought," she says, "that children must have what are commonly called 'children's epidemics,' 'current contagions,' &c., in other words, that they are born to have measles, whooping-cough, perhaps even scarlet fever, just as they are born to cut their teeth, if they live."

"'Now, do tell us, why must a child have measles?' 'Oh, because,' you say, 'we cannot keep it from infection—other children have measles—and it must take them—and it is safer that it should.' 'But why must other children have measles? And if they have, why must yours have them too? If you believed in and observed the laws for preserving the health of houses which inculcate cleanliness, ventilation, white-washing, and other means, and which, by the way, *are laws*, as implicitly as you believe in the popular opinion, for it is nothing more than an opinion, that your child must have children's epidemics, don't you think that upon the whole your child would be more likely to escape altogether?'"

I fear that this is going a little too far. Though the means suggested would, and do, no doubt, tend to diminish the malignity of certain types of disease, it can hardly be predicated that the most complete attention to carry out such means to the very utmost would necessarily have the effect of eliminating the diseases in question; and for this reason, among others —that we do not find immunity from them, or anything like it, even among those classes, the sanitary state of whose dwellings, and whose position generally, is as nearly perfect in such respects as it possibly could be.

While cordially recommending Miss Nightingale's work to the attention of the readers of the *Assurance Magazine*, and thanking her for a contribution to sanitary science of no mean value, I have felt called upon to express conscientiously in what respect I dissent from her opinions.

Miss Nightingale is, perhaps, a little enthusiastic in the subject which she has taken up so warmly; but, when we remember the extraordinary devotion with which this Lady has sacrificed herself to the interests of suffering humanity, we can hardly expect that she would be otherwise than an enthusiast in relation to any matters connected with the sacred calling she so nobly adopted at her Country's sorest need.

I am, Sir,

Your obedient servant,

Alliance Assurance Office,   H. W. PORTER.
    12th December, 1860.

THE

ASSURANCE MAGAZINE,

AND

JOURNAL

OF THE

INSTITUTE OF ACTUARIES.

On the Rates of Premium required to provide certain Periodical Returns to the Assured. By ROBERT TUCKER, ESQ., *Actuary to the Pelican Life Insurance Company, and one of the Vice-Presidents of the Institute of Actuaries.*

[Read before the Institute the 28th January, 1861, and printed by order of the Council.]

A WELL-KNOWN contributor to the *Assurance Magazine* has drawn attention, at pp. 167 and 168 of the Number for April, 1859, to " the incongruity existing between the rates of premium charged at certain ages on bonus policies, and the benefits to which they entitle the holder."

This question, " H. A. S." remarks, " hardly meets with the consideration it deserves among Actuaries." It has, nevertheless, not been altogether lost sight of; for, so far back as the year 1851, Mr. Jellicoe noticed this "incongruity";* and, so recently as the month of March, 1857' a paper was read before the Institute of Actuaries by Mr. Sprague, and published in the July Number of the *Journal* for that year, " On certain methods of dividing the surplus among the assured in a Life Assurance Company, and *on the rates of premium that should be charged to render them equitable.*" In this paper various plans are investigated and examples

* *See* that gentleman's paper on the determination and distribution of surplus, vol. i., p. 161; also his paper on the conditions which give rise to surplus in Life Assurance Companies, and on the amount of return or " bonus" which such conditions justify, vol. ii., p. 333.

given—amongst them a formula almost identical with that by "H. A. S."

I propose to discuss this subject a little more in detail, and to introduce illustrations of some of the plans adopted by Assurance Institutions for the distribution of what, as it appears to me, cannot with propriety, in all cases, be called their *surplus profits,* but which may be designated as periodical returns made to the assured.

The example given in the letter referred to supposes a reversionary bonus of P per £1 per annum declared every t years, which is added to the principal sum and forms the capital upon which the next bonus is computed, and so on.

The sum insured being £1, the annual bonus for the first period is $P = B_1$; for the second period $P(1 + tP) = B_2$; for the third period $P(1 + tP)^2 = B_3$; and, generally, for the nth period $P(1 + tP)^{n-1} = B_n$. Whence, by the ordinary commutation tables, the annual premium for such a benefit at age x is—

$$\frac{M_x + B_1 R_x + (B_2 - B_1)R_{x+t} + (B_3 - B_2)R_{x+2t} + \&c.}{N_{x-1}} \quad \ldots \quad (1)$$

If we substitute for B_1, B_2, B_3, &c., their values, we have—

$$B_1 = P,$$
$$B_2 - B_1 = P(1 + tP) - P = tP^2,$$
$$B_3 - B_2 = P(1 + tP)^2 - P(1 + tP) = tP^2(1 + tP),$$
$$B_4 - B_3 = P(1 + tP)^3 - P(1 + tP)^2 = tP^2(1 + tP)^2,$$
$$\&c. = \&c.;$$

and by dividing the annual premium into two parts, we obtain for the sum assured—

$$\frac{M_x}{N_{x-1}},$$

and for the bonus—

$$\frac{PR_x + tP^2.\{R_{x+t} + (1 + tP)R_{x+2t} + (1 + tP)^2 R_{x+3t} + \&c.\}}{N_{x-1}}.$$

Mr. Sprague commences with R_{x+1}, which, by the ordinary notation, is the value of an increasing assurance at age $(x + 1)$ years, and implies that the additions are made according to the number of years of life completed by the assured, and not upon the number of premiums paid, which is the assumption in the other case. Subject to this alteration, the expression last obtained would be identical with that given by him.

The following examples are added by " H. A. S.," showing, according to the Carlisle 3 per cent. Table— 1st, the annual premium to insure £1,000, at the ages stated; 2nd, the annual premium required to meet the increasing benefit above described; and 3rd, the ratio of the first to the second.

Age.	Premium for £1,000.	Premium for £1,000, with Bonus.	Ratio.
20	14·9358	24·3651	1·63
30	19·5192	29·6399	1·52
40	25·9932	36·8245	1·42
50	36·2236	47·7889	1·32
60	57·8955	70·1644	1·21

It is then noticed that if an addition of 30 per cent. were made to the ordinary premium, the entrants at age 50 " get a reasonable equivalent for their payments," whilst at other ages the benefits are greatly in favour of the younger lives. It is worthy of remark that those who advocate a constant addition to the net premium, in preference to a percentage on it, may here obtain a confirmation of their views; for it is observable that an addition of £10 to the annual premium, without bonus, would very nearly provide for this reversionary bonus of £15 per annum, with its accumulations.

If we make the comparison by the Northampton 3 per cent. Table—which, perhaps, more correctly represents the premiums charged by Offices adopting this mode of apportionment—the following table shows that the inequality above noticed nearly disappears. It also proves that members entering at the earlier ages of life do not contribute unduly in the shape of premium, as has been often alleged, when the ultimate benefit insured to them is taken into account.

Age.	Premium for £1,000.	Premium for £1,000, with Bonus.	Ratio.
20	21·794	24·365	1·12
30	26·672	29·640	1·11
40	33·975	36·825	1·08
50	45·301	47·789	1·05
60	63·661	70·164	1·10

In order to obtain a practical result from a comparison of our theoretical conclusions with the bonuses prevailing according to the practice of Assurance Offices in their declarations of profit, it is necessary in our illustrations to endeavour to adapt them to such practice. For instance, the reversionary addition of P per £1 per annum $= B_1$ does not usually date from the commencement of the assurance, unless the life assured be living at the end of the first period of t years; in like manner the subsequent annual increments $B_2 - B_1$, $B_3 - B_2$, &c., do not vest unless the assured be living at the end of $2t$, $3t$, &c., years respectively. Thus, in fact, the assured has an *interval bonus* at the same rate as that he was entitled to on completing the last period of t years.

The value, therefore, of the periodical additions, and of the prospective or interim bonuses, will be

$$t \text{P.} M_{x+t} + \text{P.} R_{x+t} + (B_2 - B_1) \cdot (t M_{x+2t} + R_{x+2t}) +$$
$$(B_3 - B_2) \cdot (t M_{x+3t} + R_{x+3t}) + \&c.;$$

and the annual premium for the whole benefit

$$\frac{M_x + t \text{P.} M_{x+t} + \text{P.} R_{x+t} + (B_2 - B_1) \cdot (t M_{x+2t} + R_{x+2t}) + \&c.}{N_{x-1}} \quad . \quad . \quad (2)$$

After the expiration of t years, a bonus of P per £1 per annum being added for every future premium paid, the value of these additions is evidently $\text{P.} R_{x+t}$; and the values of the subsequent increments are $(B_2 - B_1) \cdot R_{x+2t}$, $(B_3 - B_2) R_{x+3t}$, &c.

This alteration will not materially affect the results already given, and, as the computations are somewhat laborious, it seems scarcely necessary to ascertain the exact difference at the ages enumerated.

In some Offices the practice of paying bonus upon bonus does not exist, the additions being made upon the original sum assured only. Suppose these additions to be uniform, and at the same rate as before, then $B_1 = B_2 = B_3$, &c. $= \text{P}$, and the last expression becomes—

$$\frac{M_x + t \text{PM}_{x+t} + \text{PR}_{x+t}}{N_{x-1}} \quad . \quad . \quad . \quad . \quad . \quad . \quad . \quad (3)$$

Here the assured has an *interval bonus* after surviving the first t years.

When the bonus is declared at each period of t years, and no addition is made in the event of death occurring between any two periods of division, the annual premium will be—

$$\frac{M_x + t \text{P}(M_{x+t} + M_{x+2t} + M_{x+3t} + \&c.)}{N_{x-1}} \quad . \quad . \quad . \quad . \quad (4)$$

and when *each periodical bonus is reckoned from the date of the policy,* and an addition is made at the uniform rate when death occurs between two periods of division for the number of years so completed, the annual premium will be—

$$\frac{M_x + t\mathrm{P}M_{x+t} + \mathrm{P}.\,\mathrm{R}_{x+t} + t\mathrm{P}(M_{x+2t} + 2M_{x+3t} + 3M_{x+4t} + \&\text{c}.)}{N_{x-1}} \quad . \quad (5)$$

This assumes that no bonus is allowed if death take place before the expiration of the first period of t years.

If we suppose no prospective bonus to be insured, but an uniform sum of $t\mathrm{P}$ added every t years from the commencement, the premium will be—

$$\frac{M_x + t\mathrm{P}(M_{x+t} + 2M_{x+2t} + 3M_{x+3t} + \&\text{c}.)}{N_{x-1}} \quad . \quad . \quad . \quad . \quad (6)$$

which is also given by Mr. Sprague.

I now propose to add some examples of the premiums deduced from the formulæ (3), (4), and (5), according to the Carlisle 3 per cent. Table.

EQUATION (3).

Sum Assured, £1,000; Bonus, £1. 10s. per cent. per annum for life, subject to the assured living 5 years from the date of entrance.

Age.	NET ANNUAL PREMIUM TO INSURE £1,000.		
	WITHOUT BONUS.		WITH BONUS.
	Carlisle.	Northampton.	Carlisle.
20	14·9358	21·794	21·841
30	19·5192	26·672	27·292
40	25·9932	33·975	34·703
50	36·2236	45·301	45·991
60	57·8955	63·661	68·383

From this example it appears that annual premiums computed according to the Northampton 3 per cent. Table will just provide an addition to the sum assured of £1. 10s. per cent. per annum, at least, up to the age of 50.

EQUATION (4).

Sum Assured, £,1000; Bonus, £75; added every 5 years.

Age.	Annual Premium.
20	21·2398
30	26·5262
40	33·7082
50	44·5876
60	66·4635

The difference between these premiums and those resulting from equation (3) being inconsiderable, shows how very small is the value of the prospective or interim bonus.

EQUATION (5).

Sum Assured, £1,000; *Bonus,* £1. 10s. *per cent. per annum,* reckoned from the date of policy, *every* 10 *years, subject to the assured living* 10 *years; Prospective Bonus,* £1. 10s. *per cent. per annum.*

Age.	Annual Premium.
20	30·380
30	34·662
40	40·420
50	49·394
60	68·995

These premiums show, still more forcibly, the advantages given to younger members when the bonuses date from the commencement of the policy.

The following examples from equations 3 and 5 are added, showing, according to the Carlisle 4 per cent. Table—

1. *The Annual Premium to insure* £1,000 *at Death, with Addition of* £1, £1. 10s., *or* £2 *per Cent. per Annum every Five Years ; and a Prospective or Interim Bonus at the same Rate in each case.*

Age.	ANNUAL PREMIUM.			
	WITHOUT BONUS.	WITH BONUS OF		
		£1 per Cent. per Annum.	£1. 10s. per Cent. per Annum.	£2 per Cent. per Annum.
20	13·18	16·773	18·569	20·366
30	17·55	21·744	23·842	25·939
40	23·75	28·626	31·063	33·501
50	33·64	39·327	42·170	45·013
60	55·31	61·607	64·756	67·905

2. *The Annual Premium to insure* £1,000 *at Death, with Addition of* £1, £1. 10s., *or* £2 *per Cent. per Annum every Ten Years, calculated from the Date of the Policy ; and a Prospective or Interim Bonus at the same Rate in each case.*

Age.	ANNUAL PREMIUM.			
	WITHOUT BONUS.	WITH BONUS OF		
		£1 per Cent. per Annum.	£1. 10s. per Cent. per Annum.	£2 per Cent. per Annum.
20	13·18	20·773	24·569	28·365
30	17·55	25·359	29·264	33·168
40	23·75	31·546	35·444	39·342
50	33·64	41·081	44·801	48·521
60	55·31	61·792	65·033	68·274

Hitherto we have considered only the ratio which the annual premium for an assurance, with bonus additions, at a given rate bears to the net premium for the same assurance without bonus. Let us now examine the effect of a reduction of premium after a given number of years.

Suppose π_x to be the annual premium to insure £1 at age x, and ρ the reduction per cent. after t years—

$$\text{then } \pi_x\left(1 + a_{x+\overline{t-1}|} + \overline{1-\rho} \cdot a_{x}{}_{\overline{t-1}|}\right) = A_x,$$

$$\text{whence } \pi_x = \cfrac{A_x}{1 + a_{x+\overline{t-1}|} + \overline{1-\rho} \cdot a_{x}{}_{\overline{t-1}|}} = \cfrac{A_x}{1 + a_x - \rho a_{x\overline{t-1}|}}.$$

If π'_x be the premium actually paid, $\pi'_x = (1+\kappa)\pi_x$; π_x being the net premium and κ the addition to it;

$$\text{whence } \rho\, a_{x\overline{t-1}|} = 1 + a_x - \frac{A_x}{\pi'_x} = (1+a_x) \cdot \left(1 - \frac{1}{1+\kappa}\right) = \frac{\kappa(1+a_x)}{1+\kappa},$$

$$\text{and } \rho = \frac{\kappa(1+a_x)}{(1+\kappa)a_{x\overline{t-1}|}}.$$

Suppose, as before, the sum assured to be £1,000, and the reduction of premium to be 50 per cent. after 5 years, the following examples show, according to the Carlisle 3 per cent. Table, the ratio which such premiums bear to the net premium payable during life, according to the Carlisle and Northampton Tables respectively.

Age.	Carlisle, without reduction. 1.	Carlisle, with reduction of 50 per cent. after five years. 2.	Northampton, without reduction. 3.	Ratio of 1 to 2.	Ratio of 3 to 2.
20	14·936	24·796	21·794	1·66	1·14
30	19·519	31·868	26·672	1·63	1·19
40	25·993	41·483	33·975	1·59	1·22
50	36·224	55·736	45·301	1·54	1·23
60	57·895	83·706	63·661	1·44	1·31

Again: suppose the reduction to be 80 per cent. after 7 years, and we obtain the following results :—

Age.	Annual Premium, with 80 per cent. reduction after seven years.	Ratio to	
		Carlisle.	Northampton.
20	35·420	2·37	1·62
30	44·105	2·26	1·65
40	55·085	2·12	1·62
50	69·462	1·92	1·53
60	96·058	1·66	1·51

It will be seen that, while the values are greatly disproportionate at 20 and 60, they are more uniform at the intermediate ages; and that at all ages, particularly in the last examples, the premiums required are so much in excess of those usually charged, that it is natural to ask, not only how such reductions can be made, but how they can be maintained for any length of time.

Let us take another view of this case, and let us suppose an Office to charge certain rates of premium, with an implied obligation to reduce the same by 50 per cent. after 5 years, and certain other rates with a reduction of 80 per cent. after 7 years; the following examples show the single premiums corresponding to these rates, also the single premium according to the Carlisle 4 per cent. Table, which is considered to represent very nearly the net value, or prime cost, of an assurance for life. On comparing columns 3 and 5 with 6, it appears that the values are greater in 3 than in 6, except at the advanced ages, and that in 5 they are less at all ages. The premiums in column 2 are somewhat higher up to the age of 40 than are usually charged.

Age.	2. Annual Premium per cent., with reduction of 50 per cent. after five years.	3. Corresponding Single Premium.	4. Annual Premium per cent., with reduction of 80 per cent. after seven years.	5. Corresponding Single Premium.	6. Single Premium per cent., Carlisle 4 per cent.
20	2·25	30·766	2·50	23·925	25·532
30	2·75	34·625	3·00	27·293	31·338
40	3·50	39·788	3·75	32·104	38·178
50	4·50	44·753	5·25	41·895	46·658
60	6·50	51·662	7·50	51·945	58·987

The two plans of augmenting the sum originally assured by an annual percentage at stated periods, and of materially diminishing the annual premium after a certain number of years, are, I think, held most in favour by the public—probably because they are more clearly defined, and therefore better understood and appreciated, than any of the other modes of division adopted by Assurance Companies. For the same reason, it may be easier to estimate the value of these periodical additions and annual reductions than the values under any other plan, and thus to point out what the assured gain over and above their contributions, for these differences really constitute the actual bonuses realized, and not that portion of them for which a consideration is paid in the original contract.

It is the custom with some Offices to apportion their profits according to the amount received upon each policy, less its value at the period of division. This difference is called the "proportional bonus;" and the method of determining the sum in ready money to be allotted to each policy, is, to compare the surplus to be divided with the total amount of these differences—that is, with the total in the proportional bonus column. Thus, if the divisible surplus is found to be 30 per cent. of the total amount of "proportional bonus," the ready-money bonus upon each policy will, in like manner, be 30 per cent. of the "proportional bonus" appertaining thereto—or, expressed in official language, the common multiplier will be ·3.

If s^t represent the amount of £1 per annum in t years, the "proportional bonus" will then be—

$$(s^{t+1}-1)\pi'_x-(\pi'_{x+t}-\pi'_x)\cdot(1+a'_{x+t});$$
$$\text{or, } (s^{t+1}-1)\pi'_x-A'_{x+t}+\pi'^{'}_x\cdot(1+a'_{x+t});$$
$$\text{or, } (s^{t+1}+a'_{x+t})\pi'_x-A'_{x+t}.$$

At the second investigation, the premium is calculated according to the age of the life assured at the previous valuation; and not only upon the sum originally assured, but also upon the addition then made to the policy. So that the "proportional bonus" at the end of $2t$ years will be

$$\{(s^{t+1}+a'_{x+2t})\pi'_{x+t}-A'_{x+2t}\}\cdot(1+B'_1),$$

£1 being the amount insured, and B'_1 the first bonus.

At the end of $3t$ years, the "proportional bonus" will be

$$\{(s^{t+1}+a'_{x+3t})\pi'_{x+2t}-A'_{x+3t}\}\cdot(1+B'_1+B'_2);$$

and, generally, at the end of nt years the "proportional bonus" will be

$$\{(s^{t+1}+a'_{x+nt})\cdot\pi'_{x+\overline{n-1}.t}-A'_{x+nt}\}\cdot(1+B'_1+B'_2+B'_3\ldots.B'_{n-1}).$$

It does not appear that the bonuses resulting from this mode of apportionment follow any order of progression, and therefore it will be necessary, in assuming a common multiplier, to calculate the bonus at each period of division, if we wish to ascertain the corresponding premium.

The following table shows, according to the Northampton 3 per cent. premiums, the bonus accruing to a policy for £1,000 at the end of 5, 10, 15, &c., years, assuming the "divisible surplus" to be 25 per cent. of the "proportional bonus."

| No. of Years. | Ages. | | | | | No. of Years. |
	20.	30.	40.	50.	60.	
5	42·703	44·874	49·323	58·830	73·524	5
10	45·775	48·134	54·925	68·606	89·948	10
15	48·899	53·911	64·953	81·766	124·011	15
20	52·396	60·034	75·758	101·316	184·734	20
25	58·685	70·994	90·289	139·685	305·083	25
30	65·349	82·804	111·877	208·082	543·335	30
35	77·279	98·687	154·245	343·641	747·651	35
40	90·134	122·283	229·772	612·003		40
45	107·423	168·591	379·462	842·143		45
50	133·108	251·143	675·797			50
55	183·517	414·755	929·927			55
60	273·377	733·208				60
65	451·473	1014·664				65
70	804·045					70
75	1106·402					75

If we measure the value of these benefits by the Carlisle 3 per cent. Table, and compare the corresponding annual premiums with the net premiums deduced from the Carlisle and Northampton 3 per cent. premiums respectively, without bonus, we obtain the following results:—

| Age. | Premium to insure £1,000, With Bonus. | Ratio to | |
		Carlisle.	Northampton.
20	20·814	1·39	·955
30	26·495	1·36	·993
40	34·452	1·33	1·01
50	43·660	1·31	·964
60	70·833	1·22	1·11

From which it appears that, according to the Carlisle 3 per cent. pure premiums, a loading of 39 per cent. at age 20, and of 22 per cent. at age 60, is necessary to secure these periodical additions; and that the Northampton 3 per cent. premium at age 20 is in excess about 5 per cent., and is deficient about 11 per cent. at age 60, of the rates required for the same purpose. At 30, 40, and 50, the results, as in previous examples, exhibit more uniformity.

If the " proportional bonus" at each investigation were reckoned upon the " sum assured only," and not upon " the sum assured and previous additions," then every person of the same age would receive the same rate of bonus. This is evident from the general expression—

$$(s^{t+1} + a'_{x+nt})\, \pi'_{x+\overline{n-1}.t} - A'_{x+nt}.$$

The same relative result, or nearly so, would be obtained from some rates of premium by making the " proportional bonus" depend upon the improved amount of the total number of premiums paid upon a policy, less the accumulation at the previous valuation— that is, by deducing the " proportional bonus" from the expression

$$\left(s^{nt+1} - s^{\overline{n-1}.t+1}\right)\pi'_x.$$

Another mode of making periodical returns to the assured, and the last which I propose to notice, is that of adding at each valuation a sum to the policy bearing a certain proportion to the number of premiums paid upon it.

The annual premium for an insurance with such benefits will be—

$$\pi_x = \frac{M_x + t\phi\pi_x.(M_{x+t} + 2M_{x+2t} + 3M_{x+3t} + \&c.)}{N_{x-1}},$$

$$= \frac{M_x}{N_{x-1} - t\phi\,(M_{x+t} + 2M_{x+2t} + 3M_{x+3t} + \&c.)}.$$

If we take $t = 5$, and suppose the addition at each quinquennial period to be equal to 20 per cent. per annum on the premiums paid, then $\phi = \cdot 2$ and $t\phi = 1$; whence π_x becomes

$$\frac{M_x}{N_{x-1} - (M_{x+5} + 2M_{x+10} + 3M_{x+15} - \&c.)}.$$

The following examples are added, showing the net annual premium to insure £1,000 at death according to the Carlisle 3 per cent. Table, with an addition of one year's premium at the end of 5, two years' premium at the end of 10, three years' premium at the end of 15 years, and so on; also the ratio which these premiums bear to the Carlisle and Northampton 3 per cent. Tables respectively, without addition.

Age.	Premium to insure £1,000, with Bonus.	Ratio to Carlisle.	Northampton.
		Without Bonus.	
20	24·454	1·64	1·12
30	31·271	1·60	1·17
40	39·938	1·54	1·17
50	52·135	1·44	1·15
60	76·807	1·33	1·21

It is scarcely necessary to remark, that the examples given in this paper may be readily worked out from the published tables of annuities and assurances—immediate, deferred, and increasing. For instance, the formula last given,

$$\pi_x = \frac{M_x}{N_{x-1} - (M_{x+5} + 2M_{x+10} + 3M_{x+15} + \&c.)},$$

may be put under the form—

$$\pi_x = \frac{\dfrac{M_x}{D_x}}{\dfrac{N_{x-1}}{D_x} - \dfrac{M_{x+5} + 2M_{x+10} + \&c.}{D_x}}$$

$$= \frac{A_x}{1 + a_x - (A_x + 2A_x + 3A_x + \&c.)},$$

and the results readily obtained from Mr. Thomson's Actuarial Tables 1 and 2 (single lives and single deaths). I have not thought it necessary to show the net premiums required to provide *guaranteed bonuses*. These are simply another name for so many deferred assurances, not entitling the policy-holders to participate in the profits of the Company.

On taking a general review of the "bonus question," one cannot help being impressed with the advantage of selecting the participating in preference to the non-participating plan of assurance. This is apparent, if we examine with attention the large bonuses declared by some Societies, and compare the premiums charged to their members with what they ought to pay to ensure such benefits. The public are not slow to perceive that, however low the non-participating rates of an Office may be, the immediate gain of an addition to the sum assured which a given payment would secure, when compared with the amount which the same annual sum would assure according to the participating rates of premium, is not so tempting as the ultimate prospect of a much larger addition. We accordingly find that most Offices now conduct their business on the principle of admitting their policy-holders to share in their profits.

Confining our observations to reversionary bonuses, and omitting all further consideration of the plan of making a large annual reduction in the first premium after a fixed period, which, although apparently very simple, has never received a satisfactory solution, we have seen that an addition of £1. 10s. per cent. per annum, subject to the assured living 5 years, is the amount of bonus expected to accrue to members paying the Northampton 3 per cent. premium. This is without making any allowance for commission and charges of management. Against these items we may fairly set the various sources of profit realized by Assurance Companies—such as that arising from excess of interest over 3 per cent.—the rate upon which their calculations are usually based— profit from lapsed and surrendered policies, from suspended mortality, &c. &c. Suppose these various elements of profit to be sufficient to sustain the working of an Assurance Company, and that a return is made to the policy-holders equal in value to the loading or addition to the net premium for the risk undertaken, it would appear that such a Company fulfils all that can be reasonably expected from it.

Since this paper was written, my attention has been called to Mr. Scratchley's *Treatise on Life Assurance Societies*, in which some remarks are made on the errors of the bonus system, and some formulæ are given in an appendix for estimating the value of a bonus.

It would have been an omission on my part not to have referred to the labours of so popular a writer as Mr. Scratchley. The only formula, however, in his work, bearing on the particular cases which I have introduced, is

$$\frac{(\mu+1)\,\mathrm{M}_x + \cdot 02\,\Sigma\,\mathrm{M}_x}{\mathrm{N}_x},$$

the annual premium for an assurance with a guaranteed bonus of 2 per cent. per annum. The other formulæ appear to be illustrations of the author's peculiar views "As to Bonuses."

Whether, as Mr. Scratchley asserts, *the only case in which the payment of a bonus to an assurer is really proper or desirable, is when he has paid up in Office premiums, with interest, the amount of his policy,* be true or not, depends very much, I submit, on the rates of premium charged. Apart from this consideration, no satisfactory conclusion can be arrived at; and the statement must be received as the mere expression of an individual opinion.*

The truth is, that the whole bonus system is a matter of agreement. The public are invited to pay certain premiums, manifestly higher than are necessary for the risks undertaken, and for which the assured are to receive certain periodical returns—not, be it observed, " guaranteed," but dependent on the success of the particular Office they have selected.

There is clearly nothing unfair in practice, nor unsound in principle, in such arrangement, provided the returns are made with a due regard to the premiums paid. That some plans are more attractive than others, without being really better, may be readily imagined; but that unfair means should be resorted to by *respectable Offices* to make them so is not so easily credited.

I confess I have read with regret Mr. Scratchley's assertion, that *various respectable Offices have taken to declaring bonuses so large as to be obviously not justified by their financial condition nor consistent with security.*

It is difficult to believe that any *respectable Office*, properly so called, would do anything of the kind; and it seems to me unjustifiable that so severe a censure should be passed upon *various respectable Offices*, when the author only produces one solitary example in support of his declaration, and this is given on the authority of an experienced actuary whose name is withheld.

The extent of the error committed when the premiums paid are represented as creating large profits, is pointed out by Mr. Scratchley in a table *showing the amount that a net annual premium of £1 will assure at the death of a person of any given age, or the amount to which an annuity due of £1 will accumulate by the end of the year of his death.*

Mr. Scratchley appears to have committed the same error that some other writers have done, by treating these questions as identical. Professor De Morgan has shown, in the *Companion to the British Almanack for* 1842, that the answer to the first question is

$$\frac{N_{x-1}}{M_x},$$

$$\text{or,} \quad \frac{(1+r) \cdot (1+A)}{1-rA};$$

A being the value of a life annuity and r the rate of interest; and that the answer to the second is

$$\frac{a_{x+1}}{a_x}(1+r)+\frac{a_{x+2}}{a_x}(1+r)^2+\frac{a_{x+3}}{a_x}(1+r)^3+\ \ldots$$

a_x denoting the number of persons living at any age x.

Mr. De Morgan also adds the following examples by the Northampton Table at 4 per cent., showing the average sums obtained in the two cases :—

Age.	1st Case.	2nd Case.	Age.	1st Case.	2nd Case.
20	49·4	98·3	45	27·2	38·0
25	44·7	82·2	50	23·2	30·7
30	40·2	68·6	55	19·7	24·5
35	35·7	56·8	60	17·0	19·2
40	31·3	46·7			

And Mr. Hardy enters into an elaborate investigation of *the proper method of determining the amount of an annuity forborne and improved at interest during the existence of a given life,* in a paper read before the Institute on the 26th January, 1857, and published in the April number of the *Journal* for that year, clearly demonstrating that the expressions above given are identical only when *the annuity is forborne and improved for a term of years certain,*. or *when money bears no interest*—in fact, that one is the average of *present values,* the other of *amounts.*

Of Compound Interest. By Dr. Edmund Halley, *Royal Astronomer, Savilian Professor of Geometry, Oxford, and* F.R.S.*

A PRINCIPAL use of logarithms is to solve all the cases of compound interest which are not, without great difficulty, attainable by the rules of common arithmetic. But before we proceed to the practical part, it may, perhaps, not be improper to say something of the foundation or demonstration of the rules we are to give.

Therefore, let p be any sum of money forborne t times, r the rate of interest or produce of £1 and its interest in one time—that is, as 1 to r, so £1 to its amount after one year or other space of

---

* We republish this paper from *Sherwin's Mathematical Tables,* printed for W. and J. Mount, 1761; and, considering the celebrity of the writer, the ability displayed in the paper itself, and the comparative scarcity of the work from which it is taken, we believe we do no more than consult the wishes of our readers in causing it to reappear in the pages of this *Journal.*—Ed. A. M.

time; and let m be the amount of the sum p forborne t times. Now, because in one year or time unity becomes r, by the same reason r will in another time become rr, and rr in a third time become r^3, &c., it appears that r^t, or r raised to the power whose index is the number of times, will be the amount of £1 forborne t times; and, therefore—

I. The amount $m = pr^t$; therefore multiply the logarithm of r by t, and to the product add the logarithm of p; the sum shall be the logarithm of m.

Example.—What is the amount of £15. 17s. 6d. forborne 12 years, at 6 per cent. per annum compound interest?

The number $1·06 = r$ . . . . its log. 0·0253059

Which log. multiplied by $t = 12$, the years, produces . 0·3036708
The principal sum £15. 17s. 6d. = £15·875 = p, its log. 1·2007137

The amount £31. 18s. 10½d. = £31·94362 = m, its log. 1·5043845

II. $r^t = \dfrac{m}{p}$; therefore, if from the logarithm of m the logarithm of p be subtracted, and the remainder be divided by t, the quotient is the logarithm of r.

Example.—What is the rate of compound interest when the sum of £15. 17s. 6d., forborne 12 years, amounts to £31. 18s. 10½d.?

The amount £31. 18s. 10½d. = £31·94362 = m, its log. 1·5043845
The principal sum £15. 17s. 6d. = £15·875 = p, its log. 1·2007137

The remainder is log. of r^t . . . . . 0·3036708

Which, divided by $t = 12$, quotes log. of $(1·06 = r)$. 0·0253059
                   Therefore the rate is 6 per cent. per annum.

III. Because $r^t = \dfrac{m}{p}$; divide the difference of the logarithms of m and p by the logarithm of r; the quotient is t, or the time wherein the sum p will amount to m at the rate r.

Example.—In what time will the principal sum of £15. 17s. 6d. amount to £31. 18s. 10½d. at 6 per cent. per annum compound interest?

The amount £31. 18s. 10½d. = £31·94362 = m, its log. 1·5043845
The principal sum £15. 17s. 6d. = £15·875 = p, its log. 1·2007137

The remainder is log. of r^t . . . . . 0·3036708
Which, divided by 0·0253059 = log. of r, quotes 12 years for the time.

IV. $p = \dfrac{m}{r^t}$; therefore multiply the logarithm of r by t, and subtract the product from the log. of m; the remainder shall be the log. of p, the principal sum.

Example.—What is the principal sum that in 12 years, at 6 per cent. per annum compound interest, will amount to £31. 18s. 10½d.?

The amount £31. 18s. 10½d.=£31·94362=m, its log.   1·5043845

The number 1·06=r .   .   .   .   its log.   0·0253059

Which log., multiplied by t=12, the years, produces . 0·3036708
And subtracted from the log. of the amount—
The remainder is log. of (£15·875=£15. 17s. 6d.=p)=1·2007137

The four preceding rules are also readily deduced from the consideration of the rebate of money in this manner.

For if, in any time, r becomes 1, in the same time 1 becomes $\frac{1}{r}$, and in the second time $\frac{1}{r}$ becomes $\frac{1}{rr}$, and in the third $\frac{1}{rrr}$, &c.; so that, putting p the value or present worth of any sum m payable after t times, at the rate of r to 1:

I. The sum $m=pr^t$; therefore multiply the log. of r by t, and to the product add the log. of p, the sum shall be log. of m sought.

II. $r^t=\dfrac{m}{p}$; therefore from the log. of m subtract the log. of p, and divide the remainder by t; the quotient will be the log. of r.

III. Since $r^t=\dfrac{m}{p}$, divide the difference of the logs. of m and p by the log. of r; the quotient shall be t, the number of years.

IV. $p=\dfrac{m}{r^t}$; therefore multiply the log. of r by t, and subtract the product from the log. of m; the remainder will be the log. of p, which finds the value of any sum of money payable after any time assigned.

The logarithms are also serviceable to resolve all questions concerning the amount or present worth of annuities not paid as due, or purchased to be paid for in time to come.

Let a be an annuity or yearly pension whose successive amounts for times past are ar^t, and whose present values are $\dfrac{a}{r^t}$ successively, by what goes before; and the series, &c., ar^5, ar^4, ar^3, ar^2, ar, $a, \dfrac{a}{r}, \dfrac{a}{r^2}, \dfrac{a}{r^3}, \dfrac{a}{r^4}, \dfrac{a}{r^5}$, &c., will be a rank of mean proportionals continued infinitely in the ratio of r to 1. Now, the sum of all the consequents, or of the whole infinite series, will be to the said

sum increased by the next greater term (or the sum of all the antecedents) as 1 to r, by *Euclid*, v. 12 ; wherefore, putting y for the
said sum of the consequents, ry will be equal to $y + ar^t$, the sum
of the antecedents ; and $ry - y = ar^t$; and, therefore, $\dfrac{ar^t}{r-1}$ will be
equal to y, the sum of all our mean proportionals whereof ar^{t-1} is
the greatest ; and, by the same rule, $\dfrac{a}{r-1}$ will be the sum of all
the terms whereof $\dfrac{a}{r}$ is the greatest. So that if we subtract $\dfrac{a}{r-1}$
from $\dfrac{ar^t}{r-1}$, the difference will be the sum of all the terms whereof
ar^{t-1} is the greatest and a the least, their number being t, which
sum we will call z ; therefore z (the amount of the annuity of a
forborne t times at the rate r) $= \dfrac{ar^t - a}{r-1}$.

I. *The annuity* (a), *rate of interest* (r), *and time* (t), *being given,*
to find (z) *the amount.*

From the log. of a subtract the log. of $r - 1$, and to the
remainder add the log. of r^t; from the number answering to this
last sum, subtract the number answering to the remainder; the
difference shall be z, the amount sought.

Example.—What will an annuity of £34·4, forborne $12\frac{1}{2}$ years, amount
to at 6 per cent. per annum?

$$a = 34\cdot4 \qquad \text{L. } a = 1\cdot5365584 \qquad \text{L. } r = 0\cdot0253059$$
$$r - 1 = 0\cdot06 \qquad \text{L. } \overline{r-1} = 8\cdot7781513 \qquad t = \ \ . \ \ . \ \ 12\cdot5$$

$$\frac{a}{r-1} = 573\cdot3333 \qquad \text{L. } \frac{a}{r-1} = 2\cdot7584071 \qquad \begin{array}{r} 1265295 \\ 506118 \\ 253059 \end{array}$$
$$\text{L. } r^t = 0\cdot3163237$$

$$\frac{ar^t}{r-1} = 1187\cdot7660 \quad \text{L. } \frac{ar^t}{r-1} = 3\cdot0747308 \qquad \text{L. } r^t = 0\cdot3163237\cdot5$$

$$z = £614\cdot4327 = \frac{ar^t}{r-1} - \frac{a}{r-1} = \text{the amount.}$$

II. *The annuity* (a), *its amount* (z), *and the rate of interest* (r),
being given, to find (t) *the time.*

By the foregoing, $\dfrac{ar^t - a}{r-1} = z$, therefore $z + \dfrac{a}{r-1} = \dfrac{a}{r-1} \times r^t$;
wherefore, from the log. of a subtract the log. $\overline{r-1}$; to the
number answering to the remainder add the given amount, and
from the log. of the sum subtract the afore-found remainder; this

second remainder, divided by the log. of r, will quote the time required.

Example.—In what time will an annuity of £34·4 amount to £614·4327 at the rate of 6 per cent. per annum?

$$a = 34·4 \qquad\qquad \text{L. } a = 1·5365584$$
$$r - 1 = 0·06 \qquad\qquad \text{L. } \overline{r-1} = 8·7781513$$
$$\frac{a}{r-1} = 573·3333 \qquad \text{L. } \frac{a}{r-1} = 2·7584071$$
$$z = 614·4327$$
$$\frac{ar^t}{r-1} = 1187·7660 \qquad \text{L. } \frac{ar^t}{r-1} = 3·0747308$$
$$\text{L. } r = 0·0253059)0·3163237(12·5 \text{ years} = t.$$

III. *The amount* (z), *rate of interest* (r), *and time* (t), *being given, to find* (a) *the annuity.*

♦The former equation being reduced, $a = \dfrac{z \times \overline{r-1}}{r^t - 1}$; wherefore, to the log. of the amount z, add the log. of $\overline{r-1}$, and from the sum subtract the log. of $\overline{r^t - 1}$; the remainder is the log. of a.

Example.—An annuity forborne $12\frac{1}{2}$ years amounts, at 6 per cent. per annum, to the sum of £614·4327, how much is that annuity?

$$z = 614·4327 \qquad \text{L. } z = 2·7884743 \qquad \text{L. } r = 0·0253059$$
$$r - 1 = 0·06 \qquad \text{L. } \overline{r-1} = 8·7781513 \qquad t = . \ . \ 12·5$$
$$r^t = 2·071685 \qquad\qquad 1·5666256 \qquad \text{L. } r^t = 0·3163237$$
$$r^t - 1 = 1·071685 \quad \text{L. } \overline{r^t - 1} = 0·0300671$$
$$\text{L. } £34·4 \text{ the annuity} = \text{L. } a = 1·5365585$$

IV. *The annuity* (a), *time* (t), *and amount* (z), *being given, to find* (r) *the rate.*

In order to find r, the former equation is reduced to $\dfrac{z}{a} - 1$
$= \dfrac{z}{a} r - r^t$, or, in our present case, $16·8614 = 17·8614 r - r^t$, which is so affected as not readily to be resolved by the general method for resolution of equations, unless we can first approach it by some other means; for which purpose take the following rule, which will suffice where great exactness is not required.

Let $\left.\dfrac{z}{at}\right|^{\frac{t-1}{2}} = 1 + y$, and let $\dfrac{6}{t+1} = b$. I say that $\overline{bb + 2by}|^{\frac{1}{2}} - b$ is exceeding near the increase of the rate, or $r - 1$.

Wherefore take the log. of the amount, and the complements of the logarithms of the time and annuity; the sum (abating 2 in the tens place of the index) divide by $\frac{1}{2} \times \overline{t-1}$; the quotient shall be the log. of $1+y$. Then divide 6 by $t+1$, and to b (the quotient) add twice y, and to the log. of b add the log. of $\overline{b+2y}$; half the sum shall be the log. of $\overline{bb+2by}|^{\frac{1}{2}}$, from which square root subtract b; the residue will be very near the increase, or $r-1$; and, adding 1, r is found. If great exactness be desired, let r thus found be assumed, and $\dfrac{z}{a}r-r^t$, compared with $\dfrac{z}{a}-1$, will always be greater than it; and dividing the excess by $tr^{t-1}-\dfrac{z}{a}$, the quotient added to r shall verify as many more figures in the rate as were true in the assumed r.

Example.—An annuity of £34·4, forborne $12\frac{1}{2}$ years, amounts to £614·4328, required the rate of interest allowed.

$z=614\cdot4327$ L. $z=2\cdot7884743$
$t=12\cdot5$ co. L. $t=8\cdot9030900$
$a=34\cdot4$ co. L. $a=8\cdot4634416$

$$\tfrac{1}{2}\times\overline{t-1}=5\cdot75)0\cdot1550059(0\cdot0269575 = \text{L.}\ \overline{1+y}, \text{ and } 1+y=1\cdot064039$$

$b=\dfrac{6}{t+1}=\dfrac{6}{13\cdot5}=0\cdot4444444$ $2y=0\cdot128078$

$2y$ . . . . . $0\cdot1280710$ L. $b=9\cdot6478175$

$b+2y$ . . . . $0\cdot5725224$ L. $\overline{b+2y}=9\cdot7577936$

                                      $19\cdot4056111$

$\overline{bb+2by}|^{\frac{1}{2}}$ . . . $0\cdot5044353$ L. $\overline{bb+2by}|\frac{1}{2}=9\cdot7028055$

b . . . . . $0\cdot4444444$

$r-1$ . . . . . $0\cdot0599909$

Therefore the rate of interest sought is 6 per cent. per annum.

After the same manner the four cases relating to the purchase of annuities are readily solved by logarithms, and the theorems discovered with the same ease; for $a, \dfrac{a}{r}, \dfrac{a}{rr}, \dfrac{a}{r^3}, \dfrac{a}{r^4}$, &c., being a scale of mean proportionals in the ratio of r to 1, put y for the sum of all the consequents infinitely continued, whereof $\dfrac{a}{r}$ is the first, and that sum will be to the sum of all the antecedents whereof a is the first, as 1 to r (that is, $1 : r :: y : ry$), so that $ry=y+a$, and $\dfrac{a}{r-1}$ will be equal to y, the value of the fee, or the sum of all the mean proportionals less than a. And, by the same rule, $\dfrac{a}{r^t \times \overline{r-1}}$ will be the sum of all the means less than $\dfrac{a}{r^t}$, or the value of the reversion; and, subtracting the one sum from the other,

$\dfrac{a}{r-1} - \dfrac{a}{r^t \times r-1}$ will be equal to z, the sum of all the means,

whereof $\dfrac{a}{r}$ is the greatest and $\dfrac{a}{r^t}$ the least.

I. *The annuity (a), time (t), and rate of interest (r), being given, to find (z) the present value.*

The present value $z = \dfrac{a}{r-1} - \dfrac{a}{r^t \times r-1}$; therefore, from log. of the annuity subtract the log. of $\overline{r-1}$, and from the residue subtract the log. of r^t; the difference of the numbers answering to the two remainders is the present value sought.

Example.—What is £70 per annum, to continue 59 years, worth in present money, at the rate of 5 per cent. per annum?

$a = 70$	L. $a = 1\cdot8450980$	L. $r = 0\cdot0211893$
$r-1 = 0\cdot05$	L. $\overline{r-1} = 8\ 6989700$	$t = \ \ldots\ 59$
Fee $= £1400$	L. fee $= 3\cdot1461280$	1907037
Reversion, £78·6972	L. $r^t = 1\cdot2501687$	1059465
	L. rever. $= 1\cdot8959593$	L. $r^t = 1\cdot2501687$

$z = £1321\cdot3028$, the present value sought.

II. *The annuity (a), present value (z), and rate of interest (r), being given, to find (t) the time.*

Now, r^t will be equal to $\dfrac{a}{r-1}$ (or the fee) divided by the value of the reversion—that is, by $\dfrac{a}{r-1} - z$; wherefore, from the log. of the annuity subtract the log. of $\overline{r-1}$; the number answering to the remainder will be the value of the fee. From the fee subtract the present worth, the residue is the value of the reversion. Take the log. of the reversion from the log. of the fee, and divide the residue by the log. of r; the quotient will be t, the number of years sought.

Example.—In what time will an annuity of £70 per annum pay off a debt of £1321·3028, allowing the creditor 5 per cent. per annum?

$a = 70$	L. $a = 1\cdot8450980$
$r-1 = 0\cdot05$	L. $\overline{r-1} = 8\cdot6989700$
Fee $= 1400$	L. fee $= 3\cdot1461280$
$z - 1321\cdot3028$	
Reversion, 78·6972	L. rev. $= 1\cdot8959593$

L. $r = 0\cdot0211893)1\cdot2501687(59$ years $= t.$

III. *The present value* (z), *rate of interest* (r), *and time* (t), *being given, to find* (a) *the annuity.*

The former equation may be reduced to this proportion, as $1 - \dfrac{1}{r^t}$ to z, so is $r-1$ to a, the annuity sought.

Wherefore, to the complement of the log. of $\overline{1-\dfrac{1}{r^t}}$, add the logs. of z and of $\overline{r-1}$, the sum shall be the log. of a.

Example.—What annuity, to continue 59 years, can be purchased for £1321·3028, at the rate of 5 per cent. per annum?

$$\frac{1}{r^t} = 0 \cdot 0562123$$

$$1 - \frac{1}{r^t} = 0 \cdot 9437877 \qquad \text{co. L. } \overline{1 - \frac{1}{r^t}} = 0 \cdot 0251257$$

$$z = 1321 \cdot 3028 \qquad \text{L. } z \ \ldots \ 3 \cdot 1210023$$

$$r - 1 = 0 \cdot 05 \qquad \text{L. } \overline{r - 1} = 8 \cdot 6989700$$

L. $r = 0 \cdot 0211893$
$t = \ \ldots \ 59$
L. $r^t = 1 \cdot 2501687$
L. $\dfrac{1}{r} = 8 \cdot 7498313$

L. (£70 = a the annuity sought) = $1 \cdot 8450980$

IV. *The annuity* (a), *present value* (z), *and time* (t), *being given, to find* (r) *the rate of interest.*

This problem being more difficult than appears at first sight, and requiring the solution of this equation, $\dfrac{a}{z} = \dfrac{z+a}{z} r^t - \dfrac{t+1}{r}$, to which it is reduced, there must be applied some method of approaching the root r, which is by no means evident; and that approximation, as the number of years and rate are greater or less, cannot properly be obtained by one general rule, but rather by two, according as the value of the reversion is greater or less.

If the number of years be great (as suppose 40 or upwards), and especially if the rate of interest be high, $1 + \dfrac{a}{z}$ will be nearly the rate; or, more accurately, $\dfrac{z+a}{z} - \dfrac{z^t}{z+a} \times \dfrac{a}{z}$. Call it r, and $\dfrac{a}{r^t \times r - 1}$ will be exceeding near the value of the reversion, which let be x; then $1 + \dfrac{a}{z+x}$ shall approach the true rate sufficiently. But if greater exactness be desired, by repeating this process it will be obtained. Hence this rule:—from the log. of a, and also from the log. of $z+a$, take the log. of z; this latter remainder shall be nearly the log. of the rate. Multiply that log. by t, and the complement of the product add to the first remainder; the decimal fraction answering to the sum taken from the former rate shall give a more correct rate. With this rate seek x, the reversion after the time given, which add to z; then to the complement of the log. of $\overline{z+x}$ add the log. of a; the sum shall be the log. of the increase, or of $\overline{r-1}$, sufficiently near.

Example.—If £1321·3028 is paid for an annuity of £70 per annum for 59 years to come, what is the rate of interest allowed the purchaser?

$a=70$

$z=1321\cdot3028$

$\dfrac{z+a}{z}=1\cdot052978$

L. $a=1\cdot8450980$

L. $z=3\cdot1210023$

$8\cdot7240957$

co. tL. $\dfrac{z+a}{z}=8\cdot6772554$

Log. of $0\cdot00252$ . . . . $7\cdot3013511$

First $r=1\cdot050458$ L. $r=0\cdot0213787$

$t=$ 59

$z=1321\cdot3028$

$x=\quad 76\cdot002$

$z+x=1397\cdot3048$ the fee

1924083

1068935

L. $r'=1\cdot2613433$

co. L. $\overline{z+x}=6\cdot8547088$

L. a . . $1\cdot8450980$

$r-1=0\cdot05096$ L. $\overline{r-1}=8\cdot6998068$

L. $\overline{z+a}=3\cdot1434217$

L. z . . 3·1210023

L. $\dfrac{z+a}{z}=0\cdot0224194$

t 59

2017746

1120970

tL. $\dfrac{z+a}{z}=1\cdot3227446$

co. L. $r'=8\cdot7386567$

co. L. $\overline{r-1}=1\cdot2970700$

L. $a=1\cdot8450980$

L. $x=1\cdot8808247$

So that the rate of interest sought is 5 per cent. per annum.

If the number of years be small, the aforesaid rule will avail little. In this case it will be requisite to approach the rate thus:—
Let $\dfrac{t+1}{2}$ be the index of a root of $\dfrac{at}{z}$; from which root subtract 1, and the remainder call y, and let $\dfrac{6}{t-1}$ be called b. I say that $\overline{b-\overline{bb-2by}}\rvert^{\frac12}$ is sufficiently near to $r-1$, and will be still nearer the truth as the number of years is smaller; and that the error will be always in excess. Hence the rule:—divide the log. of $\dfrac{at}{z}$ by $\dfrac{t+1}{2}$, and from the number answering to the quotient subtract 1; double the remainder, and subtract it from b—that is,

from the quotient of 6 divided by $t-1$; to the logarithm of this remainder add the log. of b; then the number answering to half the sum of those logarithms taken from b will leave $r-1$, the increase of the rate sought.

Example.—An annuity of £20 per annum, to continue 21 years, is sold for £220: required the rate of interest allowed the purchaser.

$a= 20$　　　L. $a=1\cdot3010300$
$t= 21$　　　L. $t=1\cdot3222193$
$z=220$　　co. L. $z=7\cdot6575773$

$\frac{1}{2}\times\overline{t+1}=11)0\cdot2808266(0\cdot0255297 =$ L. $\overline{1+y}$, and $1+y=1\cdot060546$

$$2y=0\cdot121092$$

$$b=\frac{6}{t-1}=\frac{6}{20}=0\cdot3$$

L. $b=9\cdot4771213$
L. $\overline{b-2y}=9\cdot2526298$

$2y$ . . . . . . $0\cdot121092$

$$18\cdot7297511$$

$b-2y$ . . . . . $0\cdot178908$
$\overline{bb-2by}|^{\frac{1}{2}}$ . . . . $0\cdot231673$

L. $\overline{bb-2by}|^{\frac{1}{2}}=9\cdot3648755$

Now, $b-\overline{bb-2by}|^{\frac{1}{2}}=0\cdot068327=r-1$

Therefore $r=1\cdot068327$, the rate sought.

The rate r thus found is always some small matter too big, the true rate being $1\cdot06814$; but as the number of years are fewer, the error becomes insensible. If greater exactness be required, it will be easy by the general method for the resolution of equations having so near an approximation, to prosecute this inquiry as far as you please; but this seems abundantly sufficient for use, which is our principal design in this place.

Lastly, by way of corollary to the former: let it be required to find the rate of interest allowed the purchaser when he pays a sum z, for an annuity a, wherein he has already a term t, to have it prolonged for a certain time x.

Example.—An annuity of £20 per annum being in possession for the term of 21 years, and for £40 paid down it can be prolonged for 10 years more, or to 31 years; what is the rate of interest required?

Put $T=2t+x+1$, and $\frac{1}{2}T$ shall be the index of a root of $\frac{ax}{z}$. Let $\frac{ax}{z}\Big|^{\frac{1}{2}T}$ be equal to $1+y$, and $\frac{6T+6}{xx}=b$. I say $r-1$ is very near to $b-\overline{bb-2by}|^{\frac{1}{2}}$.

$a=20$ L. $a=1\cdot3010300$
$x=10$ L. $x=1\cdot0000000$
$z=40$ co. L. $z=8\cdot3979400$

$\tfrac{1}{2}$T$=26\cdot5)0\cdot6989700(0\cdot0263762=$L. $\overline{1+y}$, and $1+y=1\cdot062616$

$$2y=0\cdot125232$$

$$b=\frac{6\mathrm{T}+6}{xx}=3\cdot24$$

L. $b=0\cdot5105450$
L. $\overline{b-2y}=0\cdot4934257$

$2y$ . . . . $0\cdot125232$

$$1\cdot0039707$$

$\overline{bb-2by}|^{\frac{1}{2}}=3\cdot176767$ L. $\overline{bb-2by}|^{\frac{1}{2}}=0\cdot5019853$

$b-\overline{bb-2by}|^{\frac{1}{2}}=0\cdot063233=r-1$

Therefore $r=1\cdot063233$, the rate sought,

as will be readily proved by seeking the value of the reversion of an annuity of £20 per annum for 10 years after 21, at the rate of $1\cdot063233$ per cent. per annum. (*See* the Work.)

$a=20$ L. $a=1\cdot3010300$ L. $r=0\cdot0266284$
$r-1=0\cdot063233$ co. L. $\overline{r-1}=1\cdot1990562$ 21
                          co. L. $r^t=9\cdot4408036$ ————
                                                            266284
Rever. £87·275   L. rev.$=1\cdot9408898$ 532568
                          co. L. $r^z=9\cdot7337160$ L. $r^t=0\cdot5591964$

Rever. £47·2722   L. rev.$=1\cdot6746058$

£40·0028, the value sought.

Thus it appears that £40 and about three farthings is the true value of the difference of the reversions at the rate of interest before found; by which it may be judged how near an approximation the foregoing rule affords towards finding the rate of interest, when the value of an annuity for a term of years to commence after such a distant time is proposed.

Sixth Annual Report of the Insurance Commissioners of the Commonwealth of Massachusetts.*

IN no year since the business of life assurance commenced in this country has it advanced so rapidly as in the one just closed. The nineteen Companies now making their returns to this Office show over 10,000 new policies issued, insuring more than 30,000,000 dollars, and the whole amount insured by them exceeds 150,000,000 dollars.

Ninety-four per cent. of all this business consists of whole-life policies, kept up by equal annual premiums, which are largely in excess of the annual risks during the earlier years of the insurance, but will be in defect on all the lives that continue through the later years. Hence, to cover the entire mass of risks which increase year by year, the earlier excess of the premiums, which do not increase, must be carefully accumulated to provide for the future deficiency. To apply a test to this accumulation, that the public may know, in regard to each Company, how near it comes to the happy medium between a too costly excess and a ruinous deficiency, was the object of the law requiring us to make an annual valuation of the policies of all the Companies doing business in the Commonwealth. Though it, in effect, prescribes a net valuation by not requiring of the Companies data sufficient for any other, it does not prescribe the rate of mortality or of interest that is to be assumed as the basis of the work. In adopting the "Combined Experience" or English "Actuaries" rate of mortality, and 4 per cent. interest, we supposed we were travelling most safely between extremes, and we have the pleasure to know that the fault found with our choice has been about equally divided, as much of it making our valuation too low as too high. Four per cent., as the proper rate of interest, was fully approved by the Convention of Officers of Life Insurance Companies which met in New York last May, and we believe no objection has been made by any of the American Companies to the rate of mortality adopted. It is worthy to be remarked, however, that Prof. C. F. McCay, of Georgia, in a series of valuable articles in *Hunt's Merchants' Magazine*, on the valuation of life insurance policies, has expressed the opinion that our method of valuation requires too small a premium reserve. His argument is certainly conclusive as a reply to the mathematical gentlemen, Prof. Pierce, of Cambridge, and Messrs. Woolhouse and Neison, the English actuaries, on whose authority we were charged with doing great injustice to the International Assurance Society of London, by requiring too high a reserve! It is doubtless true, as Prof. McCay proves by example, that by the combined experience mortality, the value of policies entered below the average age of insurance will be smaller for a few years than by Dr. Farr's table, and this, so far as it has any effect, makes our valuation rather more favourable to new Companies than to old ones. But the aggregate value of policies entered at all ages will not differ much from Dr. Farr's, and especially when the Company has attained a few years' standing. Prof. McCay's own figures tend to show this. To show how the "Actuaries" rate does not probably require less reserve than Dr. Farr's—the only rate which has any claim to a higher

---

* We have been favoured by Mr. Elizur Wright with a copy of this Report, just published, and we give a portion of it, which we have no doubt will interest our readers. It has been necessary to omit some pages of tabular matter, and some few passages of less importance than the rest.—ED. *A. M.*

scientific authority—we give below the values of five policies of 100 dollars each, entered at five different ages, at the time when the sixteenth premium falls due, according to four different rates of mortality, at 4 per cent. interest, and also add the values by using Carlisle premiums discounted by Neison's life annuities.

Age of Entry.	Com. Ex.	Dr. Farr.	Carlisle.	Neison.	Neison and Carlisle.
20........	$11 85	$12 14	$11 99	$11 10	$12 33
30........	17 64	17 14	15 39	15 96	16 33
40........	25 57	24 51	23 48	23 14	23 36
50........	34 41	34 05	32 89	32 18	33 02
60........	43 21	42 03	41 49	42 06	41 12
Totals....	$132 68	$129 87	$125 24	$124 44	$126 16

This may serve to illustrate what is undoubtedly true, that the "Combined Experience" rate, as a means of ascertaining the proper premium reserve of an established Company, corresponds more nearly with Dr. Farr's than the other rates. For the purposes of our official valuation, it is not open to Prof. McCay's objection. And considering it as a means of adjusting the rights of different members in the same Mutual Company, it has perhaps an advantage which he does not take into account, by virtue of the probable error which he points out. When, in computing the premium, we assume, as we intentionally do, a rate of interest lower than that which will be actually realised, the effect of this reduction is to increase the premiums, but to increase those of the younger more than those of the older ages, because there is more of *time*—the food that compound interest lives on—in the case of the former. This injustice to the younger ages, from assuming an interest below the actual, may be happily compensated by a rate of mortality which makes the risk of those ages a little too small, the flexibility of one assumption correcting the excessive rigidity of the other.

It must not be inferred, from our adopting this rule of valuation as a test of the sufficiency of the actual premium reserve, that we disapprove of keeping a larger reserve. We only think it need not be left much larger at the time of distributing the surplus, or declaring dividend. The strongest believers in the deterioration of life in old Companies, by the lapse and surrender of policies on the healthier lives, have never pretended, so far as we know, that the reserve need be more than about 14 per cent. And their observations included Companies of poor credit as well as long standing, in which losses and surrenders had had a more damaging effect on the quality of the residual life than can ever occur in Companies well managed and kept in good credit. In the latter, we do not believe the deterioration of life is any reason for reserving even 3 per cent. beyond the reserve at 4 per cent. of the "Combined Experience" rate, which is founded on the mortality of Companies averaging a greater deterioration from the causes mentioned than is ever likely to occur in American Companies. Still, as a matter of general prudence and financial force, there may be ample reason for keeping a reserve 15 or 20 per cent. higher than is really needed, so that the institutions may be felt, even by the most timid, to be high and dry on the rock above any flood-tide of chance.

The valuation for the past year has been made with all possible care to avoid error, and the results are arranged in tables similar to those of last year. No doubt considerable inaccuracies have crept in among the calculations, extending to 55,360 single policies, not to speak of several thousand bonuses or reversionary additions; but we believe they do not appreciably affect the results. These show a most gratifying increase, both of the business and stability of the Companies. The ratio of the aggregate actual premium reserve to the computed is considerably greater, and the ratio of expenses to receipts is hardly 1 per cent. greater, though competition has been greatly increased by the creation of several new Companies. New Companies planted within the shadow of flourishing old ones, cannot be expected to get into successful operation without expending on their machinery more than the premium receipts for one, two, or perhaps three, of the first years. And it is, therefore, no wonder that some of the new Companies show the necessity of a guarantee capital.

This capital has in all cases been paid in cash, and invested in good securities, and if the new Companies' terms of insurance are just and liberal to the insured, there is no reason why energy, fidelity, and perseverance on its part should not soon place it beyond the need of guarantee funds. It is very true that the struggle may be arduous, and the present risk of the capital is considerable. The chance for the public-spirited capitalist to escape without loss is not such as to encourage the creation of new Companies. The worst, however, that can happen to the policy-holder with such a guarantee is, that he will not, quite so soon as he might, begin to receive back surplus from his premium; for if the capital should become too much impaired, its owners will save it from utter loss by getting their risks transferred to a strong Company. But the money spent in these experiments is by no means thrown away, even if the Companies do not succeed. The old Companies, since they probably catch the greater part of the birds started up when the bush is vigorously beaten, no matter who beats it, ought to treat the new ones with great kindness and even gratitude.

We have this year, as last, taken pains to give the ratio of our computed premium reserve to the amount insured for each year; because, by observing this ratio, it will be seen that an empirical rule, very simple and easy of practical application, may be established for ascertaining a *safe* premium reserve. It will, at any rate, do practically for all Companies that are large enough to dispense with a guarantee fund, and it is this:—

Divide the policies into groups, according to the number of annual premiums paid on each, then reserve of the amount insured in each group a percentage of twice the number of premiums paid on it. That is, on all the policies that have paid one premium, reserve 2 per cent. of the amount insured; on those that have paid two, 4 per cent.; on those that have paid three, 6 per cent., and so on.

This rule would give for all the Companies combined an aggregate reserve about 20 per cent. higher than our computed reserve, and somewhat less than the actual reserve, as will appear from the table. We do not by any means recommend this rule as a substitute for accurate arithmetical calculation, but as better than guessing without any guide.

In former Reports we have dwelt on an imperfection of life insurance, as usually practised, growing out of the nature of the contract. The insured in a whole-life policy pays in advance of the progress of the risk, and is bound to continue to pay the same premium annually during life, under the penalty not only of releasing the other party to the contract, but of forfeiting all that he has advanced beyond the risk up to the time of non-payment. Theoretically, this bargain appears disadvantageous to the insured, for he may, among the uncertainties of life, find it desirable to retire from the contract.   Practically, it has been made an objection to this system of insurance; and many practical minds have been employed to find a remedy for it as an evil.   We have endeavoured heretofore to expose the nature of this evil, and suggest a remedy; but not till now have we been in a condition to ascertain and show its magnitude.   The policies which were in force in 1858, and were returned as forfeited for non-payment of premium in 1859, might be reinstated in 1860, so that we could not last year fairly state them as forfeitures.

The returns of 1860 show how many were restored, and those that were not, will probably never be; or, at any rate, a very trifling proportion of them will be.   We give the number and amounts forfeited in thirteen Companies, dividing them into classes according to the number of premiums paid.   The net value of each class, when the premium was due, and not paid, is given, and shows what the Company gained—in addition to the policy-holder's share of accrued divisible surplus—by the forfeiture, supposing the life to have been of the average vitality.   The additions or bonuses attached to policies as reversionary dividends, in some of the Companies, and forfeited with the policies, are also given in amount separately, and with their present value at the time of forfeiture.

The total of what all the Companies gained—exclusive of the relinquished share of the excess of the Companies' actual premium reserve over our net valuation—was 234,000 dollars.   This, it is true enough, is a very inconsiderable sum compared with the annual income of the Companies, being only 3 per cent. additional to the receipts of that year; and it is to be reduced by the premium notes outstanding in some of the Companies against the forfeiting parties, which, of course, will never be collected.

Only two of the Companies—the Berkshire and the Massachusetts Mutual—return the amount of these notes, cancelled by the forfeiture of the policy.   In the case of the Berkshire, the amount of cancelled notes was 4,002·06 dollars, reducing the gain by forfeiture to 2,119·09 dollars; in that of the Massachusetts, 5,351·34 dollars, reducing the gain to 1,429·12 dollars.   In individual cases, the note cancelled was larger than the value of the policy, so that the retiring policy-holder got off without paying for more insurance than he received, or perhaps not paying for quite so much. For this reason the others lost the more.   In other and older premium note Companies, where the amount of notes has been more reduced by dividend, the cancelled notes cannot bear so high a ratio to the value forfeited. We think it a very large allowance to estimate the premium notes on all these forfeited policies at 90,000 dollars, reducing the whole amount of profit from that source—aside from the proportion of surplus—to 144,000 dollars, or about $2\frac{1}{2}$ per cent. addition to the revenue of the year.

This average gain, however, it will be seen at a glance, was shared very unequally by different Companies.   Some got nothing to speak of, or, say one-fifth of 1 per cent., others increased their revenue more than 16 per cent. by this means, or eighty times as much as the former.   We do not wish to make invidious comparisons, but those who seek information can easily cipher for themselves, between the present Report and the last, the degree of cohesion which exists in these several institutions.   Whatever may be its strength in any, it plainly does not depend on the magnitude of the penalty in case of non-payment. Forfeiture occurs where many premiums have been fully paid in cash, and large reversionary additions have been made to the policy. One all-cash Company has had nearly 8 per cent. of its liability annihilated by forfeiture, while the half-note Company, which shows the largest amount of forfeiture, has had its liability diminished by only 2·18 per cent., and since the assets are diminished by the cancelled notes, the relative gain is still less. There is every reason to believe, from the study of this table, and an acquaintance with the officers, agents, and modes of management of the various Companies, that the strength of cohesion depends rather on the fairness and honour of the agents who solicit business, and the intelligence of the people on whom they operate.

Though the evil we have been endeavouring to measure is of no alarming magnitude, when viewed in its proportion to the whole business of the vast beneficent institutions to which it adheres with such wonderful tenacity, it is still large enough to be worthy of profound consideration. Successful motion depends on taking advantage of friction in the right place and avoiding it in the wrong, and friction depends on the minute quantities of matter which make the difference between a rough surface and a smooth one. It is often a slight filing and polishing which determines the question whether a well-proportioned and powerful machine shall work well and produce immense results, or run unprofitably and wear itself out. Smoothness of finish, at any rate, always saves oil and motive power.

Hence we cannot but recal the attention of the legislature to the legal remedy which we have heretofore explained at large, and which we need not here repeat. It proposes no interference with past contracts, but simply a provision of law that in the case of future ones nothing shall be forfeited beyond the policy-holder's share of accrued divisible surplus and the right to be insured beyond the term already fully paid for in cash,—the establishment of the principle, in fact, that the policy-holder or his representatives shall be entitled to all the insurance which he pays for, whatever may be the terms of the contract.

Corporations are the creatures of the legislature, and must undoubtedly conform the contracts they make to its will, when that will is once expressed. In this case, by expressing its will against a bargain, which, in point of morals, is no better than a bet, and an unfair one at that, we believe it will benefit the Life Insurance Companies no less than those who would otherwise become the victims of their peculiar mode of obtaining pay for service never to be performed. We do not in the slightest degree question that this is done with the best possible intentions. But we have all read of a bad place paved with those good materials. Practically the law is not needed against the best Companies, which are altogether better than their bargains. But Companies, such as have been and may yet be, under dishonest, reckless, and mercenary management, can and will, under such bargains, make a good thing, in a financial sense, of their bad credit, by sending out highly magnetic and glib-tongued agents into quarters where their standing is not well known—and the world will always be too large to have it known everywhere—and alluring men to take policies, who, after several premiums are paid, will discover their error, and forfeit what they have overpaid as the best mode of escape from greater loss. By such gains, in the mother country, Companies of virtual swindlers, under the name of Life Insurance, wasting, in profligate expenditures, a full third of all the funds entrusted to them, have managed to exist through perhaps an entire generation, and make a show of solvency and respectability. They always court secrecy as to the ratio of their premium reserve to the net value of their policies, yet its insufficiency cannot be exposed without really enriching them by frightening the old policies into forfeiture, and when this has made their assets again equal to their liabilities, they will be able to recover from the temporary check given to their new business, and go on as before. In reality, for ingenious rogues, a Life Insurance Company, with the forfeiture clause in its policies, seems to be an engine for plunder with a principle of immortality in its very constitution. It seems incapable of death except by great destitution of intelligence on the part of the operators.

It is very pleasant to believe that none of the men who are charged with the management of the Companies now under our supervision, are of the sort to make any dishonourable use of this fault in the structure of their system. They all have the disposition, but they are unable, altogether to prevent it from transferring one or two hundred thousand dollars per annum, actually paid by some two thousand men for the benefit of the first fifty or sixty of their own widows, into the pockets of some forty or fifty thousand other probably more fortunate men to help them provide for their widows. We ask the State to help them prevent it.

Claims by Death against Sixteen Life Insurance Companies doing business in Massachusetts, for the year ending November 1, 1860.

Companies.	No. of Claims.	Amount.	Ratio of Loss to Amount Insured.	Ratio of No. of Claims to No. of Policies.
Massachusetts Hospital ..	4	$15,000 00	11·75	8·33
New England	18	74,302 00	0·57	0·45
State Mutual	9	12,107 54	0·42	0·50
Berkshire	8	21,000 00	1·17	0·93
Massachusetts Mutual	14	36,400 00	0·86	0·69
Mutual Life, N. Y.......	105	342,438 19	0·92	0·90
Mutual Benefit, N. J.	66	227,000 00	1·01	0·98
Connecticut, Ct.	94	235,700 00	1·31	1·02
National, Vt.	8	14,635 02	0·84	0·71
Union Mutual, Me......	25	61,900 00	1·42	1·35
Manhattan, N. Y.	20	75,698 00	0·73	0·62
Charter Oak, Ct...	23	47,317 51	0·74	0·69
American Temperance, Ct.	9	14,500 00	0·58	0·50
Knickerbocker, N. Y.....	3	8,600 00	0·43	0·42
Equitable, N. Y........	3	14,000 00	1·73	1·72
Guardian, N. Y.	1	5,000 00	2·39	1·33
Totals........	410	$1,204,998 26	1·09	0·84

It is pleasant also to turn from this view of the short-coming of the Life Insurance Companies, to their magnificent well-doing, as exhibited in the above table of the termination of policies by death. Here we see in a single year more than 400 families, smitten by death of their natural protectors and providers, shielded against the most dreaded consequences of such a bereavement, by the distribution of more than a million of dollars.

This substantial aid to the ones they most loved, in the hour of their greatest need, cost the deceased a comparatively small sum of money. It was the result of a little prudence and self-denial, availing itself of a great law that governs human life, and really gives to what is called chance the calculable certainty of granite rocks and iron bars. It will be seen, from the figures elsewhere, that very few claims remain unsettled. The whole amount is usually paid without dispute, as soon as proper evidence of the death is submitted.

Want of time has prevented us from ascertaining how this mortuary experience has been distributed among the various ages, and what is its significance as illustrating the law of mortality among the lives at risk.

When the registration has attained greater age, and the constantly increasing number of policies has been submitted to observation for a considerable series of years, the results that may be deduced will be of a practical value far more than to compensate all the expenses of this office. And it is our design, so long as we are connected with it, to use every exertion to make the registration as accurate and fruitful of good results as the nature of the subject and the powers conferred by the law admit.

The following table, prepared from the returns of the year preceding the last, may serve to show how nearly for that year the actual mortality corresponded to the rate we have adopted as the basis of our valuation.

Mortality compiled from the Returns of Sixteen Life Insurance Companies doing business in Massachusetts, for the year 1859, *compared with the Combined Experience, Dr. Farr's English Table, and the experience of the Gotha Life Insurance Company, of Germany.*

Periods of Age.	No. of Losses.	No. of Years of Life exposed.	Ratio per Cent.	Com. Ex.	Farr.	Gotha.
Under 26	7	2,086·70	0·3354	—	—	—
26 to 30.....	45	4,998·96	0·9002	0·81	0·97	0·87
31 to 35......	54	7,664·45	0·7045	0·89	1·10	0·92
36 to 40.....	52	9,198·71	0·5653	0·99	1·25	1·00
41 to 45......	68	8,561·59	0·7942	1·13	1·42	1·04
45 to 50......	65	6,356·66	1·022	1·43	1·62	1·45
51 to 55......	58	3,923·19	1·478	1·91	1·87	1·82
56 to 60......	45	2,032·76	2·214	2·65	2·71	2·77
61 to 65......	15	908·12	1·651	3·79	3·95	3·83
66 to 70......	12	355·49	3·375	5·55	5·75	6·08
70 to 75......	3	105·66	2·839	8·13	8·32	9·04
75 to 80......	5	39·08	12·790	11·88	11·94	11·35
Over 80	1	5·00	20·000	17·22	16·90	23·94

Respectfully submitted.

ELIZUR WRIGHT.
GEO. W. SARGENT.

Boston, February 20, 1861.

CORRESPONDENCE.

MR. FINLAISON'S "REPORT AND OBSERVATIONS ON THE MORTALITY OF THE GOVERNMENT LIFE ANNUITANTS." *

To the Editor of the Assurance Magazine.

Sir,—At this period, when the decennial census is on the point of being taken, it may not be out of place to call attention to the above Report, just issued, which happens to touch upon certain points that have a peculiar interest at present, as bearing upon facts which may affect the correctness of any future life tables that may be computed upon the basis of the present census.

The *Assurance Magazine* is, therefore, an appropriate organ for directing the attention of actuaries, and through them of the public, to an interesting and most valuable document, which, being published in the shape of an unpalatable Blue-book, is not likely to come under general notice.

I propose to refer to such portions of the Report only as have a bearing on the census question, in relation to the false returns which the experience of previous censuses has shown to have been made with respect to the ages of females, to an extent probably hitherto little suspected, and to point out how the advance of science may be imperilled and retarded, and the pecuniary interests of the public sacrificed, as well by the false census returns as by the loose system which very generally obtains in the method adopted in practice of arriving at the ages of deceased persons, with the view to their being recorded in the books of the District Registrars.

With reference then to the first question, Mr. Finlaison, who from the nature of his appointment is peculiarly competent to form a correct opinion on such points, states that " it is the universal tendency of persons supplying under no particular responsibility information of their own age or of that of their relations, to understate the truth when the party concerned is young, or not past middle age. On the other hand, where the person to be accounted for is old, there is a disposition to claim or attribute an exaggerated longevity." He explains that though errors in the reported ages of the dead may possibly to some extent counterbalance each other, this is never true of the reported ages of the living.

Mr. Finlaison quotes from the Report of the last census returns, vol. i., part ii., p. 24, to show the extent of the results produced by individual mis-statements.

It is mentioned in that Report that " persons of the age of 20 in

* *Report and Observations on the Mortality of the Government Life Annuitants.* By Alexander Glen Finlaison, Actuary of the National Debt, 1860. Ordered by the House of Commons to be printed, 25th August, 1860.

1851 must have been 10 years of age in 1841, and persons of the age of 25 in 1851 must have been of the age of 15 in 1841; and as there is a certain number of losses by death, it is evident that, excluding the effects of migration, the numbers at age 20 to 25 in 1851 must be less than the numbers living at the age 10 to 15 in 1841, of whom they (20 to 25) are the natural survivors.

" What are the statements which the abstract of ages express?

" 1841—The number of girls aged 10 to 15
was   1,003,119
1851—The number of young women aged
20 to 25 was, as stated in the returns   .   1,030,456

" Now, as the first number could never have swollen in 10 years to the magnitude of the second, we are driven to the hypothesis, that in 1841 and 1851 the heads of families returned several thousand ladies of the higher ages at the age of 20 to 25; and the hypothesis is confirmed by comparing the diminished numbers returned at the age 30 to 35 in 1851·with the numbers returned at 20 to 25 in 1841, where it is evident that the latter number is in deficiency as much as the former number is in excess.

" 1841—The number of young women of the age
20 to 25, as stated in the returns, was   .   973,696
1851—The number of women of the age 30 to
35, as stated in the returns, was   .   .   768,711 "

The Report goes on to show that notwithstanding an acknowledged disturbing influence caused by the immigration of Irish into Great Britain during the 10 years 1841–51, " upon comparing the above numbers with those for males at the corresponding ages, the conclusion appears to be inevitable, that some 35,000 ladies, more or less, who have entered themselves in the second age, twenty to forty, really belong to the third age, forty to sixty. Thousands have allowed themselves to be called twenty, or some age near it, which happens to be the age at which marriage is commonly contracted in England."

If we could succeed in giving extensive publicity to such facts as these, and in bringing home to the minds of those who are weak enough, under the influence, no doubt, of some misplaced feeling of shame, to act in this foolish manner, and, if we could convince such people of the fact that statistical acumen is unerring in its power of detecting the falsities in such returns, we might succeed in checking the practice to which females are so prone of concealing, or of attempting to conceal, their true ages. And though it would be difficult, in *individual* cases, to prove the truth or falsehood of the statement as to a lady's age, let it be remembered that power of discovery *in the gross* undoubtedly exists, and that, by this power of detection, not only is discredit thrown upon the returns, but the whole female population is made the subject of ridicule, while a certain proportion only is to blame; and further—how unlikely it is, that any real concealment of age can ever take place among intimate friends; and as regards the census-enumerators, can they, by possibility, care one *iota* whether any particular lady of the thousands, the records of whose ages pass through their hands, is 'sweet eighteen,' or 'fair, fat, and forty'? Let such ladies read Horace, and learn that even flowers of spring do not for ever retain their pristine glory, and that the blushing moon does not always shine with unchanging aspect.

> " Nec Coæ referunt jam tibi purpuræ,
> Nec clari lapides tempora, quæ semel
> Notis condita fastis
> Inclusit volucris dies."

"Notis fastis," I may observe, may fairly be translated, "in the records of the Registrar-General."

We should, however, endeavour to make clear to the minds of the fair sex—who, in general, do not take much interest in statistical details—the exact nature and extent of the evils that are likely to result from the falsification of returns on which so many monetary schemes, in which they are more particularly interested, depend.

For whose benefit has the greater proportion of the enormous sum of two hundred millions sterling been assured upon lives in this country? Certainly for the benefit of the wives and daughters of England.

Of the 28,367 Government annuitants—which form the subject of Mr. Finlaison's Report, now under discussion—16,538 were females; and it is reasonable to suppose, that in the lives of the 11,829 males, which make up the first-mentioned number, several thousand more wives and daughters, and other female relations of the males, were pecuniarily interested.

Now, what must be the necessary result, as regards the use of recently constructed tables based upon what should be the proper elements for the calculations of Life Assurance and Annuity Companies in this country—that is to say, upon the mortality and censuses of the country in which the transactions take place? Why, clearly this, that if, census after census, females persist in making false returns as to their ages, such Companies must secure themselves against the risk of any loss which may arise from the use of incorrect *data*, and they must accordingly, in self-protection, take a sufficient margin for their own safety in offering terms for the different descriptions of benefits in which they deal.

The effect of this will be, that such increased premiums in the one case must be charged, and such reduced annuities granted in the other, in all cases where female lives are concerned, as shall compensate for all risk of errors in the returns which form the basis of the computations, upon the correctness of which the power of granting life assurances and annuities absolutely depends, and thus the public will fail to derive any advantage from improvements in statistical science, and the Companies will be compelled to continue the use of notoriously unscientifically constructed tables.

I may point out that, even if the result of the class of errors under observation should be actually *in favour* of the Companies *in the gross*, it would nevertheless be only just that, so long as such serious errors are found to exist, public Companies should protect themselves in the manner described against loss in granting either life assurances or annuities. As statistical returns come, in course of time, to be more correctly made, and accordingly more to be depended on, there can be no reason why females should be deprived of any possible advantage that they might derive from the use of tables which represent the mortality among their own sex.

If it should be clearly proved that the duration of life among females is actually greater than among males, a less premium for life assurance should be charged where the female sex is concerned; on the other hand, the charges for annuities depending on the lives of females should be higher than those for corresponding benefits on the lives of males.*

It is shown by the Government Life Tables that the mortality among males is higher than that among females throughout the whole of life.

Mr. Finlaison refers to a remarkable peculiarity shown by those tables—which were computed by the late Mr. Finlaison in 1823—"a peculiarity at the age of 23 which is present in so many other observations, as to appear to be a law of nature. This feature consists in a marked increase of the mortality, at the particular age mentioned, which mortality afterwards diminishes to again re-augment to the same extent at an age long subsequent, viz., the age of 48. The occurrence of a similar climax in the life of the mariner at the age of 32, has also been shown by separate observations drawn by different observers from independent sources. This peculiarity, however, does not exist in the female table of mortality. Nor is this remarkable variation," adds Mr. Finlaison, "altogether unimportant, for it militates against a theory supported by one or two eminent

* The only existing Company that quotes special rates for assurances on the lives of females is, I believe, the National Provident Institution. It appears from the prospectus of this Society that for assurances on the lives of females a small *addition* is made to the premiums charged for male lives, while the rates for annuities depending on the lives of females are *less* than those for similar benefits on the lives of males—as is the case with the Government Office and with many other Companies granting annuities. It would appear, therefore, that this Institution considers female vitality to be at once both inferior and superior to that of males—an apparent anomaly. The practice, however, of thus "working the oracle" might probably be defended.

mathematicians, to the effect that mankind dies in the order of a symmetrical curve. Mankind, however, does not do so, as far as observation informs us."

Any opinion on this subject will naturally be interesting to your readers just now, considering the important discussion that has so recently taken place among the members of the Institute of Actuaries on Mr. Gompertz's law of mortality.

The result of the superiority of female over male life, as shown by the Government Life Table, is stated briefly as follows, viz.,—" In childhood—that is, at eight years of age—the female life is computed to possess $5\frac{1}{2}$ years greater mean duration of life than the male; in womanhood—that is, at 28 years of age—$4\frac{1}{2}$ years; at the age of 58, $3\frac{1}{2}$ years; and in old age—viz., 78 years of age—$1\frac{1}{2}$ years." In drawing any inferences, however, from tables constructed upon the experience of Government as regards annuities and tontines, we must always bear in mind the well-known effect of the benefit of self-selection exercised by the nominees against the Government.

On this point Mr. Finlaison remarks—" It may be doubted, indeed, whether any step in the way of selection will ever surpass the intuitive perception which reigns in the mind of the self-nominated annuitant at the time of purchase."

This influence, however, affects male no less than female lives.

Notwithstanding, then, the indications given by these tables, of the superiority of female over male life, and notwithstanding the fact that the mean future lifetime, or expectation of life, of females is greater than that of males, according to Dr. Farr's tables, still, considering the doubts that have been shown to exist as to the correctness of the *data* on which the latter tables were constructed, it will perhaps be admitted that the question of the superiority of female vitality is not yet set at rest, particularly, too, since the table founded on the experience of the seventeen Life Assurance Companies shows, that " the mortality amongst assured females, taking all ages together, is greater than amongst assured males."

It is probable, however, that even in the event of the vitality of females being positively proved to be superior to that of males, the female sex might not, *on the whole*, obtain any advantage from the use of tables of female mortality, since, no doubt, a considerably greater number of annuities than of assurances are granted upon the lives of females; but this is no reason why equitable charges should not be made in *individual* cases.

So long, however, as the records, on which the calculations of such descriptions of benefits depend, continue to be wilfully falsified by the return to the census-enumerators of incorrect ages, this will be impossible.

In connection with the question of the superiority of female over male life, Mr. Finlaison institutes a comparison between the English Life Table and the Government Life Annuitants' Table for both sexes. At the younger adult ages according to the English Life Table, the mortality among females is shown to be greater than that among males. " This conclusion," he says, " is contrary to most previous experience. It is a result which is contrary also to nature. The sexes are not created in equal numbers. For every 20 females there are produced 21 males. But no fact is more thoroughly established than that, whenever the population is counted, the females are present in considerably greater number. Making every allowance for the temporary absence of a part of the male population, such a result could not take place unless the stronger sex were subjected to a higher rate of mortality, and died off much faster than the females. Were it not so, and did not the males by their more rapid departure from the world subsequently compensate for their appearance in greater numbers in the first instance, it is evident that they would more and more preponderate at every successive enumeration."

I may mention that at corresponding ages the mortality of the female Government life annuitants appears, according to the table, to be considerably less than that of the males.

From what we have seen with respect to the census returns, there is probably great reason to doubt the correctness of the English Life Table as regards female lives; and though the same cannot be said of the male lives, still an Assurance Company would hardly adopt a table of mortality except in its entirety. If, therefore, this table came to be used at all, it would probably be a table based upon a combination of the two sexes—unless indeed the object should be to quote distinct rates for female life, and in this case the English Life Table could not probably with safety be made use of at present. It is to be regretted, however, that any circumstances should exist to prevent the use of tables founded upon the mortality of the country in which the grant of life assurances and annuities is carried on to so great an extent.

The English Life Table, constructed by Dr. Farr some years subsequent to the establishment of the Registrar-General's department for the registration of births, deaths, and marriages, naturally came to be considered peculiarly adapted for the use of Companies dealing in life contingencies—in the first place because the greater proportion of the business transacted by such Companies depended upon the lives of persons resident, for the most part throughout life, in the district embraced by the returns—and in the second place, because the table was based upon the records of the mortality experienced at a very recent period, and upon the previous decennial census; the table in question, moreover, was due to a computer of the highest eminence.

Of course, as far as respects correctness of computation, the table may be fully relied on; but minuteness of calculation, and accuracy of records—supposing the returns of the Registrar-General can be depended upon—is of no avail, unless the census-enumerators succeed in obtaining exact information; for on the numbers living at each age, according to the periodical enumerations of the people, no less than upon the number of deaths, do our mortality tables depend.

I shall presently recur to the question of the accuracy of the mortuary records.

To show, practically, the effect in actual money of the use of the English Life Table in the grant of life annuities and assurances, Mr. Finlaison has computed that if 1,000 persons of each sex, all aged 55 years—the average age of entry at the Government Annuity Office—were to purchase life annuities, there would be found to be living, according to the English Life Table, at the expiration of 20 years, 368 males and 410 females—or, in all, 778 persons—for the future payment of whose annuities it would be necessary to provide; but according to the experience of the Government Tables, at the expiration of the 20 years, the number surviving out of the 2,000 persons of both sexes would be as follows, viz., 403 males and 498 females—or, in all, 901 persons. There would therefore be, in fact, 123 persons to provide for in excess of the 778 for whom provision had been made. Assuming the annuities to have been £60 each—which Mr. Finlaison states to be the usual average—the Annuity Office would be liable to the extent of £7,380 per annum during the remainder of the lives of the surviving annuitants, for which no provision would have been made.

Again: supposing 1,000 males, all aged 27 years—which Mr. Finlaison considers to be a fair average age to assume*—were mutually to assure one another for the sum of £1,000 each, according to the mortality of the Government male annuitant lives, taking interest at 4 per cent., Mr. Finlaison computes that the annual premium payable, to assure the sum of £1,000,000, would be £17,447; while another 1,000 males, of the same age, would, according to the English Life Table for males, at the same rate of interest, have to provide the sum of £16,885 only annually, or £562 per annum less than the first set. At the expiration of 25 years there would be living, according to the Government Table, 712, and according to the English Life Table, 724, persons; and, in the case of the first society, Mr. Finlaison goes on to show that the sum of £193,920 would, according to its own table, buy up the 712 existing policies, while, in the case of the second society, the sum necessary to buy up the existing policies, according to the English Life Table, would amount to £213,877, to provide for which the second society would have been receiving annually, for 25 years,

---

* The experience of Life Offices would probably warrant the assumption of a much higher average age—possibly 40.

the sum of £562 a year less than the amount requisite to provide for the liability under the Annuitants' Mortality Table.

Upon reference to the first of these two illustrations, it will be seen that the differences between the numbers living at the age 75, according to the English Life Table and Government Tables, were respectively as follows, viz., males, 368 and 403; females, 410 and 498. The differences between these numbers are respectively 35 and 88. The variation between the tables was, therefore, considerably more than twice as great in the case of females as in that of males.

Now, without attributing the whole of the blame to false census returns—for, to do so, we must necessarily predicate the absence of all other errors and disturbing causes, as well as the absolute correctness of the Government Life Annuitants' Table—there can be no doubt, from the foregoing considerations and illustrations, that the most mischievous results might readily be caused by use of the tables in question.

The latter illustration is less apposite to the purposes of this communication, since the mortality of male life only, according to the English Life Table, is made the subject of comparison with the Annuitants' Table; and it is to the effect of the errors in respect of the ages of females upon the safety of the table, that I desire particularly to confine my attention on the present occasion.

It is referred to, however, with the object of showing how important, in a monetary sense, is the choice of a proper table for any particular description of business.

As regards that illustration, too, it may be mentioned that, in the case of an Assurance Company, many elements must necessarily enter into the question of the choice of a mortality table that need not be considered by an Annuity Company, still less by the Government, whose object in granting life annuities is simply to exchange perpetual for terminable annuities, and not to make a profit of the transactions. A table, therefore, which might be perfectly suitable for the former, would be quite unfit for the latter, whatever be the direction in which the errors tend. I am now assuming what must probably always be the case in practice, viz., that all tables of mortality necessarily contain errors. A table that might be perfectly correct, would, I apprehend, be equally suitable for an Assurance or for an Annuity Company.

It would seem that if we could arrive at such an impossible result, and if we could moreover predict the rate of interest that would be realized during a long period of years, it would be sufficient for actuaries to make such additions to the net charges for the different risks undertaken as would be required to provide for the necessary profit on the several transactions, without, as they are now obliged to do, being compelled either to make an additional charge to compensate for errors in the *data* on which they base their computations, or to make use of tables which give a far too unfavourable view of life, and derive the required profit from the mortality falling short of that which the tables showed might be anticipated.

In the case of Assurance Companies, considering the unfortunate system of speculation in bonuses which now obtains to so great an extent, of course the further necessary addition would have to be made to provide for the bonuses which the cormorant-public now so eagerly demands. This abuse has crept into the system of assurance, owing, in a great measure, to the inaccuracy of the old mortality tables, and is one which, from what we have now seen of the possible flaws in Dr. Farr's *model* table, is not likely soon to be remedied.

Probably sufficient has been said to show how much it is in the interest of the entire community to assist in checking the tendency to falsify statistical records—for, in proportion as such returns become, in course of time, more correct, so will the public be more likely to obtain, on equitable terms, the benefits conferred by Life Assurance and Annuity Companies.

Another source of error, however, exists, which tends to jeopardise the value of mortality tables.

There is reason to believe that the returns of ages at death, made to the District-Registrars of Births, Deaths, aud Marriages, are in many cases far from correct.

It is a matter fully within the cognizance of those engaged in the business of life assurance, and who have constantly occasion to test the truth of these records, by means of proper documentary evidence of the dates of birth and baptism, that the ages of persons at death, as stated in the certificates of the Registrar, are often one, two, or more years greater or less than the correct age. Upon inquiring into the cause of such discrepancies, we discover how loose the system of arriving at the age at death really is. Upon a death occurring, it becomes the duty of the medical attendant of the deceased to fill up the certificate prescribed by the Act. This certificate, which must be signed by the gentleman who last attended the deceased, states the age last birthday—when last seen—the date and the cause of death. The time from the attack till death—each form of disease, or symptom, being reckoned from its commencement *till* death—is also required to be stated. Both the primary and secondary cause of death has also to be inserted. From this certificate, the entry is made in the books of the District-Registrar. Upon the truth, therefore, of the information furnished to the medical attendant, the correctness of the return entirely depends. Probably, in the majority of cases, the medical man has no intimate acquaintance with his patient, and in very many cases he is called in towards the close of life, and after the patient has been ill for some time. His power, therefore, of forming a satisfactory independent judgment as to the probable correctness of the age, as stated to him by the relatives, is very much diminished, owing to the change in the appearance of the patient which disease has caused. The medical attendant must therefore, in most cases, be wholly at the mercy of his informants, who have often but a very imperfect knowledge themselves of the fact they are required to certify, and in a great number of cases, therefore, the age is assessed altogether at hazard.

The ideas of medical men differ very much as to the importance of statistical returns. One class will set down, without exercising any judgment in the matter, any age that may be given to them. Another will endeavour to arrive at the truth, should it seem likely that the deceased were older or younger than the age asserted. Now both these classes aid in assisting the registration of incorrect ages— the first by taking no trouble in the matter, and the second by self-deception as to the age of a person they have perhaps only seen upon a bed of sickness.

We all know how difficult it is to judge correctly even of the ages of persons in health, whom we are in the habit of seeing every day, in consequence of the varying effect produced upon the constitution by the wear and tear of life in the ever-changing circumstances in which different people are placed. Although, therefore, medical men of experience are, no doubt, better qualified than non-professional men to form a correct estimate of the age of persons at death, and to make the necessary allowances for the effect produced by the antecedent illness, I think it will be admitted that according to the system now in operation for recording the age at death, the returns are liable to be often very far from the truth.

The simple remedy for this evil would be to require that the medical attendant should give information as to facts only of which he is actually cognizant—that is to say, of the date and cause of death and duration of the illness, leaving it for the legal representative or next of kin of the deceased to furnish the Registrar with the age at death, the statement of which should be supported by documentary evidence to be furnished within a certain time after the date of death.

It would clearly be inconvenient to require that this evidence should be procured within a few days of the event; and the execution of the medical certificate could not be deferred, because the clergy are very properly not allowed to proceed to the interment of a body without the production of the Registrar's certificate, which, as previously mentioned, is based upon that of the medical attendant.

In connection with this subject, and referring to the remark of Mr. Finlaison, that "where the person to be accounted for is old, there is a disposition to claim or attribute an exaggerated longevity," I may observe that in very many of those cases where the frequent remarkable statements as to the ages which persons are registered as having attained come to be subjected to the test of actual documentary evidence, it has been found that such ages are for the most part ridiculously overstated.

It is a curious fact, and one which should influence us in denying credence to extravagant statements of this kind, unless supported by satisfactory evidence, that it is not found that members of the Peerage, whose ages at death we have instant means at hand for ascertaining, ever attain to fabulous old age. If, therefore, this favoured class do not present cases of extreme longevity, it is reasonable that we should receive with great caution such statements as to the ages of members of other classes of the community.

Probably, with very rare exceptions, the maximum age obtained in this country never really much exceeds 100 years.

The object I have had in view in this letter, has been to direct attention to certain facts, with the view to the remedy of any possible flaws that may be admitted to exist in the system of the census-enumeration and the registration of deaths in this country, as affecting the accuracy of any future tables of mortality which may be computed upon such *data*—tables on which all calculations connected with life contingencies depend.

The principal heads are as follow, viz. :—

1. The inordinate extent to which the ages of females between the ages of—say, 20 and 40—are understated in the census.

2. The over-estimate of age in very advanced life, both in the census and in the registers of death.

3. The loose system adopted in ascertaining the ages at death of all classes of people, at all ages.

4. The direct pecuniary loss sustained by females in the purchase from the Government, or from such Annuity Companies as adopt special scales for female lives, on the present assumption of the vitality of females being so much in excess of that of males.

5. The advantage the female sex might derive, were Life Assurance Companies to adopt special rates of premium for female lives.

6. The general discredit thrown upon the English Life Table, both for males and females, and the necessity that exists for Life Assurance and Annuity Companies to abstain from using it, in consequence of the incorrect returns, which both as regards the numbers living and the numbers dying, at each age, have been, no doubt, habitually made to the census-enumerators and to the District Registrars of Deaths.

The limits of a letter like the present will not allow of my even touching upon many other points which might affect the question of the suitability or otherwise, for the purposes of Life Assurance and Annuity Companies, of any life tables based upon the census returns, one of which is the fact that the labouring classes, as well as paupers, vagrants, criminals, and dissolute persons, comprise, by far, the larger proportion of the whole population; and it is not upon such lives that assurances and annuities are granted.

The mortality, too, among these classes is undoubtedly greater than that among those who rank higher in the social scale—as I have recently had an opportunity of showing, in a paper published in the Transactions of the Institute of Actuaries, in the course of which this question was entered into at considerable length.   On the other hand, it might be urged that a table containing so large a number of lives exposed to an undue rate of mortality would be, for that very reason, more suited to the requirements of public Companies.

A discussion, however, even of this one question, would lead to an inquiry into the respective merits and demerits of the existing tables of mortality, and would cause me to engross far too much of the available space in the pages of the *Assurance Magazine.*   The effect, too, of the exclusion from the census returns—notwithstanding all the care to prevent it—of a considerable proportion of the nomadic population of the country, as well as of numerous other influences, would have to be considered.

In conclusion, I may state that Mr. Finlaison—who, in connection with his recent Report, has evidently investigated and considered this subject most fully—is clearly of opinion, that "at the present day, the census, and registration of births, deaths, and marriages, cannot with prudence be adopted as the bases of the true measure of the value of life."

I am, Sir,
Your obedient servant,

Alliance Assurance Company,   H. W. PORTER.
   16*th March,* 1861.

ON MR. GOMPERTZ'S LAW OF HUMAN MORTALITY, AND MR. EDMONDS'S CLAIMS TO ITS INDEPENDENT DISCOVERY AND EXTENSION.

To the Editor of the Assurance Magazine.

SIR,—The remarks which appeared from the pen of Professor De Morgan in the number of the *Assurance Magazine* for last July, must have attracted the notice, not only of those who are interested in the history of the theory of life contingencies, but of all your readers who wish to see improvements in science attributed to their actual originators. It is no unusual thing for the title to a discovery to be contested; and it not uncommonly appears that different persons have made the same discovery independently.   This is sometimes an extremely difficult point to decide; but in the present case, the means of arriving at a satisfactory decision are unusually ample.   It must be clear to all who have read Professor De Morgan's remarks and Mr. Edmonds's rejoinder, that the charge brought by the former against the latter has been completely substantiated: viz., that Mr. Edmonds, following in the footsteps of Mr. Gompertz, and familiar with his writings, "has adopted his ideas without anything approaching to a sufficient acknowledgment." On this point, nothing further remains to be said. But Mr. Edmonds, in the course of his defence, as it may be termed, has introduced many other matters, which seem to me to call for further notice, although dismissed by Professor De Morgan as having no bearing on the question raised by him.

Mr. Edmonds, then, has made a minute comparison of his processes and results with those of Mr. Gompertz; and· has made many reflections to the disparagement of the latter (*vide* pp. 174–178), which, if allowed to remain unanswered, may have more weight than they deserve with some of your readers, and lead them to undervalue the labours of Mr. Gompertz. Having been led by Professor De Morgan's remarks to make, for my own satisfaction, a somewhat close comparison of the writings of Mr. Gompertz and Mr. Edmonds, I am induced by the above consideration to lay before your readers the conclusions at which I have arrived.

In the first place, I notice that, curiously enough, Professor De Morgan, who appears as the voluntary champion of Mr. Gompertz, does not do him full justice. Mr. Gompertz's paper in the *Transactions of the Royal Society* is full of misprints—many of them being of the most serious character. Two such are corrected by Professor De Morgan in the extract he has made (pp. 87, 88); but he has left several uncorrected, which give rise to the "obscurity" to which Mr. Edmonds refers (pp. 178, 181). In order to demonstrate this, and with the secondary object of making Mr. Gompertz's important theorem more generally known,* I subjoin its demonstration. It will be seen, that besides correcting these misprints, I have expanded the reasoning in some of the steps, and have in several instances substituted for the notation of Mr. Gompertz, that more generally used at the present time.

Let the rate of mortality at the age x be denoted by aq^x, where a and q are constant quantities, and let L_x be the number living at the age x; then the number dying in the small time dx will be $aL_x \times q^x dx$; so that—

(1) $$d L_x = -aL_x \times q^x dx;$$

(2) $$\therefore \frac{1}{L_x} \cdot \frac{d L_x}{dx} = -a \times q^x.$$

(3) Integrating both sides, $\log_e \dfrac{L_x}{d} = -\dfrac{aq^x}{\log_e q} = \dfrac{aq^x}{\log_e \dfrac{1}{q}}$;

d being the constant introduced by integration.

Now, $\log_e \dfrac{L_x}{d} = \log_{10} \dfrac{L_x}{d} \cdot \log_e 10$, $\log_e \dfrac{1}{q} = \log_{10} \dfrac{1}{q} \cdot \log_e 10$; so that the last equation gives

(4) $$\log_{10} \frac{L_x}{d} = \frac{aq^x}{\log_{10} \dfrac{1}{q} \cdot (\log_e 10)^2} :$$

---

* Since writing the above, I have found that I have been anticipated in this object by Mr. Peter Gray, who has fully explained the nature and application of Mr. Gompertz's method of interpolation in a paper (*Assur. Mag.*, vol. vii., p. 121), which is well worthy of every student's attention. In comparing Mr. Gray's form of the demonstration with that given in the text, it must be borne in mind that the constant added in the integration being arbitrary, may either be written $+ C$, as Mr. Gray has it, or $- \log_e d$, which is Mr. Gompertz's form.

(5) or, putting $\log_{10}\dfrac{1}{q} \cdot (\log_e 10)^2 = \dfrac{a}{c}$,

(6) $\log_{10}\dfrac{L_x}{d} = cq^x.$

(7) Hence $\dfrac{L_x}{d} = 10^{cq^x}.$

And if $10^c = g$, or $g = \log_{10}^{-1}c$, *i.e.* the number whose common logarithm is c,

(8) $\dfrac{L_x}{d} = g^{q^x},$

(9) and $L_x = dg^{q^x}.$

On comparing the above with Mr. Gompertz's demonstration, as quoted by Professor De Morgan, it will be seen that the mysterious constant (b), which has caused so much perplexity, is simply a misprint for the sign of multiplication $(\times)$. This will be at once admitted, when it is noticed that Mr. Gompertz writes down the *equation*

$$aL_x \times q^x x = -(L_x)^{\cdot};$$

and then concludes that $abq^x = -\dfrac{L_x}{L_x}.$

Professor De Morgan seems to have overlooked this circumstance, which proves that a constant b could not have been introduced in the manner he suggests.

It will also be noticed that the constant d, which does not appear in Mr. Gompertz's process till step (6), correctly enters on the integration in (3). Again, in (5), Mr. Gompertz's c should be $\dfrac{a}{c}$.

These corrections having been made, all " obscurity and ambiguity" disappear from Mr. Gompertz's process. Nor does that process contain, as Mr. Edmonds asserts (p. 181), "two superfluous and useless indeterminate constants." I have already disposed of b, and I shall presently show that d is neither superfluous nor indeterminate.

The preceding corrections occurred to me on reading carefully Mr. Gompertz's process, as quoted by Professor De Morgan (p. 88). In order, however, to make the comparison already mentioned of Mr. Edmonds's writings with Mr. Gompertz's, I referred to the copy of Mr. Gompertz's papers in the library of the Institute of Actuaries. That copy, which was presented to the Institute by Mr. Gompertz, exhibits the precise corrections I have already indicated, in addition to those pointed out by Professor De Morgan, made in the margin of the volume—as I presume, by Mr. Gompertz himself. (There is, however, in step (5), an *over-correction* made.) A single glance at this copy will convince any person of the accuracy of my statement, that Mr. Gompertz's paper, as it appears in the *Transactions of the Royal Society*, is full of misprints. It is to be presumed that that gentleman, like so many others of scientific eminence, does not write a very legible hand. Perhaps also the Royal Society, at the time the papers in question were printed, followed the course adopted at the present time by a few Societies—happily not by the Institute of Actuaries—and did not allow the contributors to its *Transactions* the opportunity of correcting the proofs of their papers.

We have thus seen that there is no error in Mr. Gompertz's demonstration of the theorem that "if the rate of mortality follows the law aq^x, then the number living at any age x will be given by the formula dg^{q^x};" and that Mr. Edmonds's doubts as to its correctness have arisen from the fact of its being obscured by numerous misprints. Let us now examine some of Mr. Edmonds's remarks upon that demonstration, and it will be seen that they contain errors equally grave with those falsely imputed to Mr. Gompertz, and which cannot be similarly explained. In the first place, then, Mr. Edmonds speaks (p. 180) of "b being introduced as arising in the process of integration." It will be noticed that b makes its appearance in step (2), while the integration is not performed till (3); also that d is the constant introduced by integration: so that the only conclusion appears to be that the remark just quoted was written under a misconception as to the real nature of the process of integration. This conclusion is strengthened by the perusal of the remarks (p. 178) on the limits of integration, and by the suggestion made (p. 184) of "an error committed in the process of integration, of which b represents the correction." Here a confusion of ideas is betrayed, of which it is difficult, if not impossible, to trace the origin. A similar confusion of ideas pervades the remarks (p. 181) on the constants a, b, a. The truth seems to be that Mr. Edmonds, having concentrated his attention for a long time upon a particular problem, with which it must be confessed he has shown a considerable degree of familiarity; and having arrived at a correct solution in a particular form, in which he has been aided by a knowledge of Mr. Gompertz's paper (*vide* p. 177, bottom), is yet unable, in consequence of want of familiarity with the principles of the integral calculus, to follow Mr. Gompertz in his somewhat more general way of treating the same question.

We next come to Mr. Edmonds's remarks upon the constant d. He terms this (p. 178) a "superfluous quantity," and speaks of it as constituting a "defect" in the formula. Instead of being such, it is really an essential part of the formula, *and as such is employed by Mr. Edmonds himself.* He uses the letter g to denote the same quantity as Mr. Gompertz's d; and in arriving at his formula, determines g so that $y=1$ when $x=0$. On the other hand, Mr. Gompertz leaves d undetermined; since for his purpose, there was nothing to be gained by determining its value. (This, I conjecture, is what Mr. Edmonds means by calling d, in p. 181, an indeterminate constant; but mathematicians mean by an "indeterminate" quantity, one whose value cannot be ascertained.) If it were required to determine the value of d in the formula $L_x = dg^{q^x}$, the process would be as follows:— Suppose the number living at the age $n*$ to be L_n; then $L_n = dg^{q^n}$, whence $L_x = L_n(g^{q^x - q^n})$, or $L_x = L_n(g^{q^n})^{q^{x-n}-1}$. It must be noticed that in Mr. Gompertz's formula x denotes the age measured from birth, while in Mr. Edmonds's x is only measured from birth during the period of infancy, as will be seen by his example (p. 179). So far from d being a defect in the formula dg^{q^x}, the formula $10^{\frac{k^2 a}{\lambda p}(1-p^x)}$ is, on the contrary, defective, because it assumes $y=1$ when $x=0$; and the defect in it is tacitly supplied by Mr. Edmonds himself in practice. Thus, in his example, he arrives at the result $y_{10} = 1 \div 1.076823$ ($= .928658$); and x being measured from

* I use n here in preference to Mr. Gompertz's a, because the letter a has already been used to denote another quantity in the expression aq^x.

the age of 12, this should be the number living at the age of 22. Instead of this, Mr. Edmonds is compelled to introduce the quantity we have denoted above by L_n: he takes the number living at 12 to be 100,000 ($=L_{12}$), and deduces the number living at 22 to be 92865·8; as will be seen on reference to his Table of Mean Mortality. The above is not a point of great importance; but it shows that whatever defect does exist, is in Mr. Edmonds's formula, and not in Mr. Gompertz's. Mr. Edmonds, however, falls into a far more serious error (p. 180) in calling d a particular value of y. It is easily shown that y cannot possibly be equal to d. For, in the formula $y=dg q^x$, if $y=d$, we have $g q^x=1$; hence $q^x=0$, and $x=+\infty$ or $-\infty$, according as q is $<$ or >1; values which are of course inadmissible.

Mr. Edmonds's words in speaking of d are as follows (p. 178):— "The defect in Mr. Gompertz's formula, caused by the addition of (d), is the same as that which would exist in a table of discount of money at compound interest if any other basis were adopted than the value of the sum of £1 receivable (x) years hence." It would be generally considered that £1 was the *basis* of such a table,* and not the value of £1 receivable x years hence; but without dwelling on this minor inaccuracy, I notice that in the passage just quoted, two very different things are confounded with each other, viz., the *formula* for the value of a sum of money due at the end of x years, and the *table* by which that value would be practically found. It is true that a table in which any other basis than £1 were assumed, would be *inconvenient* in practice (though it could not be correctly described as *defective*); but on the other hand, a formula in which the sum to be received is taken as £1 (*i.e.* $(1+i)^{-x}$), would be defective; the correct formula being $s(1+i)^{-x}$.

Again, we read on the same page that Mr. Gompertz has changed the sign of c; whereas the true state of the case is that Mr. Edmonds has changed its sign for his own purposes. The quantity c, which does not appear in Mr. Edmonds's own investigation, is taken by Mr. Gompertz to be equal to $\dfrac{a}{\log_{10}\frac{1}{q}\cdot(\log_e 10)^2}$, which in Mr. Edmonds's notation would be $\dfrac{k^2 a}{\lambda\frac{1}{p}}$, or $-\dfrac{k^2 a}{\lambda p}$. The latter, when using the letter c for the first time on p. 178, takes it equal to $+\dfrac{k^2 a}{\lambda p}$, and then says that Mr. Gompertz has changed the sign of c!

I now pass on to the remaining portion of my subject—a general comparison of the results obtained by Mr. Gompertz and Mr. Edmonds. Here at the outset I must direct attention to a grossly unfair misquotation by the latter of Mr. Gompertz's words. He wishes to give your readers the impression, that Mr. Gompertz's discovery is imperfect by reason of its not stating any limits of age, within which the formula is to be applied with the same constants; and for this purpose he professes to quote (p. 174, top) a passage from Mr. Gompertz's paper. *No such sentence is to be found in*

---

* Similar to this, is Mr. Edmonds's use of the word "formula" to denote an *equation*, as $L_z=dg q^x$, instead of the *expression* ($dg q^x$), which forms the second member of the equation.

that paper! Mr. Gompertz's words are as follows:—"*If* the average exhaustions of a man's power to avoid death were such that at the end of equal infinitely small intervals of time, he lost equal portions of his remaining power to oppose destruction which he had at the commencement of those intervals, *then* at the age x his power to avoid death, or the intensity of his mortality, might be denoted by aq^x." And again: "This equation $[L_x = dg^{q^x}]$ between the number of the living, and the age, becomes deserving of attention, not in consequence of its hypothetical deduction, which in fact is congruous with many natural effects, as for instance, the exhaustions of the receiver of an air-pump by strokes repeated at equal intervals of time, but it is deserving of attention because it appears corroborated *during a long portion of life* by experience; as I derive the same equation from various published tables of mortality *during a long period of man's life*, which experience therefore proves that the hypothesis approximates to the law of mortality *during the same portion of life;* and in fact the hypothesis itself was derived from an analysis of the experience here alluded to." I have been thus particular in quoting the above passages (in which the italics are mine), because the question of misquotation is one upon which all your readers will be able to form an opinion, whereas it requires for the full appreciation of many of the points here raised, a greater knowledge of the higher mathematics than they can all be expected to possess. By comparing together the above two passages from Mr. Gompertz's paper, any person may judge of the "fairness" or the truth of Mr. Edmonds's assertion (p. 174) that "it might fairly be inferred from the above statement that the vital force of man, measured by the ratio of the living to the dying, is in a constant state of decay *from birth to the end of life* at one and the same uniform rate." The reason Mr. Gompertz has stated no definite limits of age is obviously because he did not believe them to be fixed in the same nearly invariable manner as Mr. Edmonds does. This brings me to the next point I have to notice—the improvements the latter claims to have made on the former's discovery.

Mr. Edmonds has convinced himself that in the formula for the mortality at any age, aq^x, q has three fixed values, invariable or nearly so, under all circumstances, and for all populations; and he endeavours, by reiterated assertions, to convince his readers of the truth of the proposition. I say *assertions*, because he has not advanced a single particle of *proof* in support of his proposition. In this single distinction is summed up the difference, or rather the contrast, between the writings of Mr. Gompertz and Mr. Edmonds. All that the former advances is *rigorously proved;* all that the latter claims to have discovered is *simply asserted.* Mr. Gompertz describes all his computations so as to enable a person with only slight knowledge of the subject to repeat them and test their acccuracy: he shows how he has found the values of his constants g and q; and he arranges his results so as to exhibit at a glance the degree of accuracy with which they agree with previously computed tables of mortality. Mr. Edmonds has done nothing of this kind: he simply states the results he has arrived at, and apparently expects them to be received without demonstration.

If Mr. Edmonds's proposition of three invariable values for q were established, it would no doubt be justly entitled to the name of a *discovery*, and it would be a great advance on what Mr. Gompertz has done. But in the absence of any proof, it is impossible to treat the proposition otherwise than as an unproved hypothesis. If Mr. Edmonds possesses the means of proving his proposition, and publishes sufficient details of the extensive comparison of tables described by him on p. 172, the scientific public will then be in a position to judge of the truth or error of his proposition; and I for one shall not be slow to avow my reception of anything put forward by Mr. Edmonds when supported by sufficient proofs. As the case stands at present, it is impossible to admit the truth of Mr. Edmonds's proposition, or to countenance his use of the term "discovery" in describing the conclusions at which he has arrived.

It forms no part of my purpose to comment on the bad taste displayed by Mr. Edmonds in many of his remarks, nor do I think it necessary to allude particularly to his unjustifiable use of the phrase " *true* law of human mortality" to describe his hypothesis; for these are points which must force themselves on the notice of every intelligent reader. I will therefore conclude this part of my subject by pointing out two other inaccuracies into which Mr. Edmonds has fallen. He states (p. 177) that he believes Mr. Galloway to be the only person who has made a practical use of Mr. Gompertz's formula. This is by no means the case. Mr. Jellicoe, in the first volume of the *Assurance Magazine*, p. 166, has applied the formula to the adjustment of Mr. Neison's Table of Mortality deduced from the deaths among the officers of the Indian army. He has also used it in a modified form (*Assur. Mag.*, vol iv., p. 206) to graduate his own Table of Mortality derived from the experience of the Eagle Insurance Company; and the same method has been recently adopted by Mr. Dove to adjust the results deduced from the experience of the Royal Insurance Company. Mr. Farren has also employed it in his Life Contingency Tables. Again: Mr. Edmonds (p. 174) says that "nearly all he [Mr. Gompertz] offers to show is, how *interpolations* may be made for intermediate ages when the number of survivors at the beginning, and the number of survivors at the end, of a large interval of age are given." (*Conf.* p. 176, bottom.) Here Mr. Edmonds quite overlooks the fact, that it requires a knowledge of the number of the survivors at *three* ages at least, to apply Mr. Gompertz's method; which is an obvious deduction from the fact that the formula dg^{q^x} contains *three* disposable constants. When we remember that Mr. Edmonds erroneously rejects one of those constants as superfluous, it appears less surprising that he should have made this serious blunder.

We have seen that it is impossible to admit, on the evidence adduced hitherto, that Mr. Edmonds has *extended* the theory of Mr. Gompertz: it remains now to examine his claims to an independent discovery of what is common to them. Taking then Mr. Edmonds's own statement (pp. 172–178), it appears that he ascertained that the rate of mortality at different ages might be represented with sufficient accuracy by the formula ap^x, and he then learnt that Mr. Gompertz had shown that the number of survivors at any age would be given by the formula dg^{p^x}. In other words, Mr. Edmonds *very nearly* discovered what had been previously discovered by Mr. Gompertz; or, in his own phrase, if he had only had a little more time, he would certainly have discovered it. It is of no avail to make conjectures (*vide* pp. 175, 176) as to the course by which Mr. Gompertz arrived at his formula. The *fact* remains that he did arrive at it and publish it, and that Mr. Edmonds first became acquainted with the formula through Mr. Gompertz's writings. It was no doubt a great disappointment to Mr. Edmonds to find that he had been anticipated in this manner, but it may be a consolation to him to reflect that his position is by no means a singular one. There has probably been scarcely a single discovery of any kind made, but some unfortunate man has been on the point of making it too, and would have done so, if he had only had sufficient time allowed.

In conclusion, I beg to state that in all I have said, I have been influenced by no personal feeling towards Mr. Edmonds. I wish to allow him all the credit justly due to him. I believe that in accomplishing the object he had in view—viz., that of forming a new hypothetical table of mortality—he has shown considerable judgment and skill; and I shall willingly admit the universality of his constants whenever I see sufficient proofs of their accuracy produced. My object has been simply truth; and if, in the course of my remarks, I have been led to use language which may seem severe, I regret the necessity I have been under. I feel I have trespassed rather unreasonably on your space; but trust that what I have said will enable others who have but little time to devote to such subjects, to come to a correct conclusion as to the claims put forward by Mr. Edmonds. My task would have been much lighter if Mr. Edmonds had confined himself to an explanation of his own theory; for then I should not have thought it necessary to dwell upon the numerous errors into which he has fallen. This course has been rendered necessary by his uncalled for criticisms of Mr. Gompertz. Your readers will probably be better able to form an opinion as to the proper weight to be assigned to those criticisms, the whole of which I have not thought it necessary to review, when they learn that the writer of them has himself fallen into many serious errors.

I am, Sir,
Your obedient servant,

25, *Pall Mall*,   T. B. SPRAGUE.
    7th February, 1861.

EXPRESSION FOR THE VALUE OF A TERM ASSURANCE,
LIFE AGAINST LIFE.

To the Editor of the Assurance Magazine.

SIR,—Perhaps the following brief expression for the value of a short period assurance on (x) against (y) may be worth a place in your *Journal*.

By the combined use of the ordinary Commutation Tables in Jones, and the tables of Gray, Smith, and Orchard, for such an assurance to be current for m years and paid for in n annual premiums, the formula becomes

$$\frac{D_{xy} A_{\underset{xy}{1}} - D_{x+m,\,y+m} A_{\underset{x+m,\,y+m}{1}}}{N_{x-1,\,y-1} - N_{x+n-1,\,y+n-1}}.$$

I am, Sir,
Your most obedient Servant,

Aberdeen, 7th May, 1861.   H. AMBROSE SMITH.

x 2

MR. GOMPERTZ'S PAPERS.

To the Editor of the Assurance Magazine.

SIR,—Mr. Gompertz has done me the honour to make me the medium of communication between himself and the members of the Institute of Actuaries, and has addressed the following letter to me for publication in the *Journal* of the Institute, in reference to the recent controversy on the subject of his celebrated paper, read before the Royal Society on the 16th of June, 1825, and published in the *Philosophical Transactions* for that year, "On the Nature of the Function expressive of the Law of Human Mortality."

The question as to the undoubted claim of Mr. Gompertz to be considered the sole and original discoverer of the theorem enunciated in that paper being now so definitively set at rest, by the complete analysis of the whole question by Mr. Sprague, in his able paper, read before the Institute of Actuaries on the 25th of February last, I may, perhaps, be allowed to express an opinion—in which I know I shall be joined by every gentleman who has the pleasure of Mr. Gompertz's acquaintance—that he would have been the very last person to fail to acknowledge any claim to an independent discovery, had such been made; and I cannot, perhaps, conclude better than by quoting Mr. Gompertz's own words on this point, which happen to occur in the course of a previous paper, read before the Royal Society on the 29th of June, 1820,—" To a true philosopher, it will ever be much more pleasing to grant even more praise than is actually due, than to pluck the laurel from the deserving brow."

This, Sir, is a sentiment, which I think you will agree with me is no less elegant in expression than it is indicative of the known amiability of disposition which has always characterised this distinguished mathematician and actuary.

I have the honour to be, Sir,

Your obedient servant,

Alliance Assurance Company,   H. W. PORTER.
   12th March, 1861.

LETTER FROM BENJAMIN GOMPERTZ, ESQ., F.R.S.

Kennington Terrace, Vauxhall,
6th March, 1861.

MY DEAR SIR,—Not having been sufficiently in health since the gentlemen who are members of the Institute of Actuaries did me the honour to elect me an honorary member of their excellent establishment, to attend the meetings, nor to have been active in adding my mite to the papers which it publishes; with your sanction, I wish you to be the medium to express my thanks to several of the members for their frequent kind mention of my name in their valuable papers; but, in particular, I wish to allude to the distinguished and highly talented Professor De Morgan, for the two papers, cleverly written, to prevent the subject of my paper on the "Law of Human Mortality," published in the *Transactions of the Royal Society for* 1825, being lowered in the estimation of scientific men, in consequence of a claim made by a gentleman, who, with rather sharper criticism than I believe ought to have been directed either to me or my paper, of having, I think about the year 1832, made the discovery independently of me; but I am not concerned in that claim; because, whether he did, or did not, make a discovery of a theorem, seven years after it had been honoured by the approbation of scientific men, after my publication of it, cannot be an injury to me. Still, I think it was not wise of him to try to use arguments to establish his claim, which would, I think, by their close resemblance to my own words, rather prove the reverse. I have no wish to rob him of any claim, which either he, or his friends for him, may consider his due, but, on the contrary, I wish well to everyone who may, virtuously and without jealousy, feel inclined to tread in a new path to promote the objects of science, or who, with a view to add information to knowledge acquired, should modestly follow the steps pointed out by an earlier labourer, who may have ploughed in the same field in which he may hope to grow his wheat, or who may have planned a garden in which myrtles may grow, and laurels thrive, to adorn the brow of some future labourer, who may, by arduous labour joined to humble pretensions, merit praise.

I am willing to own that there are many typographical errors which disfigure the paper, which may lead a school-boy from the direct meaning of the information I meant to convey; or which may so act on the mind of a lazy student as to confuse his judgment; but in all the mathematical papers I have published, I have found such errors constantly to occur, notwithstanding all the pains I have paid to prevent them; but I observe, that as long as mathematical papers are printed, without having a sufficiently clever mathematical superintendent of the press, that annoyance will intrude.

It is my intention, or at least my wish, to publish, either in the *Transactions of the Royal Society*, or elsewhere, a notice to correct those errors; yet, I observe, such errors may have the advantage of causing a reader to go more deeply into the meaning of an author, by obliging him to dig into the ground, instead of flying over the surface. I have, for two or three years, been endeavouring to add matter of importance, in my own opinion, to the subjects of my papers of 1820 and of 1825; but the state of my health has so far interfered with my object as to prevent my getting up my intended paper, to present to the Royal Society. And in a paper I wrote for the International Congress, held in July last, I gave hints—which, I flatter myself, will be interesting, when published by the Commissioners— relative to the subject of the paper I was writing for the Royal Society, on " Mortality and Invalidity," with some striking tables, corroborative of my new views on the subject, and hints of the very extensive service of the law of mortality I had further discovered in calculating values, with respect to all sorts of intricate cases and complications of intricacies; and in the paper which I am now endeavouring to induce my health to proceed with, I have improved parts of the paper I was then writing, but the hints do not much go into the abstruse part of my paper, as I did not consider that portion adapted to the intention of the Congress, and because I preferred also to make the Royal Society, if I were allowed that honour, the medium of the mathematical essence of the paper through which my views went forth to the public. In my paper of 1825, I showed, from tables and the law of mortality for portions of life, means of calculating, without much difficulty, the value of annuities on many joint lives; and the ingenious and worthy and highly talented Professor De Morgan has shown, in a paper he wrote, published among the papers of the Institute, that if my formula $a \cdot \overline{b}|^{q^x}$ were absolutely true throughout, it would give the means of calculating the values of annuities on any number of joint lives; but as with the same constants a, b, q, the term of its applicability is limited, the method will not invariably correctly apply: and I think it proper to remark, that my investigation, published in the *Transactions*, ought not to have led the gentleman, who considers himself the second independent discoverer, to consider them absolutely constant, and my having pointed out that, at the ages between 60 and 100, they had different values, proves what my views with respect to those elements were; and I further observe, that my subsequent researches, hinted at in the paper I wrote for the Congress, and further to be explained in the paper which I am about writing, show the curious nature of the variability of those elements, and explain by what manner they put on the appearance, during long periods, of constancy. And the consequence of such slow variability of the elements which may be considered, for long periods, constant, together with the law of their variability, give the command of estimations with accuracy of most complicated cases of intricacy, which will be found useful. But by my improved statement of the law of mortality, by the methods which will be pointed out in the paper I am about to present, when sufficiently advanced, to the Royal Society, a great variety of complicated cases of contingencies will be very easily grappled with.

The tables of two joint lives, published by the Society of Useful Knowledge, are extremely useful. The famous tables of Barrett, on the Northampton Tables of Mortality of every combination of three lives, are too voluminous and too rare (if at all to be met with) to be generally useful, even where they apply; but my method, hinted at in the paper I wrote for the Congress, and which I hope shortly to lay before the Royal Society, will apply to any number of lives, whether they be all subject to the same law or to different given laws, pertaining to different individuals of the whole and through a very wide complication of conditions and contingencies. I therefore flatter myself, should my health last to enable me to proceed with, at least part of, my subject, the paper will be well received.

I wish you, my dear Sir, to present this letter to the Institute, as I wish in it to notify my thanks to Mr. Sprague, whom I have not the honour of personally knowing, for his kind and able paper in vindication of my claim to be the sole independent publisher of a theorem which, I am gratified to say, appears of importance to the scientific world — I say, the sole independent publisher, though the fact of my being the first independent discoverer has not been denied me, even by the gentleman who claims to be the second discoverer, because that claim, should it be ever proved to be a just one, would not interfere with me.

I am, my dear Sir,

Yours, with regard,

To H. W. PORTER, ESQ.   BENJ.ᴺ GOMPERTZ.

AN APPROXIMATE EXPRESSION FOR THE VALUE OF AN ASSURANCE, LIFE AGAINST LIFE.

To the Editor of the Assurance Magazine.

Sir,—The following short investigation may, perhaps, be useful to some of your readers. It will be observed that the resulting formula is much more simple than the ordinary one for the single premium for a contingent assurance; and it has been found to be sufficiently accurate for practical purposes.

PROBLEM.

To find the present value of £1 to be received at the end of the year in which a life aged x may fail, provided that such event happen during the lifetime of another, aged y; the chance of both dying in the same year being neglected.

$$\text{Let } S = \text{value required;}$$

$$\text{then } S = \Sigma v^n (p_{x,\;n-1} - p_{x,\;n}) p_{y,\;n}$$

$$= \Sigma v^n (p_{x,\;n-1}\, p_{y,\;n} - p_{xy,\;n});$$

$$\text{but } \Sigma v^n p_{xy,\;n} = a_{xy}$$

$$\text{and } \Sigma v^n p_{x,\;n-1}\, p_{y,\;n} = \frac{l_x \cdot l_{y+1}}{l_x \cdot l_y} v + \frac{l_{x+1} \cdot l_{y+2}}{l_x \cdot l_y} v^2 + \frac{l_{x+2} \cdot l_{y+3}}{l_x \cdot l_y} v^3 + \&c.$$

$$= \frac{l_{y+1}}{l_y} v \left(1 + \frac{l_{x+1} \cdot l_{y+2}}{l_x \cdot l_{y+1}} v + \frac{l_{x+2} \cdot l_{y+3}}{l_x \cdot l_{y+1}} v^2 + \&c. \right)$$

$$= v p_y (1 + a_{x,\;y+1})$$

$$\therefore S = v p_y (1 + a_{x,\;y+1}) - a_{xy}.$$

I am, Sir,

Yours truly,

Equity and Law Life Office,   ARTHUR H. BAILEY.
13th March, 1861.

INSTITUTE OF ACTUARIES.

PROCEEDINGS OF THE INSTITUTE.

First Ordinary Meeting, Session 1860-61.—Monday, 26th November, 1860.

CHARLES JELLICOE, President, in the Chair.

The minutes of the last ordinary meeting were read and confirmed.
The Secretary announced various donations to the library.

Messrs. Alexander Burnett, George Thomas Ruck, James Stark, Jun., and Mark Symons, duly nominated at the last ordinary meeting, were unanimously elected Associates of the Institute.

Mr. J. Hill Williams read a paper—"On the theory of probabilities," by Robert Campbell, Esq., M.A.

Thanks were voted to Mr. Williams and to Mr. Campbell, and the meeting adjourned to the 31st December, 1860.

Second Ordinary Meeting, Session 1860-61.—*Monday,* 31*st December,* 1860.

CHARLES JELLICOE, President, in the Chair.

The minutes of the last ordinary meeting were read and confirmed.

The Secretary announced various donations to the library.

The undermentioned gentlemen, duly nominated at the last ordinary meeting, were unanimously elected members of the Institute :—

Official Associate—Henry D. Davenport, Esq.

Associates.

Mr. R. Clarke.	Mr. C. R. Saunders.
„ W. R. D. Gilbert.	„ T. Y. Strachan.
„ A. W. Mackenzie.	„ C. J. Wilkins.

The Chairman announced that out of eleven candidates for the matriculation examination (1860), nine passed, in the following order of merit :—

	Marks.		Marks.
Mr. C. R. Saunders	440	Mr. James Stark	332
„ R. P. Hardy	364	„ C. J. Wilkins	293
„ J. R. Knowles	344	„ T. Y. Strachan	278
„ Fredk. Harper	335	„ Jas. Henderson	271
„ H. W. Manly	331		

Mr. Hodge then read a paper—"On the stability of results based on average calculations," by Robert Campbell, Esq., M.A.

Thanks were voted to Mr. Campbell and Mr. Hodge, and the meeting adjourned to Monday, 28th January, 1861.

Third Ordinary Meeting, Session 1860-61.—*Monday,* 28*th January,* 1861.

CHARLES JELLICOE, President, in the Chair.

The minutes of the last ordinary meeting were read and confirmed.

The Secretary announced various donations to the library.

The undermentioned gentlemen, duly nominated at the last ordinary meeting, were elected members of the Institute :—

Official Associate—Frank McGedy, Esq.

Associates.

Mr. F. Harper.	Mr. J. R. Knowles.
„ Jas. Henderson.	„ Henry Wm. Manly.

Mr. Tucker read a paper—"On the rates of premium required to provide for certain periodical returns to the assured."

Thanks having been voted to Mr. Tucker, the meeting adjourned to Monday, 25th February, 1861.

Fourth Ordinary Meeting, Session 1860-61.—*Monday,* 25*th February,* 1861.

CHARLES JELLICOE, President, in the Chair.

Read and confirmed the minutes of last ordinary meeting.

Various donations to the library were announced.

The undermentioned gentlemen, duly nominated at the last ordinary meeting, were elected members of the Institute, viz. :—

Fellow—W. S. B. Woolhouse, Esq.

Associates.

Mr. C. F. Haycraft. | Mr. S. J. Shrubb.

Mr. Sprague read a paper " On Mr. Gompertz's law of mortality."

Thanks having been given to Mr. Sprague, the meeting adjourned to Monday, the 25th March, 1861.

ANNUAL EXAMINATIONS.

The members of the Institute will, no doubt, recollect that early in the last year some alterations were suggested by the examiners in the syllabus published by the Institute; and that their suggestions being approved by the Council, the amended syllabus was adopted and ordered to be thereafter acted upon.

The changes then made have necessarily affected, to a certain extent, the character of the questions; and as it is obviously desirable that the way in which these changes have operated should be understood, we now publish the questions given on the last occasion for the second and third years' examinations — those for the first year being but very slightly influenced by them. It will be seen that the third year's questions are those mainly affected by the alterations.

SECOND YEAR'S EXAMINATION, 1860.

1. Assuming the formula for $\log_e (1 + x)$, prove that

$$\log_e 3 = 1 + \frac{1}{3} \cdot \frac{1}{4} + \frac{1}{5} \cdot \frac{1}{4^2} + \frac{1}{7} \cdot \frac{1}{4^3} + \dots$$

$$\tfrac{1}{2} \log_e 10 = \log_e 3 + \frac{1}{19} + \frac{1}{3} \cdot \frac{1}{19^3} + \frac{1}{5} \cdot \frac{1}{19^5} + \dots$$

2. Explain what is meant by the modulus of a system of logarithms. By means of the formulæ in question, show that

$$\log_e 3 \ = 1 \cdot 098612$$
$$\log_e 10 = 2 \cdot 302585$$

Also show that the modulus of the common system of logarithms is ·434294.

3. When an event can happen in more ways than one, show that the probability of its happening is equal to the sum of the probabilities in respect of the different ways.

4. Determine (1) the probability that two persons now aged x and y respectively will both die in the nth year from the present time; (2) the probability that one or both of them will die in that year.

5. Find the present value of a sum of money due at the end of any time, assuming the operation of compound interest.

Prove that the value of £1 due at the end of $p + q$ years = value of £1 due in p years × value of £1 due in q years.

6. Describe a practical method of calculating a table of the value of £1 due at the end of any number of years from 1 to 100; and show how the

results may be employed to calculate a table of the values of annuities certain.

Ex.—The value of £1 due at the end of 10 years, at 4 per cent., is ·67556417: deduce the first ten terms of the value of an annuity for any number of years.

7. Explain what is meant by a table of mortality, and state the different methods in which the facts embodied in it may be exhibited.

8· Give some account of the origin and relative merits of the tables of mortality known as the Carlisle, Equitable, Experience, and English Life Tables.

9. State accurately what is meant by the expectation (or mean duration) of life at any age according to a given table of mortality.

Prove the formula for calculating the chance of living a year at any age from the expectation—

$$p_x = \frac{e_x - \frac{1}{2}}{e_{x+1} + \frac{1}{2}}.$$

10. Prove that if $e_{x-1} = e_x = e_{x+1}$, then will $p_{x-1} = p_x$; but that if $p_{x-1} = p_x = p_{x+1}$, e_{x-1} will not be equal to e_x unless $e_{x+2} = \frac{1}{2} \cdot \frac{1 + p_x}{1 - p_x}.$:

11. The value of an annuity on the life of a person of a given age is frequently supposed to be equal to that of an annuity certain for a number of years equal to the mean duration of life at that age. Explain why this is not the case, and state upon what hypothesis it would be true.

12. If B represents the present value of a benefit of £1 upon a given life (x), B_1 the same upon a life one year older $(x+1)$, p the probability of a payment of B being received in the first year, and Π the probability of (x) surviving a year, prove that

$$\log B = \log v\, \Pi + \log \left(\frac{p}{\Pi} + B_1 \right).$$

13. Describe the process of calculating a table of annuities by means of the formula $a_{m-1} = (1 + a_m) p_{m-1,1} . r$; and show that this formula is a particular case of the one in the last question.

14. Explain fully the construction and use of columns D, N, and M, in the columnar method of calculating the values of annuities and assurances; also state the superior advantages this method possesses over the old one.

15. Prove that the value of an annuity of £1 during the joint lives of x and y, and for t years afterwards, should x survive so long, is

$$a_x - \frac{D_{x+t}}{D_x} (a_{x+t} - a_{x+t.y}).$$

16. Give an expression for the annual premium for a contingent annuity to commence at the death of A and to continue as long as either B or C is living.

17. Find, in a convenient form for computation, the single and annual premiums for an annuity to commence at the death of y and continue payable during the remainder of x's life, but to be payable only if y dies within t years.

18. Prove that the single premium for an assurance payable on a life now aged n years attaining the age of $n+t$, or dying previously, may be

represented by the formula (Jones's notation), $\dfrac{1-ia_{\overline{n_{i-1}}|}}{1+i}$; and show that

this formula is equivalent to $\dfrac{M_n-M_{n+t}+D_{n+t}}{D_n}$.

19. Give an expression for the single premium for an assurance on the life of A provided he die after B.

20. Prove that $A_{(x\,y)_{\overline{1}|}^1 t}=r^t . p_{(x\,y)t} . A_{(x+t.y+t)}^1$.

21. Determine the value of an assurance on a life of 30, payable at the age of 60 or previous death, commencing at £a, increased by £p at the end of 5 and 10 years respectively, and thereafter increasing by £q *per* annum.

22. Prove the formulæ for the value of a policy of £1—

(1) $1-\dfrac{1+a_{m+n}}{1+a_m}$,   (2) $(P_{m+n}-P_m)(1+a_{m+n})$,   (3) $1-(d+P_m)(1+a_{m+n})$.

Show that the same forms apply when the policy is on the joint duration of two lives.

23. Find the value of an annuity on two successive lives, x and y, of which the second is to be nominated at the death of the first, and is supposed to be then y years of age. Also show that if I represent the value of a perpetuity of £1, the value of the annuity will be equal to $I-(1+I)A_xA_y$.

24. Find the single premium for an annuity to x after the death of y, with the condition that the premium is to be returned if x die before y.

25. A, aged x years, is entitled to the interest of £1 for life. If A die within t years, the interest is payable to B, or his representatives, till the expiration of the t years; when C, if living, is entitled to the capital. If, however, C, now aged y years, die either before A or within the t years, the capital reverts to B or his representatives. Determine the values of B's and of C's interests.

THIRD YEAR'S EXAMINATION, 1860.

1. Describe Mr. Gompertz's method of graduating tables of mortality, and give his formula.

2. Describe the method proposed by Mr. Milne for that purpose.

3. Give a brief description of the other methods which have been devised with the like object.

4. Say which of these is, in your opinion, the most effective; and give your reason for thinking it so.

5. In what way do you consider that the surplus of an Assurance Company can be most equitably distributed amongst the assured?

6. When an assurance is effected by one person on the life of another, state the effect of an untrue statement by the life assured or the referees upon the validity of the policy.

7. Explain the difference in the constitution—(1) of Companies formed under the Act 7th and 8th of Victoria, c. 110; (2) of Companies formed before the passing of that Act, and having a special Act of Parliament; (3) of Companies so formed without a special Act; giving instances of each among the existing Companies.

8. How does this difference affect proceedings in the courts of law?

may be employed to calculate a table of the values of annuities

Ex.—The value of £1 due at the end of 10 years, at 4 per cent., ·676 .... hence the first ten terms of the value of an annuity for any number of years.

7. Explain what is meant by a table of mortality, and state the different methods in which the facts embodied in it may be exhibited.

8. Give some account of the origin and relative merits of the tables of mortality known as the Carlisle, Equitable, Experience, and English Tables.

9. State accurately what is meant by the expectation (or mean duration) of life, according to a given table of mortality.

10. Find a formula for calculating the **chance** of living a year at any time ....

$$p_x = \frac{e_x - \frac{1}{2}}{e_{x+1} + \frac{1}{2}}.$$

Find that if $e_x = e_{x+1}$, then will $p_{x-1} = p_x$; but that if $p_{x-1} = p_x$, then e_x, e_{x+1} will not be equal to e_x unless $e_{x+2} = \frac{1}{2} \cdot \frac{1 + p_x}{1 - p_x}$.

11. The value of an annuity on the life of a person of a given age is frequently assumed to be equal to that of an annuity certain for a number of years equal to the mean duration of life at that age. Explain why this is not the case, and state upon what hypothesis it would be true.

12. If B represents the present value of a benefit of £1 upon a given life (x), B_1 the same upon a life one **year older** $(x+1)$, p the probability of a payment of II being received in the first year, and II the probability of surviving a year, prove that

$$\log B = \log e \, II + \log \left(\frac{p}{II} + B_1 \right).$$

13. Describe the process of calculating a table of annuities by means the formula $a_{n-1} = (1 + a_n)p_{n-1, 1-r}$; and **show** that this formula is a particular case of the one in the last question.

14. Explain fully the construction and use of columns D, N, and M, in the columnar method of calculating the values of annuities and assurances; also mention the superior advantages this method possesses over the old one.

15. Prove that the value of an annuity of £1 during the joint lives of ... for t years afterwards, should x survive so long, is

$$a_x - \frac{D_{x+t}}{D_x}(a_{x+t} - a_{x+t,y}).$$

16. Give an expression for the annual premium for a contingent annuity commencing at the death of A and to continue as long as either B or C is living.

17. Find, in a convenient form for computation, the single and annual premium for an annuity to commence at the death of y and continue payable during the remainder of x's life, but to be payable only if y dies within t years.

18. Prove that the single premium for an assurance ... now aged ... years attaining the age of $n + t$, or dying ...

represented by the formula (Jones's notation), $\dfrac{1-i\alpha}{1+}$; and show that

this formula is equivalent to $\dfrac{M_n-M_{n+t}+D_{n+t}}{D_n}$.

19. Give an expression for the single premium for a assurance on the life of A provided he die after B.

20. Prove that $A_{(x\,y)\underset{1}{\overline{}}|t} = r^t . p_{(x\,y)t} . A_{(x+t\,y+t)\underset{1}{}}$.

21. Determine the value of an assurance on a life of J, payable at the age of 60 or previous death, commencing at £a, increasd by £p at the end of 5 and 10 years respectively, and thereafter incrsing by £q *per* annum.

22. Prove the formulæ for the value of a policy of £1—

(1) $1-\dfrac{1+a_{m+n}}{1+a_m}$, (2) $(P_{m+n}-P_m)(1+a_{m+n})$, (3) $1-(d-P_m)(1+a_{m+n})$.

Show that the same forms apply when the policy is o the joint duration of two lives.

23. Find the value of an annuity on two successive ves, x and y, of which the second is to be nominated at the death of the rst, and is supposed to be then y years of age. Also show that if I resent the value of a perpetuity of £1, the value of the annuity w. be equal to $I-(1+I)A_x A_y$.

24. Find the single premium for an annuity to x afte the death of y, with the condition that the premium is to be returned if ae before y.

25. A, aged x years, is entitled to the interest of £1 fr life. If A die within t years, the interest is payable to B, or his reprentatives, till the expiration of the t years; when C, if living, is entitled to the capital. If, however, C, now aged y years, die either before A or wnin the t years, the capital reverts to B or his representatives. Determine e values of B's and of C's interests.

THIRD YEAR'S EXAMINATION, 1860.

1. Describe Mr. Gompertz's method of graduating tabs of mortality, and give his formula.

2. Describe the method proposed by Mr. Milne for tha purpose.

3. Give a brief description of the other methods wich have been devised with the like object.

4. Say which of these is, in your opinion, the most efftive; and give your reason for thinking it so.

5. In what way do you consider that the surplus of an ssurance Company can be most equitably distributed amongst the assured

6. When an assurance is effected by one person on theife of another, state the effect of an untrue statement by the life assuredr the referees upon the validity of the policy.

7. Explain the difference in the constitution—(1) of Copanies formed the ' nd 8th 'oria, c. 110; (2) of Copanies formed tl f thr having a special Act of Parliament; (forr special Act; giving itances of each or is dings in the cots of law?

9. What qualifications are mainly necessary in a deed assigning a policy of assurance?

10. What legal remedy has an assignee, should a Company decline to pay under his policy?

11. Explain the process of forming a table of mortality from the burial registers of a stationary population, and the means of correcting for increase of population.

12. Point out the errors of the Northampton Table, and show how they arose from imperfection in the means of observation.

13. Describe the progress of an Insurance Company which receives, at the beginning of each year, a fixed number of new members of a given age, and in which the members only cease to be such by death.

14. Find the number of members at the end of any given number of years, and determine how long it will be before the number of deaths in a year is equal to the number of new members admitted.

15. What is the population of the United Kingdom at the present time, and what does the annual taxation amount to per head?

16. What is the effect at home and abroad of a depreciation in the value of the metallic currency of a country?

17. What consequences arise from the quantity of the coinage in circulation being excessive?

18. In what way, if at all, can the Bank of England influence the amount of circulation?

19. To what extent is the gold in the issue department of the Bank available, at any given time, in the banking department?

20. What is the precise nature of the security which a Bank of England note constitutes?

21. Describe the manner in which you would construct the accounts, or "open the books," of a Life Assurance Company commencing operations.

22. Explain the methods of determining the market value of a contingent reversion.

23. A wishes to buy an annuity for his wife, to commence at his decease; on what terms, as regards rates of interest and mortality, would he be likely to get it, and what regulates the terms? Both are in good health.

24. A has been presented with a policy of assurance on his own life, on which all the premiums have been paid; he wants to surrender this, in consideration of an annuity while he lives. In what way would you find the annuity to be given?

25. B is entitled to an estate at the death of his father, if he survive him. He has assured his life, against that of his father, for half the value of the estate, by a single payment; but he has to pay an annuity whilst he and his father are both living. Describe the way in which you would arrive at the market value of B's interest, and say what rates of interest and mortality you would use in the determination.

THE

ASSURANCE MAGAZINE,

AND

JOURNAL

OF THE

INSTITUTE OF ACTUARIES.

On the Rate of Mortality prevailing amongst the Families of the Peerage during the 19th *Century. By* ARTHUR HUTCHESON BAILEY, *Actuary of the Equity and Law Assurance Society, and Fellow of the Institute of Actuaries; and* ARCHIBALD DAY, *Actuary of the London and Provincial Law Assurance Society, and Fellow of the Institute of Actuaries.*

[Read before the Institute the 29th April, 1861, and printed by order of the Council.]

IN a note in the introduction to Milne's *Treatise on Annuities,* the author remarks—" There can, I think, be no doubt but that the mortality is greater among the higher than the middle classes of society. They form too small a proportion of the population to have any sensible effect here; but it would be of importance to the Life Offices to determine the law of mortality among them." Since the publication of this work, forty-six years ago, some attempts have been made to test the accuracy of this assertion, and to supply the desideratum; but none with which we are acquainted are by any means conclusive.

In a tract, privately printed in the year 1832, entitled, *Observations on the Mortalities among the Members of the British Peerage,* by the late Mr. George Farren, the author investigates the average duration of the enjoyment of the title by each peer, from observations made in 740 cases. Out of these, the ages are stated to have been recorded in 447 instances,—of which 288 succeeded to the title between the ages of 10 and 40; from which the conclusion was deduced (and this is the only result stated), that, at the age of 25, the mean duration of life of the peers is very nearly 32 years, which corresponds with the mean duration by the Carlisle Table, at the age eight years older. The process adopted is not stated; but, independently of this and other objections, it will probably be considered that the number of facts observed was too small to warrant any general conclusions on the subject.

Mr. Edmonds contributed to the *Lancet* of the 10th February, 1838, and the 9th March, 1839, two papers—on the "Duration of Life in the English Peerage," and on the "Lineage of English Peers." His observations—extending, apparently, over a long interval of time—were made on 675 peers (32, whose deaths were violent or accidental, having been excluded), during the period of their possession of the title only. These 675 peers were the representatives of 109 titles, the first and last peer in the line of succession having been omitted. In 243 cases the ages were not stated, and this defect was supplied by an approximation made by observing the recorded dates of birth of other peers in the same line of descent. The results are given in three tables, from which it appears that the mortality among the peers is very much in excess of that of the Carlisle Table, but corresponds pretty nearly with a theoretical table constructed by Mr. Edmonds himself, which he designates "City Mortality," according to which the mortality at every age is 50 per cent. greater than in the Carlisle Table. The process by which these results are arrived at is nowhere stated, so that any verification of them is impossible. The number of facts seems to us altogether insufficient, and has been needlessly diminished by the exclusion of the first peer and the existing peer of every title, apparently for no other reason than that they were not required for another investigation which Mr. Edmonds had then in progress. The proportion of cases in which the ages were not recorded is very large—36 per cent. of the whole number; and the hypothetical method adopted for supplying this defect is, we think, open to considerable objection, especially as Mr. Edmonds seems to imply that, had he carried out his original intention of deducing the mortality from those peers only whose ages were recorded, the results would have been different. Also, the probable effect of confining the observations to the period of the occupancy of the title only, will have been, if hereditary tendencies had their usual influence, to exaggerate the mortality of the younger ages, because the peers who succeeded to the title when young must, as a general rule, have been the sons of short-lived parents, and they would not enter into combination for several years with the sons of the long-lived parents, as the latter, for the most part, would not have succeeded to the title until older ages had been attained. Moreover, Mr. Edmonds's plan excludes altogether female and infant mortality, and takes but little account of that of the periods of childhood and youth. For these reasons, we cannot consider his investigations satisfactory or sufficient.

More recently, Dr. Guy made some observations on a much greater number of facts, the results of which he embodied in two papers read before the Statistical Society, and published in their Journal for March, 1845, and March, 1846. Dr. Guy's process is very clearly stated, and is sufficiently simple. He extracted from a Peerage and Baronetage in his possession the ages at death of 2,291 male lives who died above the age of 21 in a period of time extending from the 13th century to the year 1830, and having obtained the number of deaths at each age, a table of the mean duration of life was calculated from these materials. The result—somewhat unexpected, Dr. Guy says—being, that the mean duration of life among these classes is nowhere greater, and at all ages under 70 is materially less, than among the general male population of the country. Entertaining, as we do, a very sincere respect for Dr. Guy's scientific attainments, we cannot, at the same time, avoid remarking that his investigations upon the subject of mortality are unsatisfactory ; all his observations having been made on the ages at death only, without any regard to the numbers living. This is not a fitting opportunity, nor can it be necessary here, to enlarge on this point. That observations made from deaths alone will always give erroneous results of the mortality, except in the case of a stationary population, is well known to all who have studied the subject : those who have not, may consult Milne's article in the *Encyclopædia Britannica* on "Mortality," and the 5th and 6th Annual Reports, with the Appendices, of the Registrar-General.

The results obtained by Dr. Guy have frequently been quoted both in newspapers and scientific publications—among others, by Mr. Neison, in his *Contributions to Vital Statistics*—and reasons attempted to be assigned for the conclusions arrived at. Now, as it is certain that, whether Dr. Guy's conclusions are correct or not, they cannot be deduced from the facts which he collected, and not being satisfied with the investigations of his predecessors in the same path, we resolved to undertake the not inconsiderable labour of determining the mortality prevailing among the families of the peerage by proper technical methods, tracing each case through every year of the period of observation, and comparing the number of deaths with the number of living in each year of age.

Before commencing this task, however, the question arose—what reliance could be placed upon the accuracy of the peerage books? Having made some inquiries on this matter, we are informed, on good authority, that the process of their compilation is some such as the following :—The editors glean from the newspapers, from time to time, occurrences in the way of births, deaths, and marriages, and make the necessary alterations in the current editions of their works. A proof is then sent annually to each peer, of the portion relating to his own family, with a request that he will be good enough to correct and return it. And although the proofs are not all returned annually, the editors express the utmost confidence that, except in a few instances, communications are always made to them whenever corrections are required, and that the several dates are accurately recorded. Having satisfied ourselves upon this important point, the next matter was to select which of the different Peerages should be made use of. The first at hand, that of Burke, was quickly discarded. With a chivalrous spirit, worthy of the Ulster King-at-Arms, he invariably omits the dates of birth of the ladies; and his work, therefore, although probably well adapted for general circulation, was altogether unsuited for our purpose. We then turned to *Lodge's Peerage;* here the facts we required were fully and clearly given, even the still-born births being accurately recorded, and the only objection was, that it did not extend sufficiently far back in point of time. To remedy this, recourse was had to an older Peerage by Debrett; and from these two sources, with occasional reference to Burke, in cases of doubt or discrepancy, all the necessary information was obtained.

We had then to consider what should be the limits of our investigation. To have extended our researches back to distant generations would have been objectionable, because, as an impression commonly prevails that the rate of mortality has progressively improved, comparisons with other tables derived from modern data might have been considered inadmissible. Also, had we followed Mr. Farren and Mr. Edmonds, in restricting the observations to peers only, the number of facts would, in our judgment, have been insufficient, independently of other objections. On the other hand, had we included all the collateral branches, the distinctive features of the class would have been lost. We therefore determined, that the observations should be limited in point of time to the present century, and should be made upon peers, sons and daughters of peers, and sons and daughters of the eldest sons of peers.

The observations commence with the anniversary of the date of birth in the year 1800 of those who were born in the last century, and with the actual date of birth for the remainder, and terminate on the 31st December, 1855; the only assumption introduced being, that when, as it occasionally happened, the year only of birth or death was given, the day being omitted, this latter was taken to be the 30th June. Those cases were altogether excluded in which the year of birth was not recorded. The number of facts obtained was as follows, viz. :—

	Males.	Females.	Both Sexes.
Deaths	1,938	1,253	3,191
Existing, 31st Dec., 1855 ..	2,283	1,999	4,282
Totals..........	4,221	3,252	7,473

When it is considered that the data from which the Carlisle Table of Mortality, which is probably the one in most general use, was derived, were two enumerations, made at an interval of about nine years, of 7,677 persons living on the first occasion, and 8,677 on the second; that the number of deaths was 1,840; and the ages obtained merely from information voluntarily given, we hope that in regard to time, numbers, and accuracy, the present materials may possess some value.

The extracts having been made separately for each sex, and afterwards combined, the results will be found in Tables I., II., and III., which are sufficiently explained by the headings of the columns. The only remarks needed being, that the number exposed to the risk was obtained by subtracting from the number who completed the age half of the number existing on the 31st December, 1855—that, as was unavoidable, the numbers at the oldest ages were somewhat arbitrarily dealt with—and that the mean duration of life was computed by means of the equation—

$$\text{Log. } e_x = \text{log. } p_x + \text{log. } (1 + e_{x+1}),$$

Where e_x represents the curtate mean duration at the age x. All the processes, both of making the extracts and the computations, were performed by each of us independently of the other.

As was to have been expected, the column representing the annual mortality exhibits numerous irregularities, but we have purposely abstained from applying any method of graduation. Indispensable as this process is for many applications of tables of mortality, and great as has been the ingenuity expended upon it, the process itself can only be regarded as a necessary evil; and, without going so far as to say, with Professor De Morgan, that " the practice cannot be too strongly condemned," the force of his remark, that, " the tables thereby lose some of their value as representations of physical facts," is undeniable. And as the process was neither necessary nor desirable for one of the chief objects of this investigation—the comparison of the results with those of other tables—we have preferred to leave the observations unadjusted.

Examples of these results, compared with those of other tables of authority, will be found in Tables IV., V., and VI. The observations selected for comparison have been—of male lives exclusively, the English Life Table, representing the mortality of the population in general; the experience of the Equitable Society (chiefly males); that of the Government Annuitants, from Mr. Alexander Finlaison's recent Report; and the mortality of healthy districts, from a paper by Dr. Farr, lately read before the Royal Society. The latter, having been derived from the records of 63 districts of the country where the average annual mortality in 1,000 did not exceed 17, the general average being 22, is considered by Dr. Farr to be " the nearest approximation we can obtain to a table representing the human race in the normal state." For female lives exclusively, we have selected for comparison the corresponding tables of the English Life Table, the Government Annuitants, and the healthy districts; and for both sexes combined, the Carlisle Table, and the general table of the healthy districts.

Referring, first, to male lives only, it will be found that the average mean duration of life among the families of the peerage is, at all ages under 73, greater than among the general population of the country, greater even than among the selected lives of the Equitable Society (with the unimportant exception of the period from 15 to 21, where the number of cases in that Society was very small) ; greater, at all ages under 62, than among the Government annuitants; and, throughout, approaches pretty nearly to the standard table of the mortality of the healthy districts. At the older ages, it very nearly coincides both with the English Life Table and the Equitable experience, but is somewhat less than that of the Government annuitants. Looking to particular periods of life, it will be observed that the advantage in favour of the families of the peerage is most remarkable in infancy and childhood—the mortality under the age of 10 years being little more than one-third of that of the general population. In the next decade, the two tables nearly coincide; but at the succeeding period, from 20 to 29, a very singular anomaly occurs. There the mortality among the families of the peerage is not only in excess of that of the English and Equitable Tables, but, contrary to our previous notions, is also materially greater than in the next decennial interval. A similar anomaly occurs in the Government annuitants, and also in the Society of Friends.* At all other ages, up to 80, the advantage is with the aristocratic class.

It is evident, therefore, that Dr. Guy's conclusions on this subject are erroneous; and, to make this more clear, we have thought it worth while to compute a table of the mean duration of life, deduced from the deaths alone, in the present observations, and have placed some examples of it in juxtaposition with the corresponding results deduced by the correct method, and also by Dr. Guy.

Males—Peerage Families.

Age.	Mean Duration of Life.	Do. deduced from the Deaths only.	
		Present Observations.	Dr. Guy.
20	41·46	36·42	38·48
30	35·51	31·63	30·88
40	28·33	25·36	24·45
50	21·40	19·38	17·92
60	14·56	13·27	12·57
70	8·77	8·03	8·15
80	4·58	4·20	5·09

The above table is, we think, deserving of some attention—partly, because the erroneous method of observation is very common with the medical profession; and partly, because the results of such

---

* In this Society, the annual mortality per cent. among the males has been found to be at the ages 20–29, ·881; and at the ages 30–39, ·782. Our information on the subject is derived from a most complete and interesting paper " On the Vital Statistics of the Society of Friends," by Joseph John Fox, read before the Statistical Society, 21st December, 1858.

observations are received with implicit confidence in some quarters where it could least have been expected.

Turning next to the tables for female life exclusively, it will be observed that the average mean duration of life among the families of the peerage is, throughout, materially greater than with the general population, coincides in a very remarkable manner, up to the age of 55, with that of the Government annuitants, and also of the inhabitants of the healthy districts, but surpasses them both for the remainder of life—indicating a more favourable mortality at the older ages than any table whatever with which we are acquainted. It will be seen that, as with the males, the contrast with the general population is most marked under the age of 10, but that in the next decade the mortality is somewhat in excess; the abnormal feature noticed in the males thus appearing among the females, but not to so great an extent, and occurring at a somewhat earlier period of life. The singularly favourable mortality above the age of 70 will not escape attention.

We believe that no previous attempt has been made to investigate the mortality prevailing among females of the higher classes in this country, and are not without hope that the present one may, owing to the trustworthiness of the data, help to elucidate the subject of female mortality generally. The known difficulty of obtaining accurate information on so delicate a subject as ladies' ages has, undoubtedly, tended to throw some degree of suspicion upon all results obtained from any general records of the female population. Even the compilers of the Census of 1851, laudably eager as they were to break a lance in defence of the veracity of their countrywomen, were reluctantly compelled to admit that this part of the returns required some awkward and troublesome corrections, and cited the case of an eminent French statist, who, after many persevering but fruitless attempts, had abandoned in despair a philosophical inquiry which had for its object to determine the ages of his wife and his cook.

Table VII. will illustrate the comparative mortality of the sexes. As this is a subject of some interest, and as the evidence respecting it is conflicting, we have thought it deserving of some further examination.

The superior longevity of the female sex having long been well known, one of the most striking results of the combined experience of the Assurance Offices was that the mortality amongst assured females was greater than amongst assured males. Several explanations of this apparent anomaly have been offered, none of them very satisfactory, and some altogether absurd; but, after a careful consideration of the matter, we cannot help thinking that the anomaly does not really exist at all. According to Dr. Farr's Tables, the mortality in the period of infancy is greater among boys than girls, from 2 to 40 years of age the mortality of females is in excess, while for the remainder of life the females have the advantage over the males in a marked degree. And we find that the remarkable tenacity of female life in old age so far outweighs the greater mortality of the sex shortly before and during the childbearing period, as to give to the female sex an average mean duration of life greater than that of the male throughout the whole of the table. These results the present observations confirm in every respect—they exhibit the superior vitality of females in infancy, their remarkable tenacity of life in old age, and their greater mortality in youth and the prime of life; for, notwithstanding the abnormal mortality of the males between the ages of 20 and 30, it will be found that, taking the entire period from 5 to 45 years of age, the female mortality is in excess. Bearing these peculiarities in mind, it was certainly somewhat startling to find, on referring to the Combined Experience Tables, that the female mortality in the decade 70–79 was both so great absolutely, and so much in excess of the male, especially as between the ages of 45 and 70 the female mortality is less. But on looking more closely, and remarking that the objection often urged, that these tables represent the experience not of lives but of policies, is serious when the numbers are small, though not, perhaps, very important when they are large, we discover that the whole number of claims above the age of 70, under policies on female lives, in all the combined Offices, was exactly 60, and only 477 at all ages. And the actuaries themselves assign the paucity of the numbers as a reason for the apparent excess of the female mortality at the older ages, admitting that no importance is to be attached to this result. But as the facts were avowedly insufficient to bring out the peculiar tenacity of life of the female sex in old age, the result is perfectly consistent with other observations. Instead, therefore, of asserting that the "mortality amongst assured females, taking all ages together, is greater than amongst assured males," we think a more accurate statement would have been, that the experience of the Offices confirms the result of other observations, which show the greater mortality of females during the childbearing period, and affording little or no information for the periods of infancy and old age, it did not possess the means of exhibiting the general mortality of the sex. In corroboration of this may be adduced the valuable information furnished by the experience of the Eagle Office, where the number of female deaths exceeded those of all the combined Offices. In the experience of the Eagle, the female mortality was lower than the male, taking all ages together; while, at the same time, under the age of 42, it was considerably in excess. And it must not be forgotten that the average period of observation of each life, although greater than with the combined Offices, did not exceed 8½ years; so that there is reason to believe that, if the observations were continued to the present time, the greater general vitality of the female sex would be still more apparent.

On the other hand, the late Mr. Finlaison, whose experience and authority on this subject were very great, stoutly contended that "the fact is undoubtedly certain, that the mortality of the female sex, at every period of life, is less than that of the male sex at the same ages, excepting only in infancy," and supported this assertion by numerous tables deduced from observations made on the nominees of certain tontines, and also on the Government annuitants from 1808 to 1825. These observations have been continued to the 31st December, 1850, by Mr. Alexander Finlaison, and the results have been recently published in a report by that gentleman. The later results confirming, in this particular, the former observations, Mr. Alexander Finlaison appears to consider that his father's views are now unimpeachable, and form a standard by which the accuracy of other tables may be tested. Applying this test to the English Life Table, he makes the following remarks:—"At the youthful and earlier adult ages, the mortality of the female is represented to be greater than that of the male. This conclusion is contrary to most previous experience. It is a result which is also contrary to nature. The sexes are not created in equal numbers. For every 20 females there are produced 21 males. But no fact is more thoroughly established, than that whenever the population is counted the females are present in considerably greater number. . . . Such a result could not take place, unless the stronger sex were subjected to a higher rate of mortality, and died off much faster than the females."

From these observations we must express our dissent. Dr. Farr's conclusion is not contrary to most previous experience. Not only in the English Life Table, but also in the observations on the inhabitants of the healthy districts, on assured lives, and on the Society of Friends, the mortality of the female in the youthful and earlier adult ages is found to be greater than that of the male;

and in *every* published table of mortality to which we have been able to refer, excepting only Mr. Finlaison's and the Swedish Tables, the mortality of the female is, at particular ages, in excess of that of the male. The peculiarity of Mr. Finlaison's results may, perhaps, be accounted for by the probability that a large proportion of the females in his observations are unmarried, and by the scantiness of his materials in middle life compared with their abundance at the older ages.

The English Life Table is not contrary to nature; for, concurrently with the greater mortality of the female at particular ages, we almost invariably observe a greater general mortality of the male. The two circumstances—the excess of male births and the greater general mortality of that sex—together cause that nearly uniform proportion of the sexes which successive enumerations disclose.

How far the long and elaborate attack on the English Life Table in Mr. Alexander Finlaison's report is consistent with the official etiquette usually observed by different departments of the State towards each other, we do not presume to determine. If the National Debt Office and the General Register Office are at variance, assuredly *non nostrum est tantas componere lites.* We only endeavour, amongst conflicting statements, to search for the truth; and the result of our inquiries on this particular subject— the comparative mortality of the sexes—has been to confirm the views of Dr. Farr, and not those of Mr. Finlaison. On the whole, if human life be divided into three great periods—of infancy, maturity, and old age—the weight of evidence is in favour of the general conclusion, that, at the two extremes, the mortality of the female sex is less, and at the intervening period greater, than that of the male; the probable after-lifetime being, at all ages, greater for the female.

The examples of the general table of mortality of the families of the peerage will suffice to dispel the previous views that have been propounded on the subject, and remove the erroneous ideas into which even so judicious and accomplished a writer as Milne has fallen, and which led Mr. Edmonds to make the unfounded assertion, that " the severest mortality is to be looked for in the poorest class of a city population, and in the highest class of the monied or nonlabouring portion of the community." It will be observed that the mean duration of life among the families of the peerage approaches nearly to that in Dr. Farr's standard table, and, with one slight exception, is throughout greater than in the

Carlisle Table. This exception occurs about the age of 80, and may be readily explained by the circumstance, that in the population of Carlisle, from which the mortality table was framed, the females were about 55 per cent. of the whole number—a much greater proportion than occurs in the general population. In our observations, on the contrary, the males outnumber the females— a greater number of cases in the latter sex having to be rejected owing to deficiency of dates. Had it not been for this circumstance, the results of the general table would have proved even more favourable.

The peculiar features of the mortality at different periods of life have already been sufficiently discussed.

One or two remarks on the application that may be made of the results of the present investigation, in the occupations in which most of us are engaged, may not be considered inappropriate. Adopting the common division of the different ranks of society— into upper, middle, and lower classes—it may be safely stated that the latter, although forming the great bulk of the community, have hardly any dealings with Life Offices. And we are inclined to think that the somewhat heterogeneous mass called the middle class does not resort to these Offices to the same extent, in proportion to its number, as the higher classes. It is, we are aware, a popular belief, that the extent of the practice of life assurance in this country affords strong evidence of the provident habits of the community, and that the success of the Offices in Great Britain is attributable to the greater degree of prudence and forethought prevailing here than among the nations of the Continent. But some of those who have had the most experience in the matter would probably be of opinion that the practice of life assurance affords quite as much evidence of improvident as of provident habits ; that the Offices obtain as many supporters from those who exceed as from those who live within their incomes ; and that their success is attributable, in no slight degree, to the extent to which the practice of making settlements of property prevails in this country, and to the consequent number and variety of life interests of a pecuniary nature arising therefrom. However this may be, there can be no doubt that those who are beneficially interested in these settlements, but, at the same time, have not usually much ready money at command (a description not unfrequently applicable to the class now under observation), are introduced to the Assurance Offices in considerable numbers. It cannot, therefore, be otherwise than satisfactory to find that the mortality prevailing among the class

in question is decidedly more favourable than any which the experience of the Offices has yet furnished.

The present investigation also indicates that the effect expected
to be produced by selection of lives is much exaggerated; at the
same time, it confirms an opinion occasionally expressed, and which
seems to be founded on experience, that the best lives are those
that are assured for large amounts.

It may, perhaps, be of some service in a department of our practice which urgently requires amendment—the system upon which
extra premiums for foreign residence are charged. The male lives
that have formed the subject of the present investigation have
been found to experience an unusually favourable mortality; yet
they enter the army and navy in large numbers, travel extensively,
and are certainly more exposed to what the Assurance Offices consider extra risks, than the middle classes. It would seem, therefore, to be no unfair inference, that differences of climate have
less effect on human mortality than differences of occupation and
position in life; and as the Offices do not attach much importance
to the latter—taking a butcher and a country clergyman on similar
terms—they might perhaps relax somewhat in their estimate of
the former. Considering the very unsatisfactory character of the
present practice in this respect, it might be worth consideration
whether any serious risk would be incurred by dispensing altogether,
in the majority of cases, with the existing restrictions on foreign
travelling and residence.

On the other hand, the results of this investigation afford some
suggestions for the exercise of caution. The most painful comparison presented by the present tables is the remarkable difference
in the mortality of the children when contrasted with the general
population. Now, as it may be tolerably safely assumed that all
children for whom endowments are purchased will be well cared
for, it would seem that both the Carlisle and English Tables are
unsafe data for the calculation of endowment premiums.

The exceptional mortality also of the period of early manhood,
confirmed as it is by the experience both of the Government annuitants and of the Society of Friends, indicates that those assurances
are not the most desirable that are effected under the age of
30 for terms of years, on an increasing scale of premium, or on
that most inconsistent and odious method called the "half-credit
system." The same circumstance will perhaps explain why, in the
Economic Office, it has been found that the highest rate of mortality
has been experienced on term assurances.

The information obtained upon the comparative mortality of the sexes leads to the conclusions, that tables of mortality for all purposes of life assurance should be derived from observations on male lives chiefly, or exclusively; that the greater vitality of the female will not justify any reduction of premium in contracts of assurance, because, on account of the large proportion of policies that are suffered to lapse, the greatest amount of risk will usually be incurred in middle life, at which particular period the female mortality is greater than the male. On the other hand, the distinction of sex is of serious moment in all contracts of annuities, immediate, deferred, and contingent; because in those cases the most important period is that of old age, where the distinction between the mortality of the sexes is most marked. The distinction is also important in another branch of our pursuits, where, we believe, it is frequently overlooked—we mean in the valuation of reversionary interests.

Finally, if this investigation should tend to encourage the belief that the mortality of each well-defined class has peculiar characteristics of its own, it must weaken the hold that the Carlisle Table has upon some of its votaries, who seem to consider that for all purposes, and under all circumstances, their favourite table is applicable. A consideration of the characteristic features, both of these and of other observations on persons in affluent circumstances, may suggest to another class of enthusiasts, that there are many other causes affecting the mortality of mankind besides the sanitary condition of their habitations; and that although ventilation, drainage, and water supply are all very necessary things, they are not " all the law and the prophets" notwithstanding. That the peculiar features of the present observations belong to the normal law of mortality of the human race, it would, we think, be very unwise either to affirm or deny. Notwithstanding all that has been written on that subject, we remain of opinion that that law is yet undiscovered, and that a much greater number and variety of observations than we at present possess will be required for its discovery. Such a law, if discovered, would be of high interest, both to the physiologist and the mathematician. But it will represent the law that really prevails among the living, moving, thinking men that inhabit the earth, much in the same way that the statue of the Apollo Belvedere represents their bodily form. Such a law will never supersede, in our pursuits at least, the exercise of that careful judgment and sound discrimination which it should be our study to cultivate, and without which the most varied talents will be useless and the greatest attainments vain.

TABLE I.—*Males.*

Age (x).	Completed the age x.	Existing 31st Dec, 1855, between the Ages x and x+1.	Died between the Ages x and x+1.	Number exposed to the Risk from the Age x to x+1.	Probability of Dying in the Year.	Probability of Surviving the Year.	Mean Duration of Life.
0	2534	31	197	2518·5	·07821	·92179	52·00
1	2358	38	38	2339·0	·01625	·98375	55·37
2	2326	27	20	2312·5	·00865	·99135	55·25
3	2328	42	9	2307·0	·00390	·99610	54·73
4	2315	42	10	2294·0	·00436	·99564	53·93
5	2302	27	11	2288·5	·00481	·99519	53·16
6	2293	45	4	2270·5	·00176	·99824	52·42
7	2276	34	6	2259·0	·00266	·99734	51·51
8	2271	49	10	2246·5	·00445	·99555	50·66
9	2250	30	8	2235·0	·00358	·99642	49·88
10	2249	42	9	2228·0	·00404	·99596	49·04
11	2238	36	13	2220·0	·00586	·99414	48·23
12	2230	26	8	2217·0	·00361	·99639	47·52
13	2231	28	11	2217·0	·00496	·99504	46·68
14	2227	36	5	2209·0	·00226	·99774	45·91
15	2225	34	9	2208·0	·00408	·99592	45·02
16	2223	33	16	2207·5	·00725	·99275	44·21
17	2211	30	14	2196·0	·00638	·99362	43·53
18	2197	27	16	2183·5	·00733	·99267	42·82
19	2188	36	17	2170·0	·00783	·99217	42·13
20	2174	48	18	2150·0	·00837	·99163	41·46
21	2135	42	35	2114·0	·01656	·98344	40·80
22	2089	32	20	2073·0	·00965	·99035	40·47
23	2077	31	24	2061·5	·01163	·98837	39·87
24	2053	36	18	2035·0	·00885	·99115	39·33
25	2035	31	23	2019·5	·01139	·98861	38·67
26	2010	38	25	1991·5	·01255	·98745	38·10
27	1985	24	15	1973·0	·00760	·99240	37·58
28	1978	40	17	1958·0	·00868	·99132	36·86
29	1949	27	18	1935·5	·00930	·99070	36·18
30	1938	36	24	1920·0	·01250	·98750	35·51
31	1899	35	11	1881·5	·00585	·99415	34·96
32	1882	30	19	1867·0	·01018	·98982	34·17
33	1867	34	14	1850·0	·00757	·99243	33·51
34	1851	31	12	1835·5	·00654	·99346	32·75
35	1835	26	10	1822·0	·00549	·99451	31·97
36	1815	33	25	1798·5	·01390	·98610	31·15
37	1773	49	11	1748·5	·00629	·99371	30·58
38	1738	31	17	1722·5	·00987	·99013	29·77
39	1720	28	15	1706·0	·00879	·99121	29·08
40	1697	30	23	1682·0	·01367	·98633	28·33
41	1657	36	16	1639·0	·00976	·99024	27·71
42	1626	26	22	1613·0	·01364	·98636	26·99
43	1604	37	17	1585·5	·01072	·98928	26·34
44	1565	38	10	1546·0	·00647	·99353	25·63
45	1537	39	22	1517·5	·01450	·98550	24·80
46	1489	26	22	1476·0	·01491	·98509	24·16

TABLE I. (continued).

Age (x).	Completed the Age x.	Existing 31st Dec., 1855, between the Ages x and $x+1$.	Died between the Ages x and $x+1$.	Number exposed to the Risk from the Age x to $x+1$.	Probability of Dying in the Year.	Probability of Surviving the Year.	Mean Duration of Life.
47	1464	37	19	1445·5	·01314	·98686	23·52
48	1419	30	19	1404·0	·01353	·98647	22·82
49	1385	26	17	1372·0	·01239	·98761	22·13
50	1355	29	21	1340·5	·01567	·98433	21·40
51	1317	29	18	1302·5	·01382	·98618	20·74
52	1280	31	15	1264·5	·01186	·98814	20·02
53	1244	33	21	1227·5	·01711	·98289	19·25
54	1201	21	27	1190·5	·02268	·97732	18·58
55	1161	28	19	1147·0	·01656	·98344	18·00
56	1128	35	24	1110·5	·02161	·97839	17·30
57	1074	27	11	1060·5	·01037	·98963	16·66
58	1042	33	27	1025·5	·02633	·97367	15·83
59	989	18	20	980·0	·02041	·97959	15·25
60	957	15	19	949·5	·02001	·97999	14·56
61	930	19	26	920·5	·02818	·97182	13·85
62	891	16	24	883·0	·02717	·97283	13·23
63	859	18	32	850·0	·03764	·96236	12·59
64	815	23	24	803·5	·02986	·97014	12·05
65	775	20	33	765·0	·04313	·95687	11·41
66	725	13	27	718·5	·03758	·96242	10·90
67	687	18	27	678·0	·03983	·96017	10·30
68	645	16	34	637·0	·05338	·94662	9·71
69	602	16	35	594·0	·05892	·94108	9·23
70	557	13	36	550·5	·06539	·93461	8·77
71	513	15	32	505·5	·06330	·93670	8·35
72	470	15	25	462·5	·05405	·94595	7·88
73	432	6	32	429·0	·07457	·92543	7·31
74	396	10	35	391·0	·08952	·91048	6·85
75	355	9	39	350·5	·11127	·88873	6·48
76	314	8	29	310·0	·09354	·90646	6·23
77	280	8	27	276·0	·09783	·90217	5·82
78	235	8	27	231·0	·11688	·88312	5·40
79	212	5	22	209·5	·10501	·89499	5·05
80	187	4	20	185·0	·10811	·89189	4·58
81	165	5	29	162·5	·17846	·82154	4·08
82	131	6	16	128·0	·12500	·87500	3·85
83	109	3	21	107·5	·19535	·80465	3·33
84	85	,,	19	85·0	·22353	·77647	3·02
85	66	2	15	65·0	·23077	·76923	2·74
86	49	3	13	47·5	·27369	·72631	2·41
87	33	1	11	32·5	·33846	·66154	2·14
88	21	3	8	19·5	·41026	·58974	1·97
89	10	,,	3	10·0	·30000	·70000	2·00
90	7	,,	3	7·0	·42857	·57143	1·64
91	4	,,	1	4·0	·25000	·75000	1·50
92	3	,,	2	3·0	·66666	·33334	·83
93	1	1	,,	·5			

TABLE II.—*Females.*

Age x.	Completed the Age x.	Existing 31st Dec., 1855, between the Ages x and x+1.	Died between the Ages x and x+1.	Number exposed to the Risk from the Age x to x+1.	Probability of Dying in the Year.	Probability of Surviving the Year.	Mean Duration of Life.
0	2152	33	127	2135·5	·05948	·94052	53·71
1	2029	30	33	2014·0	·01639	·98361	56·09
2	2003	36	14	1985·0	·00705	·99295	56·00
3	1987	35	11	1969·5	·00559	·99441	55·41
4	1979	35	11	1961·5	·00561	·99439	54·71
5	1963	36	5	1945·0	·00257	·99743	54·02
6	1945	34	11	1928·0	·00571	·99429	53·15
7	1926	33	11	1909·5	·00576	·99424	52·45
8	1912	31	8	1896·5	·00422	·99578	51·75
9	1901	35	7	1883·5	·00372	·99628	50·98
10	1889	34	16	1872·0	·00854	·99146	50·16
11	1861	34	18	1844·0	·00976	·99024	49·58
12	1835	29	11	1820·5	·00604	·99396	49·06
13	1834	40	4	1814·0	·00220	·99780	48·35
14	1822	33	12	1805·5	·00665	·99335	47·46
15	1813	35	8	1795·5	·00445	·99555	46·77
16	1794	34	10	1777·0	·00563	·99437	45·97
17	1776	30	20	1761·0	·01136	·98864	45·23
18	1745	37	11	1726·5	·00637	·99363	44·74
19	1722	38	18	1703·0	·01057	·93943	44·02
20	1698	26	12	1685·0	·00712	·99288	43·48
21	1679	34	12	1662·0	·00722	·99278	42·79
22	1656	23	11	1644·5	·00669	·99331	42·10
23	1643	32	16	1627·0	·00983	·99017	41·39
24	1621	36	14	1603·0	·00873	·99127	40·80
25	1596	29	14	1581·5	·00885	·99115	40·16
26	1575	32	11	1559·0	·00706	·99294	39·50
27	1560	29	13	1545·5	·00841	·99159	38·78
28	1538	25	18	1525·5	·01180	·98820	38·10
29	1518	27	11	1504·5	·00731	·99269	37·54
30	1495	30	9	1480·0	·00608	·99392	36·82
31	1479	27	12	1465·5	·00818	·99182	36·04
32	1461	24	13	1449·0	·00897	·99103	35·33
33	1434	32	18	1418·0	·01269	·98731	34·65
34	1405	22	10	1394·0	·00717	·99283	34·09
35	1383	22	20	1372·0	·01458	·98542	33·35
36	1358	27	10	1344·5	·00744	·99256	32·83
37	1335	20	12	1325·0	·00906	·99094	32·08
38	1315	29	11	1300·5	·00846	·99154	31·37
39	1287	31	12	1271·5	·00944	·99056	30·64
40	1257	23	20	1245·5	·01606	·98394	29·93
41	1225	32	13	1209·0	·01075	·98925	29·40
42	1190	24	8	1178·0	·00679	·99321	28·71
43	1164	28	16	1150·0	·01391	·98609	27·91
44	1123	34	15	1106·0	·01356	·98644	27·30
45	1079	28	13	1065·0	·01221	·98779	26·66
46	1049	35	14	1031·5	·01358	·98642	25·99
47	1006	16	9	998·0	·00902	·99098	25·34
48	986	33	8	969·5	·00825	·99175	24·56
49	952	25	13	939·5	·01383	·98617	23·76
50	924	29	16	909·5	·01759	·98241	23·08
51	883	21	10	872·5	·01146	·98854	22·49
52	857	19	5	847·5	·00590	·99410	21·74

TABLE II. (continued).

Age x.	Completed the Age x.	Existing 31st Dec, 1855, between the Ages x and x+1.	Died between the Ages x and x+1.	Number exposed to the Risk from the Age x to x+1.	Probability of Dying in the Year.	Probability of Surviving the Year.	Mean Duration of Life.
53	839	19	11	829·5	·01326	·98674	20·87
54	816	31	16	800·5	·01998	·98002	20·14
55	776	21	17	765·5	·02220	·97780	19·54
56	740	20	20	730·0	·02740	·97260	18·98
57	704	22	10	693·0	·01443	·98557	18·49
58	673	20	15	663·0	·02262	·97738	17·76
59	639	22	10	628·0	·01592	·98408	17·15
60	607	16	15	599·0	·02504	·97496	16·42
61	576	10	18	571·0	·03152	·96848	15·83
62	549	14	18	542·0	·03321	·96679	15·34
63	517	19	11	507·5	·02167	·97833	14·84
64	490	8	20	486·0	·04115	·95885	14·16
65	463	12	22	457·0	·04814	·95186	13·74
66	430	13	13	423·5	·03069	·96931	13·41
67	405	14	12	398·0	·03015	·96985	12·82
68	380	15	18	372·5	·04832	·95168	12·20
69	349	13	14	342·5	·04088	·95912	11·80
70	326	19	13	316·5	·04108	·95892	11·28
71	296	13	17	289·5	·05872	·94128	10·74
72	267	12	11	261·0	·04214	·95786	10·39
73	246	6	13	243·0	·05350	·94650	9·82
74	227	6	20	224·0	·08929	·91071	9·35
75	202	8	10	198·0	·05051	·94949	9·22
76	184	4	10	182·0	·05495	·94505	8·69
77	170	7	16	166·5	·09610	·90390	8·16
78	147	3	8	145·5	·05498	·94502	7·98
79	136	7	9	132·5	·06792	·93208	7·41
80	120	8	13	116·0	·11207	·88793	6·92
81	99	4	13	97·0	·13402	·86598	6·73
82	82	7	10	78·5	·12739	·87261	6·69
83	65	1	6	64·5	·09302	·90698	6·60
84	58	3	3	56·5	·05310	·94690	6·22
85	52	1	11	51·5	·21360	·78640	5·54
86	40	2	3	39·0	·07692	·92308	5·91
87	35	3	3	33·5	·08955	·91045	5·36
88	29	..	4	29·0	·13793	·86207	4·84
89	25	1	3	24·5	·12245	·87755	4·54
90	21	1	4	20·5	·19512	·80488	4·10
91	16	1	2	15·5	·12903	·87097	3·97
92	13	2	2	12·0	·16667	·83333	3·49
93	9	..	1	9·0	·11111	·88889	3·08
94	8	1	2	7·5	·26667	·73333	2·41
95	5	..	1	5·0	·20000	·80000	2·10
96	4	2	..	3·0	·33333	·66667	1·50
97	2	..	..	2·0	·50000	·50000	1·00
98	2	..	..	2·0			
99	2	..	..	2·0			
100	2	..	..	2·0			
101	2	1	..	1·5			
102	1	..	..	1·0			
103	1	..	..	1·0			
104	1	1	..	·5			

TABLE III.—*Both Sexes.*

Age x.	Number exposed to the Risk from the Age x to $x+1$.	Died between the Ages x and $x+1$.	Probability of Dying in the Year.	Probability of Surviving the Year.	Mean Duration of Life.
0	4654·0	324	·06962	·93038	52·62
1	4353·0	71	·01631	·98369	55·52
2	4297·5	34	·00791	·99209	55·44
3	4276·5	20	·00468	·99532	54·87
4	4255·5	21	·00494	·99506	54·13
5	4233·5	16	·00378	·99622	53·40
6	4198·5	15	·00357	·99643	52·60
7	4168·5	17	·00408	·99592	51·78
8	4143·0	18	·00434	·99566	50·99
9	4118·5	15	·00364	·99636	50·21
10	4100·0	25	·00610	·99390	49·39
11	4064·0	31	·00763	·99237	48·69
12	4037·5	19	·00471	·99529	48·07
13	4031·0	15	·00372	·99628	47·29
14	4014·5	17	·00423	·99577	46·47
15	4003·5	17	·00425	·99575	45·66
16	3984·5	26	·00653	·99347	44·85
17	3957·0	34	·00859	·99141	44·15
18	3910·0	27	·00691	·99309	43·52
19	3873·0	35	·00904	·99096	42·82
20	3835·0	30	·00782	·99218	42·21
21	3776·0	47	·01245	·98755	41·54
22	3717·5	31	·00834	·99166	41·05
23	3688·5	40	·01084	·98916	40·40
24	3638·0	32	·00880	·99120	39·83
25	3601·0	37	·01027	·98973	39·18
26	3550·5	36	·01014	·98986	38·58
27	3518·5	28	·00796	·99204	37·97
28	3483·5	35	·01005	·98995	37·27
29	3440·0	29	·00843	·99157	36·65
30	3400·0	33	·00970	·99030	35·96
31	3347·0	23	·00687	·99313	35·30
32	3316·0	32	·00965	·99035	34·54
33	3268·0	32	·00979	·99021	33·87
34	3229·5	22	·00681	·99319	33·20
35	3194·0	30	·00939	·99061	32·43
36	3143·0	35	·01114	·98886	31·73
37	3073·5	23	·00748	·99252	31·08
38	3023·0	28	·00926	·99074	30·31
39	2977·5	27	·00907	·99093	29·59
40	2927·5	43	·01469	·98531	28·86
41	2848·0	29	·01018	·98982	28·28
42	2791·0	30	·01075	·98925	27·57
43	2735·5	33	·01206	·98794	26·86
44	2652·0	25	·00943	·99057	26·18
45	2582·5	35	·01356	·98644	25·43
46	2507·5	36	·01436	·98564	24·77
47	2443·5	28	·01146	·98854	24·12
48	2373·5	27	·01138	·98862	23·40
49	2311·5	30	·01298	·98702	22·66
50	2250·0	37	·01644	·98356	21·95
51	2175·0	28	·01287	·98713	21·31
52	2112·0	20	·00947	·99053	20·58

TABLE III. (continued).

Age x.	Number exposed to the Risk from the Age x to x+1.	Died between the Ages x and x+1.	Probability of Dying in the Year.	Probability of Surviving the Year.	Mean Duration of Life.
53	2057·0	32	·01555	·98445	19·77
54	1991·0	43	·02160	·97840	19·08
55	1912·5	36	·01882	·98118	18·49
56	1840·5	44	·02391	·97609	17·83
57	1753·5	21	·01198	·98802	17·26
58	1688·5	42	·02487	·97513	16·46
59	1608·0	30	·01866	·98134	15·87
60	1548·5	34	·02196	·97804	15·16
61	1491·5	44	·02950	·97050	14·49
62	1425·0	42	·02947	·97053	13·91
63	1357·5	43	·03167	·96833	13·32
64	1289·5	44	·03412	·96588	12·74
65	1222·0	55	·04501	·95499	12·17
66	1142·0	40	·03503	·96497	11·72
67	1076·0	39	·03624	·96376	11·13
68	1009·5	52	·05151	·94849	10·53
69	936·5	49	·05232	·94768	10·08
70	867·0	49	·05652	·94348	9·61
71	795·0	49	·06163	·93837	9·15
72	723·5	36	·04976	·95024	8·72
73	672·0	45	·06696	·93304	8·15
74	615·0	55	·08943	·91057	7·70
75	548·5	49	·08934	·91066	7·41
76	492·0	39	·07927	·92073	7·08
77	442·5	43	·09718	·90282	6·65
78	376·5	35	·09296	·90704	6·31
79	342·0	31	·09064	·90936	5·91
80	301·0	33	·10963	·89037	5·45
81	259·5	42	·16185	·83815	5·06
82	206·5	26	·12591	·87409	4·93
83	172·0	27	·15698	·84302	4·57
84	141·5	22	·15547	·84453	4·33
85	116·5	26	·22317	·77683	4·04
86	86·5	16	·18497	·81503	4·05
87	66·0	14	·21212	·78788	3·86
88	48·5	12	·24742	·75258	3·77
89	34·5	6	·17391	·82609	3·84
90	27·5	7	·25455	·74545	3·54
91	19·5	3	·15385	·84615	3·58
92	15·0	4	·26667	·73333	3·14
93	9·5	1	·10526	·89474	3·10
94	7·5	2	·26667	·73333	2·41
95	5·0	1	·20000	·80000	2·10
96	3·0				
97	2·0				
98	2·0				
99	2·0				
100	2·0				
101	1·5				
102	1·0				
103	1·0				
104	·5				

TABLE IV.—*Males.*

Age.	MEAN DURATION OF LIFE.				
	Peerage Families.	English Table Dr. Farr.	Equitable. Morgan.	Government Annuitants. A. G. Finlaison.	Healthy Districts. Dr. Farr.
0	52·00	40·36	..	..	48·56
10	49·04	47·47	48·32	45·57	51·28
20	41·46	39·99	41·67	38·74	43·40
30	35·51	33·21	34·53	33·39	36·45
40	28·33	26·46	27·40	27·12	29·29
50	21·40	19·87	20·36	20·53	22·03
60	14·56	13·60	13·91	14·41	15·06
70	8·77	8·55	8·70	9·08	9·37
80	4·58	4·97	4·75	5·22	5·37
90	1·64	2·80	2·56	2·78	2·99

Age.	ANNUAL MORTALITY PER CENT.			
	Peerage Families.	English Table. Dr. Farr.	Equitable. Morgan.	Government Annuitants. A. G. Finlaison.
Under 5	2·227	7·072	..	..
5 to 9	·345	·926	..	·718
10—19	·536	·581	..	·742
20—29	1·046	·882	·749	1·315
30—39	·870	1·094	·928	1·216
40—49	1·227	1·487	1·243	1·368
50—59	1·764	2·275	2·111	2·269
60—69	3·757	4·654	4·304	3·971
70—79	8·714	10·012	8·994	8·685
80—89	23·836	21·370	20·786	18·600

TABLE V.—*Females.*

Age.	MEAN DURATION OF LIFE.			
	Peerage Families.	English Table. Dr. Farr.	Government Annuitants. A. G. Finlaison.	Healthy Districts. Dr. Farr.
0	53·71	42·04	..	49·45
10	50·16	47·86	50·07	50·88
20	43·48	40·65	43·27	43·50
30	36·82	34·06	36·65	36·85
40	29·93	27·50	29·91	30·00
50	23·08	20·84	22·99	22·87
60	16·42	14·49	16·17	15·69
70	11·28	9·12	10·14	9·85
80	6·92	5·34	5·69	5·64
90	4·10	3·09	2·94	3·11

Age.	ANNUAL MORTALITY PER CENT.		
	Peerage Families.	English Table. Dr. Farr.	Government Annuitants. A. G. Finlaison.
Under 5	1·882	6·037	..
5 to 9	·440	·900	·668
10—19	·716	·639	·648
20—29	·830	·917	·850
30—39	·921	1·120	·995
40—49	1·179	1·389	1·149
50—59	1·708	2·107	1·621
60—69	3·508	4·079	3·063
70—79	6·092	9·095	7·119
80—89	11·601	19·461	16·724

TABLE VI.—*Both Sexes.*

Age.	MEAN DURATION OF LIFE.		
	Peerage Families.	Carlisle.	Healthy Districts.
0	52·62	38·72	49·00
10	49·39	48·82	51·08
20	42·21	41·46	43·45
30	35·96	34·34	36·64
40	28·86	27·61	29·64
50	21·95	21·11	22·44
60	15·16	14·34	15·37
70	9·61	9·18	9·61
80	5·45	5·51	5·51
90	3·54	3·28	3·05

Age.	ANNUAL MORTALITY PER CENT.	
	Peerage Families.	Carlisle.
Under 5	2·069	7·324
5 to 9	·388	1·011
10—19	·617	·588
20—29	·951	·761
30—39	·892	1·053
40—49	1·208	1·423
50—59	1·742	1·863
60—69	3·668	4·082
70—79	7·737	8·801
80—89	17·514	17·262

TABLE VII.—*Peerage Families.*

Age.	MEAN DURATION OF LIFE.	
	Males.	Females.
0	52·00	53·71
10	49·04	50·16
20	41·46	43·48
30	35·51	36·82
40	28·33	29·93
50	21·40	23·08
60	14·56	16·42
70	8·77	11·28
80	4·58	6·92
90	1·64	4·10

Age.	ANNUAL MORTALITY PER CENT.	
	Males.	Females.
Under 5	2·23	1·88
5 to 9	·35	·44
10—14	·41	·66
15—19	·66	·77
20—24	1·10	·79
25—29	·99	·87
30—39	·87	·92
40—49	1·23	1·18
50—59	1·76	1·71
60—69	3·76	3·51
70—79	8·71	6·09
80—89	23·84	11·60

On the Law of Human Mortality; and on Mr. Gompertz's new exposition of his Law of Mortality. By T. R. EDMONDS, B.A., *formerly of Trinity College, Cambridge.*

IN the last number of the *Assurance Magazine* (April, 1861), there appears a letter from Mr. Gompertz, in which reference is made to his and my claims to the discovery of the law, or part of the law, of human mortality. In this letter, Mr. Gompertz declares that my claim cannot interfere with his claim, because I had acknowledged the priority of his discovery. I admit the truth of this declaration, and can add thereto, that I also am able to declare, with equal truth, that his claim, as now described by himself, cannot interfere with any claim hitherto advanced by me. The law of human mortality, now claimed to be discovered by Mr. Gompertz, differs materially from the law which I have declared to exist. The two laws are in accordance with one another only at one period of life (extending, say, from the age of 15 to the age of 55 years), and then only in a partial degree. It is in this partial agreement, at this particular period of life, that the priority of Mr. Gompertz's discovery consists. The description of the law of human mortality, as *now* believed by Mr. Gompertz to exist, is contained in a paper, referred to in the letter above mentioned, and since published ·(page 454) in the Appendix to the *Report of Proceedings of the Fourth International Statistical Congress, held in London in the year* 1860.

The universal law of human mortality which I believe to exist (and have so believed for the last 33 years), may be described as consisting of three parts, all of great simplicity.

The first part is, that, from birth to extreme old age, human life is divided into three well-marked physiological periods, of growth, fruitfulness, and decay, which periods differ from one another in having distinctive rates of decrease or increase of their mortality, according to age. These periods may be described as those of infancy, fecundity, and senescence. The discovery of these three periods of different progressive mortality according to age, was first made public by Dr. Price, in the year 1769, as I have already stated in the *Assurance Magazine* (page 171) for October, 1860. According to Dr. Price, human life, from birth upwards, grows *gradually stronger* until the age of 10 years, then *slowly loses* strength until the age of 50, then *more rapidly loses* strength, until, at 70 or 75, it is brought back to all the weakness of the first month.

The second part of the law of human mortality is, that, from the beginning to the end of any one of the three stages of infancy, fecundity, or senescence, the rate of mortality varies with the age in a geometrical progression, the common ratio of which may, or may not, be different for different populations, or for the same population at different times. This law of increase or decrease is the simplest conceivable law. For example, suppose it to have been ascertained, by observation of the same population, that the annual mortality at the age 25 years was 1 per cent., and at the age 45 years, 1·80 per cent., and suppose it to be required to interpolate, according to some law, the mortality corresponding to intermediate ages. In that case, the assumption first made, as being the simplest, would most probably be, that the annual proportional increase according to age was constant throughout the entire period of 20 years of age. Adopting this assumption, it would ensue, that the common ratio of the geometric progression at which the mortality increases, is equal to the twentieth root of 1·80, or to 1·03. If every term is to the next preceding term as 1·03 to 1, then the twenty-first term will be to the first term as 1·80 to 1.

The third and last part of the law of human mortality consists in the permanency of each of the three common ratios of geometric progression of mortality, according to age, in all populations. The common ratios for the several periods of infancy, fecundity, and senescence, being known for one population are known for all other populations. The respective values of these three common ratios, for annual intervals of age, have been ascertained to be the following, nearly—viz., $\frac{1}{1\cdot479108}$, 1·0299117, and 1·0796923 (say $\frac{2}{3}$, 1·03, and 1·08), these being the numbers of which the common logarithms are $(-\cdot17)$, $(+\cdot0128)$, and $(+\cdot0333)$.

The public are indebted to Mr. Gompertz for the discovery (in 1825) of a portion of the second part of the law of human mortality. This portion is, that in one of the well-marked physiological periods of human life, the mortality increases with the age in a geometric progression, the common ratio of which may, or may not, be different for different populations. The period to which this discovery applies, is that of " fecundity," extending, say, from the age of 15 to the age of 55 years. The evidence offered of the existence of this portion of the law of human mortality was indirect. For Mr. Gompertz, without saying anything as to the division of human life into three, or any other number, of physiological stages or periods, gave a general formula of geometric increase of mortality *"for a long period of man's life,"* without assigning any limits to such period. All the examples of the applicability of his formula, with one exception only, were founded upon observations of mortality made in the period of "fecundity." From these facts it may be fairly concluded, that Mr. Gompertz is entitled to the credit of the discovery, that in one of the three periods into which human life is divided, the mortality increases with the age in a geometrical progression, the common ratio of which may, or may not, be different for different populations.

In his paper, printed in the *Philosophical Transactions* of 1825, Mr. Gompertz expresses homely facts in transcendental language. Instead of supposing the mortality to increase in a geometrical progression with the age, he supposes the *power to oppose destruction* (which is inversely as the *mortality*) to decrease in equal proportions in equal infinitely small intervals of time. Instead of saying that the law of mortality, from the age of 15 to the age of 55 years, is such that the annual rate of mortality increases in a geometrical progression, of which the common ratio is (p), he states that the law is such that the survivors, or numbers left alive under the operation of this law for the term of (x) years, will be represented by the transcendental (dg^{p^x}). The law of mortality, however, is better and more simply expressed in terms of the mortality than it can be in terms of the survivors according to the same law. In the former case, we have (ap^x) to express the law; in the latter case we have the same law expressed by $y = 10^{\frac{k^2 a}{\lambda p}(1-p^x)}$, which is the quantity (dg^{p^x}), corrected and reduced to its simplest terms. Both expressions indicate the same facts, or the mathematical consequences of the same facts. In either case, where (a) and (p) are given, the only variable quantity is (x), measured from the time when the mortality was (a). It is not stated by Mr. Gompertz which of the two equivalent expressions for the law of mortality in the period of "fecundity" was first discovered by him. His expression for the survivors according to age, might have been discovered by inspection of published tables of the logarithms of survivors, at yearly intervals of age, accompanying ordinary tables of mortality. It might easily have been gathered from such inspection, that the differences between the logarithms of the survivors, at successive decennial intervals of age, are in geometric progression, and of the form (Cp^x). From this fact, the equation $y = dg^{p^x}$ might easily have been deduced.

In supplying evidence of the truth of his law of mortality, Mr. Gompertz invariably omits to compare the resulting mortality as deduced from his theoretical tables with the observed mortality according to age, which is or ought to be the foundation of such tables. In my own case, I have always supplied evidence of this complete and most simple kind. Nevertheless, Mr. Gompertz and his two officious, but accepted, advocates, concur in saying, that I have not given evidence of the truth of my theory of mortality, or of the existence of any law of mortality beyond that discovered by Mr. Gompertz, as applicable " *to a long period of man's life,*" to which no limits are assigned. This concurrence of views between Mr. Gompertz and his two advocates is the more remarkable, because it is the only instance of such concurrence to be found in the recently published papers of the three parties. His advocates advance claims on his behalf, which Mr. Gompertz has never advanced himself. The statements of the advocates are not supported by any corresponding statements of their principal. The only support which Mr. Gompertz gives his two advocates, is, to praise them for the talent which they have displayed in the advocacy of a claim rightfully or wrongfully put forward for his benefit. Mr. Gompertz appears to consider himself free from any responsibility in lending his approbation to papers containing statements, which, from his superior knowledge of the facts, he could not, with any regard to truth, have uttered himself. Except in the present instance, Mr. Gompertz has written nothing against my claim, or in support of the claim made on his behalf, which I am at all interested in contradicting. But with respect to his two advocates, I deny that either of them has said anything in depreciation of my claim which is true in substance or even in form.

Mr. Gompertz's theory of mortality (like my theory) rested greatly for support on the observations of mortality according to age of the populations of Sweden and Carlisle, made in the last century by Wargentin and Heysham respectively. The simplest evidence of the soundness of his theory would have been to show, that, taking successive decennial periods of age, the mortality deduced from either of his theoretical tables was in near agreement with the mortality at the same intervals of age observed by Wargentin or Heysham, and that both the theoretical and observed mortality were in geometric progression from the age of 15 to the age of 55 years. Instead of doing this, Mr. Gompertz has merely compared his series of survivors according to age with the series of survivors formed by Price and Milne from the observed facts. He had previously taken no pains to inquire, whether the tables of Price and Milne were in agreement with the facts in mortality on which they are professed to be founded, and whether Price and Milne had not deviated from the facts presented, in order to smoothe irregularities of decrement, according to some ideal law of mortality which they may have entertained without expressing.

In proof of the soundness of my theory of mortality, I know no fact more convincing or interesting than one which has already appeared in the pages of the *Assurance Magazine* (year 1855, page 144). It is there shown, for that part of the total male population of England which was contained in five decennial periods of age, commencing at age 25 and ending at age 75 years, that the mortality observed in these decennial intervals of age was almost in exact accordance with my theory of mortality published in 1832— being three years before there existed any complete observation of the mortality of any part of the general population of England, except that of Carlisle, in the year 1787. To show the closeness of the coincidence to the present reader, the following extract is made from the more extended table, of which it forms part.

Annual Mortality per Cent., during 10 *Years ending with* 1850, *of all the Male Population of England comprised between the Ages of* 25 *and* 75 *Years.*

Between Ages..	25-35.	35-45.	45-55.	55-65.	65-75.
Fact observed....	·97	1·25	1·78	3·14	6·61
Theory	·96	1·28	1·75	3·22	6·78

In the Appendix (page 455) to the *Proceedings of the Fourth International Statistical Congress*, there is contained a statement by Mr. Gompertz of the view which he *now* entertains of the *"one continuous and uniform law of mortality from birth to at least the age of* 100 *years."* He there states his belief that such law is expressed by the following equation, wherein (L_x) represents the numbers surviving or living at any age (x) measured from birth :—

$$L_x = \text{constant} \times A^{e^x} \times B^{e^{x}x} \times C^{q^x} \times D^{Px}.$$

The following are the remarks made by Mr. Gompertz with respect to the applicability of the above formula :—" In all the tables I have examined, all the factors but C^{q^x}, being between the ages 20 and 60, are so nearly constant that the difference from it may be neglected. This equation may be put into the form, constant $\times C^{q^x}$." Afterwards, he goes on to say that, by the disappearance of the two first transcendental factors, the equation from the age 60, and ever after, is of the form $L_x =$ constant $\times C^{q^x} \times D^{P_x}$.

From the above, it will be manifest that it is only at the interval of ages from 20 to 60 years that there is any agreement between Mr. Gompertz's theory and my theory of mortality; for, according to my theory of mortality, in either of the three stages into which human life is divided, the equation for the living or surviving at any age (x), from the beginning of that stage, is $L_x =$ constant $\times g^{p^x}$, the quantity (p) having a fixed and distinct value in each of the three stages; whilst, according to Mr. Gompertz's theory of mortality, human life is divided into four stages, and in three of the four stages no more than two out of the four transcendentals which enter into the above general formula of mortality ever disappear.

Mr. Gompertz also gives the following as his general equation of mortality for the whole of life expressed in common logarithms—

$$\lambda L_x = \text{constant} + ke^x + \underset{1}{k} \underset{1}{e^x} x - nq^x - P_x,$$

and makes the following observations thereon :—

" ke^x is at birth $= k$ and decreases regularly as x increases, and becomes, before the age of 20, and ever after, in the table I have examined, perfectly insignificant.

" $\underset{1}{k} \underset{1}{e^x}$ is of no value at birth, but increases in a very short time to a maximum of significant value; but in less than 12 months, in the table I have examined, decreases to perfect insignificance.

" nq^x is $= n$ at birth, and continually increases, with the increase of x, to the remotest age.

" P_x is perfectly insignificant at birth, and till an age but a few years below 60, and it then continually increases with the age until far beyond the age of 100, and then decreases to perfect insignificance."

In speaking of the equation $L_x = AB^{q^x}$, used in his paper of 1825, Mr. Gompertz remarks—"That AB and q were supposed to represent constant quantities—or, at least, were shown to differ very little from constants—for a very long term of years (for instance, about 50 years), but differing a little for length of term and from one locality to another; and A and B and q being so related to one another, that, supposing them to be constant, we should have $A \times B = L_0$, the number at birth. But, in making the investigation, I did not pretend that A and B were absolutely constant. They were determined from a random selection from three distant periods of age, from a statement of the number of persons who will be living at different ages out of a certain number of persons stated to have been born."

In neither of the two papers referred to as recently published in the name of Mr. Gompertz do I find any statement relative to our respective claims which impugns the correctness of any statement made in my *Life Tables*, or in my paper contained in the *Assurance Magazine* for October, 1860. I do not believe that Mr. Gompertz has ever complained of my having failed duly to acknowledge the share which he had in the discovery of the law of human mortality. Since the publication, in 1832, of my *Life Tables*, I have, at various times, been in free personal communication with gentlemen who had been, a short time previously, in free personal communication with Mr. Gompertz. From none of these gentlemen have I heard that Mr. Gompertz imagined himself to have been wronged by me. As Mr. Gompertz himself, during the space of 28 years, remained insensible to the supposed wrong, it may fairly be presumed that such supposed wrong had no existence, and that the charge of wrong brought against me by a third party (Mr. De Morgan) is entirely without foundation.

In the *Assurance Magazine* for July, 1860, Mr. De Morgan brought the charge against me of having, in the year 1832, in my *Life Tables*, "unfairly suppressed due acknowledgment to the writings of Mr. Benjamin Gompertz." After reading my paper in the next following Number of the *Magazine*, Mr. De Morgan writes as follows (page 214) in the Number for January, 1861:—
"If Mr. Edmonds had given all the description which he has now given . . . . . . there would have been no suppression." This admission appears to me to amount to the virtual abandonment of his original charge, although Mr. De Morgan denies this by saying that suppression existed in 1832 and did not cease until the year 1860. The erroneous conclusion of Mr. De Morgan is founded on the erroneous assumption, that, in 1860, I made any statement or admission more favourable to Mr. Gompertz's alleged claims than I had made in 1832. I do not think that Mr. De Morgan will find anyone (even Mr. Gompertz himself) to concur with him in such an erroneous assumption. In the year 1832 I wrote as follows:—
"The honour of first discovering that some connexion existed between tables of mortality and the algebraic expression (a^{b^x}) belongs to Mr. Gompertz." These words appear to me to convey a sufficient intimation of the high value I then attached to the discovery, and of the credit due to Mr. Gompertz for making it. The same words conveyed a suggestion to the reader to inquire and judge for himself of

the comparative merits of my complete law and Mr. Gompertz's then incomplete law of mortality. The paper which I wrote in the *Assurance Magazine* of October, 1860, was written in self-defence (against an outrageous attack of Mr. De Morgan), for the purpose of circumscribing, within just and narrow limits, the favourable admission which I had made in general terms in the year 1832.

In a question like the present, when I have to meet an apparently unfounded accusation, it appears to me to be of importance that the reader should know that Mr. De Morgan and myself were not strangers to one another in our years of student life; that previous to publication, in 1832, my *Life Tables* obtained his approval; and that, without giving me any intimation of the withdrawal of this approval, he had been secretly writing and speaking against me (in relation to Mr. Gompertz's supposed claim), for the space of 28 years. Mr. De Morgan and myself were students of the same year of admission, of the same University, of the same College, and of the same class-room of the College. We attended the same mathematical class for three years—or, rather, we should have done so if I had not omitted to read mathematics during the whole of the second year. In our third year the present Astronomer-Royal was the mathematical lecturer of our class, which, near the termination of the year, was reduced to two students—Mr. De Morgan and myself. In the year 1832, before publication, I sent Mr. De Morgan a printed copy of my *Life Tables*. A few days afterwards, at a personal interview, he expressed his general approbation of the work, but objected to one sentence only, which had no connexion with Mr. Gompertz or his supposed claim. He recommended me to have this sentence (and the page containing it) cancelled. This I declined doing; and since that time I have had no communication with Mr. De Morgan, except on one occasion, when I discovered that he had ceased to be favourably disposed towards me—for no reason, that I could imagine, except my refusal to act upon his recommendation given in 1832. It appears to me that Mr. De Morgan, after expressing to me a favourable opinion of my *Life Tables* in 1832, was not justified in changing that opinion, and acting upon that changed opinion, without previously communicating with me, and calling for an explanation. If he had done so, I should have offered the explanation which is given in the *Assurance Magazine* of October, 1860—which explanation, as Mr. De Morgan himself admits, would have exonerated me from the charge of "unfair suppression" if it had been given in the year 1832.

The formula by which Mr. Gompertz represents the number surviving according to age in one of his four periods of human life is $y = dg^{p^x}$. This formula is defective in several respects. In the correct formula the exponent of (g) is $(p^x - 1)$, and not (p^x) as stated by Mr. Gompertz. The factor (d), in the formula of Mr. Gompertz, is superfluous; and, besides being superfluous, is defective as being a composite quantity, of which the other factor (g) forms part. I will proceed to place before the reader the successive steps by which the correct formula was obtained by me 33 years ago; and I will afterwards exhibit the course of investigation pursued, or intended to be pursued, by Mr. Gompertz in arriving at his defective formula, giving Mr. Gompertz the benefit of the most recent corrections made by himself and his two advocates.

The problem for solution is this :—Given the annual mortality (a) at the beginning of a period (x) years of age, given also the common ratio (p) of annual geometrical progression of mortality according to age; required to find an expression for (L_x), or the number living and surviving at the end of the (x) years of age, in terms of (a), (p), and other known quantities.

Previous to forming the differential equation, it is essential that the nature of the quantity (a), given to represent the proportional mortality for one year when the mortality is constant, should be determined. There are two different quantities by which this annual mortality may be expressed. In one case, (a) is the finite decrement for one year, and is such that the number living at the beginning of the year of age is to the number surviving at the end of the same year as $(1 + a)$ is to 1. In the other case, the annual mortality at the age 0 is represented by a quantity a, which is the product of an indefinitely small decrement $\left(\dfrac{a}{\nu}\right)$ multiplied by an indefinitely large number (ν). If the finite decrement for one year, when the mortality is constant, is known to be equal to (a), the value of (a) when $x = 0$ may be determined from the equation $e^a = 1 + a$, which gives $a =$ hyperbolic logarithm of $(1 + a)$. If $a = \cdot0063845$, then will $a = \cdot0063643$.

In Mr. Gompertz's investigations no statement is made of any algebraical or arithmetical value of (a) or (a), although the object of his investigation was to find an expression for the number living at any age (x) in terms of the annual mortality, whether (a) or (a). This omission of separate notice of the most important quantity in the formula affords ground for doubting whether Mr. Gompertz was acquainted with the nature of the quantity used by him to denote the annual mortality when (x) was equal to 0. An independent table of mortality could not have been constructed from Mr. Gompertz's formula without the previous knowledge now indicated. It so happens that Mr. Gompertz has published no portion of a theoretical table of mortality which is independent of previously constructed tables. His examples show how interpolations may be made in existing tables—not how portions of new tables may be constructed from the mortality only. The differential equation of Mr. Gompertz is erroneous if he has used the finite decrement (a) to represent the mortality; for the differential equation is true for all values of (x) (whole number or fractional) only when the annual mortality is represented by (a), which is equal to the hyp. logarithm of $(1+a)$.

Beginning with the differential equation, we have—

$$dy = -yap^x dx;$$

integrating,

$$\log. \ y = \log. \ d - \frac{ap^x}{\log. \ p} \ ;$$

when $x = 0$,

$$\log. \ L_0 = \log. \ d - \frac{ap^0}{\log. \ p} \ ;$$

by subtraction,

$$\log. \frac{y}{L_0} = \frac{a}{\log. \ p} - \frac{ap^x}{\log. \ p} = \frac{a}{\log. \ p}(1 - p^x);$$

whence

$$\frac{y}{L_0} = e^{\frac{a}{\log. \ p}(1-p^x)} = 10^{\frac{k^2 a}{\lambda p}(1-p^x)} = 10^{c(1-p^x)} = g^{1-p^x};$$

whence

$$y = L_0 \times g^{1-p^x} = L_0 \times \frac{g}{g^{p^x}} = L_0 \times \frac{1}{g^{p^x-1}}.$$

Instead of the correct expression for (y) or (L_x), just obtained, Mr. Gompertz gives the defective equation $y = dg^{p^x}$. In deducing the latter erroneous value, Mr. Gompertz stops at the second step of the investigation, and thus fails to discover that the constant (d) is a compound quantity, including (g) as one of its factors. According to the defective process of Mr. Gompertz—

$$\log. \frac{y}{d} = -\frac{ap^x}{\log. \ p}, \ \text{so that} \ \frac{y}{d} = e^{-\frac{ap^x}{\log. \ p}} = 10^{-\frac{k^2 a}{\lambda p}p^x};$$

and, putting

$$+c = \left(-\frac{k^2 a}{\lambda h}\right), \ \text{and} \ 10^{+c} = g,$$

he gets

$$y = d \times 10^{-\frac{k^2 a}{\lambda p}p^x} = d \times 10^{+cp^x} = dg^{p^x}.$$

The latter equation is defective by reason of the constant (d) not having been determined by the aid of the equation log. $L_0=$ log. $d- \dfrac{a}{\log. p}$. If it had been so determined, (d) would have been found to be equal to $L_0 \times \dfrac{1}{g}$. If he had made the proper substitution, his equation would have become

$$y=L_0 \times \frac{1}{g} \times g^{p^x}=L_0 \times g^{p^x -1}.$$

In addition to the defect just mentioned, there is the further defect of putting (g), whose exponent is positive, equal to $10^{-\frac{k^2 a}{\lambda p}}$, a quantity greater than unity whose exponent is negative, and thus transferring to the numerator a quantity which originally formed the denominator of a fraction. If Mr. Gompertz had not made this unnecessary change in the sign of the exponent of (g), his corrected formula, on substituting g for $\dfrac{1}{g}$, would have become

$$y=L_0 \times \frac{1}{g^{p^x-1}}.$$

In addition to the defect last mentioned in Mr. Gompertz's formula, there is the further defect of using in the formula any such quantity as (L_0) at all; for (L_0) is the common multiple of all the terms of the series for the living at successive ages, and conveys no information whatever as to the proportions surviving at different ages, which is all the information sought by the formula. If, for (L_0), we substitute unity, we shall have the equation complete in its most simple terms and form, viz.:—

$$y=\frac{1}{g^{p^x-1}}=g^{1-p^x}=10^{c(1-p^x)}=10^{\frac{k^2 a}{\lambda p}(1-p^x)},$$

the last being the form in which it was published by me in the year 1832.

In my paper of October, 1860, I designated the quantity (d) in Mr. Gompertz's formula as superfluous, useless, and indeterminate. The new advocate of Mr. Gompertz objects to my use of the word "indeterminate," because the word is usually limited to the description of a quantity which cannot be determined. It appears to me, that my use of the term is in accordance with his own definition; for Mr. Gompertz and his two advocates have not hitherto been able to determine the value of (d) by decomposing it into its two constituent factors $\left(L_0 \text{ and } \dfrac{1}{g}\right)$, by doing which, they might have simplified aud rectified the original defective equation $y = dg^{p^x}$. Moreover, I would submit, that a quantity, whether intrinsically defective, as (d), or admitting an endless variety of arbitrary values, as (L_0) (none of which concern the object of the investigation), may with reason be called "indeterminate," because there is nothing to be determined, or worthy of being determined.

In the same paper I also termed (d) a particular value of (y), whilst, in strictness, I should have said that that part of (d) which did not contain (g) is a "particular value" of (y). I had previously said, that (L_0), or y when $x = 0$, was equal to $\dfrac{d}{g}$, which is the same thing as saying that (d) was a particular value of (gy). In order to avoid circumlocution, I omitted to say that the argument was the same—whether (d) was a particular value of (y), or of (gy). On account of this omission, the new advocate of Mr. Gompertz has charged me with having committed "a far more serious error" than that of calling (d) an indeterminate constant. He says nothing of the really serious error of Mr. De Morgan, in the exposing of which I committed the pretended verbal error. In the *Assurance Magazine* of July, 1860, page 88, Mr. De Morgan gives a formula, designed to express the number living at any age in terms of the mortality, in which formula he omitted the most essential part, the quantity (a), representing the mortality. It may be said, on Mr. De Morgan's behalf, that he copied the error previously committed by Mr. Gompertz. On the other hand, it is to be stated, that Mr. De Morgan reproduced the investigation and formula of Mr. Gompertz, in order to prove that they were as good as mine, designed for a similar purpose. Having regard to the task voluntarily undertaken, it is clear that Mr. De Morgan was responsible for the error adopted, whether originally committed by Mr. Gompertz or his printer.

Mr. Gompertz has recently presented to the Institute of Actuaries a corrected copy of his paper, which was printed in the *Philosophical Transactions* of 1825. The corrections with which we are now concerned, relate to errors in the printed investigation of his formula $y = dg^{p^x}$, and especially to two admitted errors repeated— abq^x and abq^x, immediately following one another in two separate equations. The correction of the first error, abq^x, is stated to be $a\dot{x}q^x$, or $adxq^x$; and of the second error, abq^x, it is stated to be aq^x. I have had the opportunity of minutely examining the printed errors and their corrections, and do not concur, with the new advocate of Mr. Gompertz, in the opinion that either of the two errors has its source in a mistake of the printer, (b) being substituted for ($\times$), the sign of multiplication. The first of the two corrections may be passed over as genuine and unobjectionable, although the (dx) occupies an unusual position, coming before, instead of after, (q^x). With regard to the second correction, I question its admissibility. For the equation, as corrected, is more defective, in form, at least, than was the equation intended to be corrected. A necessary result of the integration made, was the appearance of a new constant, (b) or (d), on one side or the other of the equation. The correction now made would leave no new constant on either side of the equation. In the absence of any satisfactory account of the second of the two admitted errors, I will now offer, for consideration, the explanation of this error which occurred to my mind before Mr. De Morgan gave it as his opinion that (b) was a constant, additional to, and independent of, the constant (d)—both constants being introduced (rightly or not) by Mr. Gompertz in his process of integration. The source of Mr. Gompertz's error, in my belief, was, that he had carelessly and erroneously concluded, that the constant (b) on one side of the equation, was the equivalent of the constant (d) on the other side of the equation.

In the same paper, I stated that the unknown constant (b) had been introduced by Mr. Gompertz through his incorrect process of integration; and that such constant was not only superfluous and useless (as Mr. De Morgan had declared it to be), but that it was erroneous also. I supported this statement by giving a new proof of my formula $y = 10^{\frac{k^2 a}{\lambda p}(1 - p^x)}$, and showed, without the aid of the differential calculus, that no quantity similar to (b) formed part of the formula. The new advocate of Mr. Gompertz does not call in question the truth of the substantial part of my statement, but accuses me of having committed an error in describing the mode in which the error was supposed to have been arrived at by Mr. Gompertz. He says, that the process of integration had not commenced when the erroneous (b) first appeared. In reply, I have to state, that there are two distinct errors of (b), both appearing after the completion of the differential equation. I have also to state, that the first error only has been satisfactorily corrected by the substitution of $(\ddot{x})$, or dx, for (b); and the second erroneous (b) is, apparently, not a printer's error, but a new and erroneous constant, introduced by Mr. Gompertz immediately after the disappearance of (dy) and (dx) from the differential equation.

It is usual to consider the process of integration as commencing at the step immediately following the completion of the differential equation.   In the present case, I will venture to offer the opinion, that the completion of the differential equation was the first and most important step in Mr. Gompertz's process of integration. For the integral sought was known, or believed to be known, by Mr. Gompertz, before the differential was known to him, as I have already shown, at page 176, in the paper before referred to.   To know the correct differential and the correct integral, generally includes the knowledge of all the intermediate steps.   It is doubtful, however, whether Mr. Gompertz possessed any exact knowledge either of the integral or differential sought.   For his integral is erroneous, in exhibiting (p^x) as the exponent of the factor (g); and his differential equation would be erroneous, if the quantity (a) is intended to represent the finite annual decrement when the mortality is constant.

In the paper before referred to, I gave, as a quotation from Mr. Gompertz's paper in the *Philosophical Transactions* of 1825, a passage which has been repeatedly quoted as the essential part of his law of mortality.   In doing this, however, I omitted the first word of the sentence, which is " if;" which word is used by Mr. Gompertz to signify "let it be assumed."   In consequence of the omission of this insignificant word, I have been charged by the new advocate of Mr. Gompertz with quoting a sentence or passage which "has no existence in the writings of Mr. Gompertz."   The writer does not describe the passage misquoted, although he refers to its place in the *Assurance Magazine* of October, 1860, nor does he state in what the alleged misquotation consisted.   If he had done so, the charge (made in italics) would have appeared too ridiculous to be seriously entertained.   I can truly say of this new advocate of Mr. Gompertz, whilst using his own expressive language, that his remarks " contain errors equally grave with those falsely imputed " to me.

The new advocate of Mr. Gompertz alleges, in opposition to the most obvious facts, that the defective quantity (d) " is really an essential part of the formula, and, *as such, is employed by Mr. Edmonds himself.*"   There is not the remotest approach to truth in this statement, enforced by *italics*.   The defective quantity (d) is obtainable from the equation $d = \dfrac{L_0}{g}$, and, when obtained, is worth less than nothing.   Instead of attempting to prove that I had ever made use of such a quantity as (d), the new advocate of

Mr. Gompertz attempts to prove that I had used the correct but superfluous quantity (L_0) as distinct from unity. This he entirely fails to do, for he reproduces the identical figures which were given by me as obtained by the substitution of unity for L_0.

On the various methods pursued in the Distribution of Surplus among the Assured in a Life Assurance Company; with a comparison of the relative merits of such methods.* By WM. POLLARD PATTISON, ESQ., *of the London and Provincial Law Assurance Society.*

[The Author of this Essay obtained for it the Prize offered by the Institute for an Essay on the subject.—ED. A. M.]

THERE is, probably, no department of an actuary's pursuits where there is such diversity of practice as that of which this paper treats. After the calculation of the premiums and the periodical valuations, the distribution of surplus is infinitely the most important duty that an actuary has to perform. The premium income of the various Life Offices now exceeds £8,000,000 a year, and probably not less than £1,000,000 is annually divided as surplus. Within the last 14 years, one Office alone has divided upwards of £1,200,000; and another, for many years past, has annually distributed as surplus more than £100,000. This will be sufficient to show the magnitude of the interests concerned; and yet, as will be seen in the sequel, these distributions are, in some cases, made in a most arbitrary manner—without reference to principles of justice and equity, and without•the basis of accurate reasoning. After what has been written on this subject by Mr. Jellicoe and Professor De Morgan, it is not to be supposed that any actuary would defend the schemes here referred to as equitable; those who *do* follow them, content themselves with stating their advantages. In many cases, however, their conservatism arises from the difficulty of disturbing vested interests; and the maxim, " *Omnis innovatio plus novitate perturbat quam utilitate prodest*," is very attractive to those in office, upon whom would fall the labour of carrying out a new system. Hence, in this, as in many other cases of old-established errors, all true reform must come from without. But the application of prin-

---

* By the title of this paper it will be perceived that the methods actually pursued are those which are the subject of investigation and comparison. I emphatically wish it to be understood that they are in no sense theoretical, but such as are adopted at the present time, and mostly, too, by Offices of high standing. The names have been carefully excluded, and everything that might lead to their identification.

ciples of reform is very difficult in the matter under consideration. It is a subject which can never become popular. The assuring public will never enter into the abstract merits of any particular scheme; but if the comparative numerical results of each one be clearly exhibited, the effect must be that the adoption of equitable methods would be forced upon the different Offices. Each would gradually lose the custom of any class of lives to which its method was known to be unfavourable, and the favoured classes would ultimately be drawing profits contributed only by themselves. In addition, therefore, to a statement of what I hold to be correct principles, and a description of the various methods in practice, I have appended a table of comparison, which, at a glance, will show the working of any particular scheme as compared with any other, in a manner that can be sufficiently understood even by persons who are not familiar with such matters.

The need for an equalization of the risks of death called into existence the plan of life assurance. The theory upon which this system depends, presupposes that the duration of life in a community can be accurately measured. Every year additional facts are acquired, and every year tends to confirm the soundness of the theory, though it modifies its application. The theory also presupposes that the rate of interest at which money will accumulate can be foreknown; a matter on which our predictions can never be more than approximately accurate. The uncertainty thus caused by modifications in the laws of mortality and the fluctuations in the rate of interest, apart from expenses of management, require that a marginal addition (technically termed a "loading") should be made to the estimate of the risk, sufficient to provide for the probable divergencies of the future from the experience of the past. Hence the premiums charged are greater than have been hitherto found sufficient for the risk; and this brings us to the consideration of the surplus, which may be traced to—

1. The loading, or margin for guaranteeing the engagements of the Company against the fluctuations of mortality and the rate of interest, for providing for the expenses of management, and for forming a bonus fund:

2. Interest—a higher rate being realized than was assumed: and,

3. Miscellaneous sources, lapses, surrenders, non-profit policies, office age over real age, fines.

The premiums charged to the public consist of two parts— (1), that which simply provides for the risk; and (2), the loading, the whole or a portion of which is at intervals found to be surplus, and accordingly distributable among the assured. In respect of the former portion of the premiums, the assured form a community of assurance; and in respect of the latter portion, a community of profit. The only principle of the division of the surplus which has been advocated, is, that each of the assured should receive in proportion to his contribution to the profit fund. I have already pointed out that, as regards one portion of the premium, the assured may be considered as a community of profit—the criterion of all partnership; and the principle just laid down is that which guides all partnership contracts, whether the contribution to profits be property or labour. Now, the error of certain systems of distribution is owing, not to the denial of this principle, which no one has, as far as I know, ever attempted to controvert, but to the almost wilful refusal to connect the sources of surplus with the contributions of the members, without which connexion the principle of proportional distribution cannot be applied.

Let p = premium charged;

p' = premium for the risk, or true premium;

ϕ = loading, or $p - p'$;

a^n = amount of an annuity of £1 for n years;

s = total amount of divisible surplus; and

Σ be the symbol of summation.

The valuation being assumed to have been made by a true table of mortality and at a true rate of interest—and if not so based, the real surplus would obviously not be known at all—the principle would be correctly applied by the formula $\dfrac{s}{\Sigma \phi a^n} \phi a^n$,* to represent the share of the assured in the divisible surplus. These assumptions apply to the valuation alone; the table on which the premiums are based has nothing to do with the question. When p is not more than p', then there would be no bonus, as should be the case. The application of the principle by the above formula includes the assumption that the expenses of management are paid out of the miscellaneous profits which have not been contributed to by those who participate in the surplus; or, if these sources of profit will not pay the whole expenses, then it assumes that the management should be borne *pro ratâ* to the loading. I shall hereafter refer to this point, and, for the present, it will be convenient to exclude it from the general question which is the subject of this paper. Since the

---

* *Vide* Mr. Jellicoe's papers in the *Assurance Magazine*, vols. ii. and iii. I wish this to be considered as a general reference, applying to other portions of this Essay.

formula depends on the rigid fulfilment of the above conditions, it is necessary to inquire how far they can be obtained.

1st. *A true table of mortality.*—The English Life Table of the general population of the country may be safely considered as representing the mortality of the majority of Offices—certainly all with a large proportion of country business; and Mr. Morgan's Equitable Tables, as representing the mortality of the better class of assured lives. The valuation surplus would probably be derived very largely at the first and second periodical divisions, and afterwards, in a less degree, on account of suspended mortality, arising from recent selection of the lives; but this portion of the surplus ought not to be divided, inasmuch as the subsequent mortality cannot safely be assumed to follow the average. To ascertain the divisible surplus, deduction from the valuation surplus ought to be made for suspended mortality. Hence, then, no profits can be said to be made from that source; should they appear, inappropriate tables have been employed, and all certainty of profit gone. For these considerations, I would disregard this source of profit from the question of the true mode of distributing profits.

2nd Condition. *A true rate of interest.*—I would almost venture to say, that this cannot be fulfilled. While, on the one hand, we find $3\frac{1}{2}$ and $3\frac{3}{4}$ realised by two Offices of the highest standing, so, on the other, there are Offices making $4\frac{1}{4}$ and $4\frac{1}{2}$, and in one first-class Office very nearly 5 per cent. is made, with the life funds, through preference of investments being granted to that portion of a large accumulation of money. Four per cent. has in some valuations been assumed as the true rate. Now, it is to be observed, that with the accumulation of funds the difficulty of maintaining the rate of interest increases; and further, it depends upon good management. An Office may have now a skilful financier as its actuary; he would, doubtless, advocate a 4 per cent. valuation. But should the management change, and a few men, with prejudices for and against particular securities, become members of the Board, $3\frac{1}{2}$ per cent. might be the rate of interest—and all this apart from the fluctuations in the value of money. Supposing £1,000,000 to be the liability by a 4 per cent. valuation, the change in the base to 3 per cent. would diminish the divisible surplus by upwards of £100,000; and this might cause considerable alarm among the Office connexions. I would, therefore, adhere to a lower rate of interest—3, or, at most, $3\frac{1}{2}$ per cent.—and introduce, as a further element in the ratio of division, the profit from this source. From the total divisible surplus, the profit so derived would be deducted, and would be composed of two parts—

1. To be divided among all that have paid premiums within the period ; and,
2. to be divided among those who participate in the previous division.

It is clear that, if at the last valuation 3 per cent. were assumed and 4 per cent. have since been realised, the sum then reserved was more than sufficient, and the difference ought to be returned, as far as practicable, to those who would then have participated in a larger amount of surplus.   This element in the distribution might be easily introduced where the premiums are based on the true table of mortality, and where the loading is wholly a percentage on the premiums or a percentage with a small constant for expenses. The following methods of distribution all omit a proportion of profit from interest to those who participated in a former division ; and all but the first, the second (where the loading is a percentage), and the third (where the loading is a percentage and a constant), with respect to the increased interest obtained within the last period.   In giving my opinion, therefore, that any one mode is equitable, it will be always subject to modification in the case of the interest obtained being more than the assumed rate.

The following are the principal methods at present pursued :—

1. In proportion to premiums paid since the last division, accumulated at compound interest to date of valuation.
2. In proportion to the loading—that is, the difference between true and Office premiums—accumulated at compound interest between the intervals of division.
3. In proportion to the difference between the pure and Office premiums, less a constant for expenses, accumulated at compound interest between the intervals of division.   (This is adopted, with some modification.)
4. In proportion to the difference between the premiums paid, accumulated with interest, and the respective values of the policies.
5. In proportion to the values of the policies.
6. A reversionary bonus in proportion to the premiums paid, without interest, always reckoning from the commencement of the policy.
7. In augmentation of the sums assured by a percentage per annum on the sums assured and the bonuses already

declared, the number of years being reckoned back to the last distribution of profits only.

8. In augmentation of the sums assured by a percentage per annum from the date of the policy.

9. In reduction of the annual premiums by an uniform percentage to all members after payment of a fixed number of full premiums.

10. In proportion to the amounts assured under the policies which have become claims within the year.

11. Amongst those of the assured only whose premiums, accumulated with interest, exceed the amounts assured under their policies.

1st method.—In proportion to premiums paid since the last division, accumulated at compound interest to date of valuation.

This is the most general mode of distribution, and if the premium is formed from the true premium for the risk, and loaded by a percentage, it is equitable; and, which is a great recommendation, is a practically good mode. By "practically good," I mean that it produces that proportion between the bonuses at different ages which will give most general satisfaction among the assured.

Using the same notation, the working of this will be shown by the formula $\frac{s}{\Sigma pa^n}pa^n$; and if ϕ be a percentage on p', and, as before, $p=p'+\phi$, then $p:\Sigma p::\phi:\Sigma\phi$. Hence this formula will give the same results as $\frac{s}{\Sigma\phi a^n}\phi a^n$.

But though this method of distribution, under the assumption of a percentage loading, is equitable, such a loading is very rare in practice; and according as the loading diverges from this formation, so does the method depart from what is just. I have known this method applied where Northampton premiums were charged. Its effect with such premiums will be readily seen from the following :—

Age.	Northampton.	English Life Table.	Δ.
	Premium per Cent. per Annum.	Premium per Cent. per Annum.	
	£ s. d.	£ s. d.	s. d.
25	2 8 1	1 15 9	12 4
55	5 6 4	4 16 9	9 7

If the English Life Table be held as the true rate, the respective shares in the surplus at ages 25 and 55 at entry would be as 24 : 53, or about 4 : 9, though age 25 contributes annually to the surplus 2s. 9d. per cent. more than age 55. Were commission taken into account, the unfairness of the method applied to such premiums would be still more manifest.

If Mr. Morgan's Equitable be taken as the true rate, we have—

Age.	Northampton.	Equitable.	Δ.
	£ s. d.	£ s. d.	s. d.
25	2 8 1	1 17 2	10 11
55	5 6 4	4 14 7	11 9

that is to say, the ratios of division would be as 4 : 9, though the contributions were nearly equal, viz., 10s. 11d. and 11s. 9d. The Northampton rates exhibit the unfairness in its highest degree, but there are many others which to a less extent are unfair. Before passing on to the next method, I would notice that the ratio of the premiums paid would be a fair ratio for dividing that portion of the profit which has arisen from excess of interest derived on the premiums paid *within the period since the last division.* The profit from interest on the reserve for liabilities made at the former division would have to be divided among those only who then participated. I have remarked upon the expediency of this method of distribution, and upon this point it is to be further noticed that the ratio of division (premiums accumulated within the intervals of division) is, after the first division, a constant. The successive bonuses being equal, the share of each in cash will also be equal, and the reversionary bonuses with advancing years will continuously diminish. Now the reversionary bonus is the measure among the public, and the reason for the diminution is not understood. This result has, of course, nothing to do with the équity of the method, whatever it may have with its expediency. If the view of the public with respect to increasing bonuses were incompatible with equitable distribution, it would be necessary to dispel the notion, difficult as that might be. This, however, is not necessary. Mr. Sprague has pointed out that the cash bonus may be applied as an increasing assurance; and there is the plan pursued by some Offices of converting the cash into reversion by the table of single premiums given in the prospectus for participating policies, and then giving bonus on bonus in future divisions. When the Office is "Mutual," no objection can be urged on the ground of principle; but when the Office is "Mixed," the effect is to give to the shareholders a portion of a former division, although they have already received their full share. This injustice is increased by the conversion of the cash bonus into reversion by the table above quoted, which is very rarely used when the assured have the option. The rates include a charge for Office expenses and commission ; so that when the assured have a cash bonus, a portion of the two-thirds or four-fifths, as the case may be, to which they alone are entitled, that portion is taxed with a charge for the management of it, with commission which is never paid, and with a contribution one-third or one-fifth of which will go to the shareholders, though such contribution be a portion of the two-thirds or four-fifths of profits, to which they have no title whatever. If this system be advisable, deduction should at least be made of that portion of the loading, about one-half, which provides for expenses and commission.

2nd method.— *In proportion to the loading—that is, the difference between true and Office premiums—accumulated at compound interest between the intervals of division.*

It has before been stated, that any profit that may arise will be from ϕ; and if this be the measure of each one's profit, this method satisfies the only principle that has been laid down as equitable.

The application of this method is extremely easy, various as may be the premiums payable under the policies. The existence of twenty different rates of premiums would not be the least obstacle to its application.

3rd method.—*In proportion to the difference between the pure and the Office premiums, less a constant for expenses, accumulated at compound interest between the intervals of division.**

This is a modification of No. 2 method. Nos. 1 and 2 both assume that the expenses of management should be paid out of the general surplus fund, so that each one's share of expenses would be in proportion to his contribution to surplus. Now, if this be a constant, his share of the expenses is also a constant. This I hold to be correct. One £100 policy costs (disregarding commission, which is a percentage) as much per annum as another, whether the premium be £5 or £2. The charge for this purpose should therefore be the same at all ages. If this principle be admitted, the second method is erroneous to this slight extent, and the third is perfectly equitable. This constant need not be sufficient for the full amount of expenses, as the miscellaneous profits may be justly

---

* The working of this is, I believe, not exactly as described. The accumulation is made by a table of temporary annuities; and the former bonuses are taken into account in determining the ratios for subsequent divisions.

set off to diminish these. Could they be found sufficient, the second method would be equitable, and, as no constant would then have to be deducted, the third method would coincide with it.

4th method.—*In proportion to the difference between the premiums paid, accumulated with interest, and the respective values of the policies.*

The application of this at the first division of profits is very simple; and, supposing x to represent the ages at entry, the formula for working would be $\dfrac{s}{\Sigma(p_x a^n - l_x)} \times (p_x a^n - l_x)$ for the share of each policy. At the second and subsequent divisions, it becomes more complicated. All the policies that participated in the former division are, for the purpose of the second division, considered as new policies effected at the first division. Thus, if a man assure at age 25, and is 32 at the first division, the accumulated premiums paid, less the value of a policy 25 at entry, after seven years, will express the ratio of his share in the surplus at the first division. If the second division be seven years after the first, seven years accumulations of premiums payable at the age of 32—his age when the first division took place—less the value of a policy on the life of 32, after seven years, will express the ratio of his share at the second division. There are modifications of this in some instances, as regards the bonus. The bonuses declared at previous divisions are added to the amount assured, and the premiums are formed upon the augmented amount assured. Thus, if in the example before given £100 were the bonus at the first division on a policy for £1,000, the ratio for the second division would be the accumulated premiums on £1,100 at age 32, less the value of a policy of £1,100 seven years after inception at that age.

The formulæ for the working of this method at the second division, without and with the modifications as regards bonus, would be—

$$\frac{s}{\Sigma(p_{x+7}a^7 - {}^7 l_{x+7})} \times (p_{x+7}a^7 - {}^7 l_{x+7})$$

and

$$\frac{s}{\Sigma\{p_{x+7}(M+B)a^7 - {}^7 l_{x+7}(M+B)\}} \times \{(p_{x+7}(M+B)a^7 - {}^7 l_{x+7}(M+B)\},$$

M being the amount originally assured, B the bonus, and in the second formula p and l are the premiums and values per £1 assured. The first of these formulæ is to be preferred, as the working of it is more simple, and it does not further augment the bonuses on policies of long standing, which are otherwise too large.

The principle on which this method seems to have been founded is this, that the accumulation of all premiums, less the value of all the policies, would represent the gross surplus, if there were no expenses and no claims, and the contribution of each to that gross surplus would be the accumulation of the premiums he had paid, less the value of his policy. The divisible surplus being the amount of such imaginary surplus, less the claims and expenses, the share of each in the divisible should be represented by the ratios in the imaginary surplus. This notion includes the fallacy, that if a man aged 55 lives over a period, say of five years, he has contributed more profit than one aged 25 living five years, apart from the question of the loading of the premium.

Now, it has been already shown that this is contrary to the very essence of life assurance. In a true table of mortality, no profit is derived from long life, except as regards that portion of the premium which expressly provides for profits and expenses. If this distribution between the two parts of the premium be not regarded, all consistency in our deductions will disappear.

Referring again to the formula $\dfrac{s}{\Sigma(pa^n-l)}pa^n-l$, if we substitute for p its equivalents $p'+\phi$, and $\therefore$ $(p'+\phi)a^n$ for Σpa^n and for Σl, the net liability on all the policies $\Sigma(p'a^n-c)$; for it is clear that if p' represents the true premiums, the amount which should be in hand at any time would be the sum of the amounts of every p' or $\Sigma p'a^n$ less the total amount in claims, or Σc; so also for l we may substitute $p'a^n-c$ (c being p' contribution to claims), we then have

$$\frac{s}{\Sigma(p'+\phi)a^n-\Sigma(p'a^n-c)}\times\{(p'+\phi)a^n-(p'a^n-c)\},$$

or

$$\frac{s}{\Sigma(\phi a^n+c)}\phi a^n+c;$$

that is to say, the profit is allotted in the proportion of the contributions to profits and the payment of claims, which has been shown to be contrary to principles of equity, and to include an assumption opposed to the fundamental principles of life assurance.

5th method.—In proportion to the values of the policies.

By this method the surplus is divided in proportion to the liability under each policy, and the working of it is expressed by the formula $\dfrac{s}{\Sigma.l}\cdot l$.

How inequitably this system works will be seen by the table of comparison appended, and this is with a well-graduated premium upon the same base as the valuation. If applied in an Office where different rates are payable—some with a large margin, others with a small or no margin—those paying a large marginal addition will have the same bonus as those paying a small one (the ages of the lives and the dates of policies being the same), and even a less bonus than a policy with a premium inadequate for the risk. It is, however, to be observed, that if all the participating policies in an Office paid according to the same scale of premiums, and this were formed on a true base, with a percentage and small constant for loading, this method would provide to the new policies compensation in future years, since the value of every policy, whether effected on a young or old life, would, if the life attained the extreme age in the table, range from 1 to 99 per cent. The ratios of division, therefore, range between the same limits for all policies.

6th method.—A reversionary bonus in proportion to the premiums paid, without interest, always reckoning from the commencement of the policy.

The clause in the deed wherein this scheme was devised runs as follows :—

" The dividend to each member shall bear the same proportion to the total sum to be divided as the total premium received of the existing members insured for life, which sum allotted to each individual shall be added to the sums originally insured."

The Office, I believe, interprets this clause to mean, that the reversionary bonus is to be in proportion to the premiums paid, not the cash bonus, as it has been reasonably considered to mean, on the ground that the sum to be divided being cash, the fourth term of the proportion must be like, and also cash. There is no trace of principle in this. The profits are given to the long livers, and in greater proportion than by any other scheme.

7th method. —In augmentation of the sums assured by a percentage per annum on the sums assured and the bonuses already declared, the number of years being reckoned back to the last distribution of profits only.

This is unscientific ; but, by a graduated loading, this plan could be made nearly equitable. It would not, however, admit of adjustment, should subsequent experience render it desirable to modify the table of mortality.

It will be seen in the table of comparison, that according to the loading there made, it favours the young lives. How far the plan of giving bonus on bonus is inequitable, depends on the way in which the cash surplus is converted into a reversionary sum, and on the Office being " Mutual" or " Mixed."

8th method.—In augmentation of the sums assured by a percentage per annum from the date of the policy.

The origin of this is well known to have been the discovery that the experimental premiums charged to the early assured were greatly in excess of the risk, and this was devised to restore the equality. It could never be adopted again, and nothing but a great reputation for wealth could sustain an Office against the inexpediency of the method. The application of the principle varies in different Offices. The Office in which this method originated converts the cash bonus into reversion, and then divides this according to the above principle. Another Office, however, applies the principle of distribution to the cash surplus, and, though it is a *cash* share of profits, it is not payable until death. The effect of this in a proprietary Office, giving only, say two-thirds, of the profits, may be readily conceived. Assuming the surplus to be £300,000, the shareholders would get £100,000, and the policy-holders £200,000; but since they would only come into this on the falling in of the policies, the present value would not be much more than £100,000. The surplus at the next division would consequently be increased by the remaining £100,000 accumulated, with interest. But here again the shareholders would get their one-third share, and of the policyholders' £100,000 accumulated as part of the general surplus.

The above-described methods are all that are included in the table of comparison before referred to. I append this table, which contains the different methods, arranged in the order I have described them.

The policies are all assumed to be effected in the commencing month of the quinquennial period in which they first participate. They are all for £1,000; and the ages at entry are 25, 40, and 55. The premiums are according to the 3 per cent. English Life Tables (published in the *Registrar-General's Report for* 1849), with loading of 10 per cent., and a constant of 5s. per cent. This loading has been adopted to obtain an average premium, and not from any reasons for that specific loading. The calculations have been made according to the same tables.

Table of Comparison.

Calculations based on the English Life Table (Males), 3 per Cent.

Age at Entry	Premium payable	Age at Valuation	No. 1 Cash	No. 1 Reversion	No. 2 Cash	No. 2 Reversion	No. 3 Cash	No. 3 Reversion	No. 4 (i) Cash	No. 4 (i) Reversion	No. 4 (ii) Cash	No. 4 (ii) Reversion	No. 5 Cash	No. 5 Reversion	No. 6 Cash	No. 6 Reversion	No. 7 Cash	No. 7 Reversion	No. 8 (i) Cash	No. 8 (i) Reversion	No. 8 (ii) Cash	No. 8 (ii) Reversion
25	22·16	30	20·	48·6	25·4	61·5	19·	46·	21·4	51·9	21·4	51·9	18·3	44·4	14·2	34·5	25·5	61·8	25·5	61·8	...	33·3
40	32·70	45	29·5	56·	31·	58·7	29·4	55·6	29·2	55·3	29·2	55·3	30·	56·9	26·8	51·	32·6	61·8	32·6	61·8	...	33·3
55	55·80	60	50·5	74·4	43·6	64·3	51·6	76·	49·4	72·8	49·4	72·8	51·7	76·2	59·	87·	41·9	61·8	41·9	61·8	...	33·3
First Period			**£100·**		**£100·**		**£100·**		**£100·**		**£100·**		**£100·**		**£100·**		**£100·**		**£100·**		**£100·**	
25	22·16	35	20·	44·8	25·4	56·9	19·	42·5	20·2	45·2	20·7	46·3	25·8	57·9	19·5	43·6	27·3	61·1	34·9	78·2	...	44·4
40	32·70	50	29·5	51·4	31·	54·	29·4	51·2	32·4	56·4	30·	52·3	41·4	72·	36·9	64·4	35·1	61·1	45·	78·2	...	44·4
55	55·80	65	50·5	69·3	43·6	59·8	51·6	70·9	62·4	85·4	60·5	83·	66·7	91·4	80·1	110·	44·5	61·1	57·0	78·2	...	44·4
25	22·16	30	20·	48·6	25·4	61·5	19·	46·	18·3	44·4	19·	46·	12·1	29·6	9·	21·8	23·7	57·6	16·	39·1	...	22·2
40	32·70	45	29·5	56·	31·	58·7	29·4	55·6	24·8	47·1	26·	49·3	19·8	37·4	17·	32·2	30·4	57·6	20·6	39·1	...	22·2
55	55·80	60	50·5	74·4	43·6	64·3	51·6	76·	41·9	62·	43·8	64·6	34·2	50·4	37·5	55·3	39·	57·6	26·5	39·1	...	22·2
Second Period			**£200·**		**£200·**		**£200·**		**£200·**		**£200·**		**£200·**		**£200·**		**£200·**		**£200·**		**£200·**	
25	22·16	40	20·	41·2	25·4	52·1	19·	39·1	20·6	42·5	20·1	41·4	30·3	62·4	22·9	47·4	29·3	60·3	40·4	83·4	...	50·
40	32·70	55	29·5	47·2	31·	49·6	29·4	47·	32·1	51·3	31·2	50·	48·6	77·8	43·	69·	37·7	60·3	52·3	83·4	...	50·
55	55·80	70	50·5	65·	43·6	56·	51·6	66·4	78·6	101·3	73·	94·	72·4	93·3	91·5	117·9	46·8	60·3	64·9	83·4	...	50·
25	22·16	35	20·	44·8	25·4	56·9	19·	42·5	17·7	39·6	18·2	40·6	19·1	42·9	14·1	31·3	25·4	56·8	24·8	55·6	...	33·3
40	32·70	50	29·5	51·4	31·	54·	29·4	51·2	25·7	44·8	26·4	46·	30·6	53·1	26·3	46·	32·6	56·8	32·	55·6	...	33·3
55	55·80	65	50·5	69·3	43·6	59·8	51·6	70·9	52·6	72·3	53·	72·7	49·8	68·2	57·2	78·8	41·4	56·8	40·6	55·6	...	33·3
25	22·16	30	20·	48·6	25·4	61·5	19·	46·	15·6	37·8	16·8	40·7	9·	21·9	6·4	15·6	22·1	53·6	11·5	27·8	...	16·6
40	32·70	45	29·5	56·	31·	58·7	29·4	55·6	21·2	40·2	22·8	43·	14·7	28·	12·1	23·	28·3	53·6	14·7	27·8	...	16·6
55	55·80	60	50·5	74·4	43·6	64·3	51·6	76·	35·9	52·9	38·6	56·8	25·5	37·6	26·5	39·4	36·4	53·6	18·8	27·8	...	16·6
Third Period			**£300·**		**£300·**		**£300·**		**£300·**		**£300·**		**£300·**		**£300·**		**£300·**		**£300·**		**£300·**	

From the results here exhibited, it will be observed—

1. That according to the first three (which I will, for convenience, call first class) methods of division, the shares of each age in cash are constant—the same at one division as at another.

2. That the remaining methods—or second class—(omitting No. 6), give generally nearly similar results with one or other of the first class methods as regards the first period.

3. That at the second and third periods a rapid augmentation in the profits of the old policies takes place under the second class methods—omitting No. 7.

4. That this augmentation mainly goes to the old *lives* in No. 4 (i. and ii.); in the remaining portion of the second class methods it goes to the old *policies*, which retain very nearly the same proportion of profit, one with another, as at the first division.

5. That the bonus on policies of five years' standing rapidly diminishes with the increase in the number of old policies.

It has been observed, as regards No. 4 (i. and ii.), that " at each septennial division the same degree, or very nearly the same degree, of injustice is perpetrated as at the first"*—a little consideration of the *tendency* of the results will show that this is not so. The cash bonus of 25 at entry, taking the results of the third period under 4 (ii.) is, after 5 years, 16·8; after 10 years, 18·2; after 15 years, 20·1; which last is nearly equal to the bonus allotted at age 40 after 5 years. If this table had been worked out to a fourth period, the bonus of 25 would have been the same as 40 after 5 years. This result follows from the application of the principle as before described. The effect of this method, if continued to more distant divisions, will give, at the 11th period, to six policies taken out at the same age 25, but of different standing, the following proportions of bonus :—

	Years standing.					Years standing.
The policy of 20 would have about		1½	as much as policy of	5		
„	30	..	.. 2	..	..	„
„	45	..	.. 4	..	..	„
„	55	..	.. 8	..	..	„

A policy taken out at an early age, and continuing in force to the extreme age, would have successively the ratios of division for all the interval ages up to the highest.

---

* *Assurance Magazine*, vol. iii., p. 191.

Of the methods not included in the table, I have but few observations to make—little more, indeed, than reasons for their omission.

From the peculiarity of the system of No. 9 (by which the annual premiums are reduced by an uniform percentage to all members, after payment of a fixed number of full premiums), no fair comparison could be instituted with the standard adopted in the table, and on this ground it was omitted. It is clear, however, that the bonuses are in proportion to the premiums, and therefore its equity mainly depends on the loading. The time of the postponement of the privilege of reduction being the same for the old as for the young lives, is also to be regarded in considering its equity. If the loading were a percentage, this rule would act very unequally, and would be unfavourable to the members entering at advanced ages.

The 10th method is neither more nor less than a pure lottery.

The 11th method, which assumes that those only are entitled to profits who have paid up their policies, is a system of distribution utterly opposed to the principles and intention of life assurance. As it would preclude profit to any policy within the period of those given in the table, it could not there be introduced.

The conditions upon which the first three methods in the table are equitable have been given. As regards the remaining methods (omitting No. 7, which I have before disposed of), if the principles laid down as to the sources of profit be borne in mind, the table of comparison establishes *that the profits contributed by the new policies are apportioned to the old, and when these new policies become old the methods provide compensation out of the contributions of the future new policies. How if these should not be forthcoming?*

I append a further table, based on Mr. Morgan's Equitable Tables. The loading in this is 15 per cent. on the pure premiums, with a constant of 5s. per cent. In the calculations of the bonuses according to No. 5 method, the values of the former bonuses have been included in the ratios of division. This has been done, as it is the practice of one of the Offices which has adopted this method. In other respects, the table explains itself.

Table of Comparison.

Calculations based on Mr. Morgan's Equitable Tables, 3 per cent.

Age at entry.	Premium payable.	Age at Valuation.	No. 1 Cash.	No. 1 Reversion.	No. 2 Cash.	No. 2 Reversion.	No. 3 Cash.	No. 3 Reversion.	No. 4 Cash.	No. 4 Reversion.	No. 5 Cash.	No. 5 Reversion.
25	21·55	30	19·5	49·4	23·9	60·5	18·5	46·9	20·2	51·15	18·6	36·8
40	32·36	45	29·2	56·7	30·6	59·3	28·9	56·	27·9	54·09	30·7	48·4
55	56·87	60	51·3	76·4	45·5	67·8	52·6	78·3	51·9	77·27	50·7	63·3
First Period ………			£100·		£100·		£100·		£100·		£100·	
25	21·55	35	19·5	45·2	23·9	55·4	18·5	42·9	19·8	45·9	26·9	49·4
40	32·36	50	29·2	51·6	30·6	54·1	28·9	51·1	28·5	50·4	44·5	64·7
55	56·87	65	51·3	70·9	45·5	62·7	52·6	72·5	62·2	86·	69·4	81·1
25	21·55	30	19·5	49·4	23·9	60·5	18·5	46·9	18·1	45·8	11·	21·7
40	32·36	45	29·2	56·7	30·6	59·3	28·9	56·	25·	48·5	18·2	28·7
55	56·87	60	51·3	76·4	45·5	67·8	52·6	78·3	46·4	69·1	30·	37·5
Second Period ……			£200·		£200·		£200·		£200·		£200·	
25	21·55	40	19·5	41·4	23·9	50·7	18·5	39·2	19·5	41·4	31·7	54·
40	32·36	55	29·2	47·2	30·6	49·4	28·9	46·5	30·7	49·6	52·3	70·3
55	56·87	70	51·3	66·4	45·5	58·7	52·6	67·8	74·6	96·5	77·3	85·1
25	21·55	35	19·5	45·2	23·9	55·4	18·5	42·9	17·3	40·1	18·6	34·1
40	32·36	50	29·2	51·6	30·6	54·1	28·9	51·1	25·	44·2	30·9	44·9
55	56·87	65	51·3	70·9	45·5	62·7	52·6	72·5	54·5	75·3	48·1	56·2
25	21·55	30	19·5	49·4	23·9	60·5	18·5	46·9	15·8	40·	7·6	15·
40	32·36	45	29·2	56·7	30·6	59·3	28·9	56·	21·9	42·5	12·6	19·9
55	56·87	60	51·3	76·4	45·5	67·8	52·6	78·3	40·7	60·6	20·9	26·1
IIIrd Period ……			£300·		£300·		£300·		£300·		£300·	

CORRESPONDENCE.

MR. FINLAISON'S REPORT AND THE ENGLISH LIFE TABLE.

To the Editor of the Assurance Magazine.

SIR,—The letter of Mr. Porter, in your last number, contains so many questionable assertions, and does such injustice to one of the most valuable tables of mortality we possess, as to require some further notice.

The information contained in the tables of Mr. Finlaison's Report is of great interest and value; the accompanying remarks, embracing a variety of different subjects, as might be expected, afford room for considerable difference of opinion. But most of your readers will probably agree in thinking, that the greater part of the eleven pages devoted to criticising the English Life Table would have been advantageously omitted. That Mr. Finlaison should look with some disfavour upon actuarial investigations proceeding from another Government department than his own, is not, perhaps, very surprising; but that Mr. Porter, passing over so much that is valuable in the Report, should have chosen this particular passage for comment, is in every way unfortunate. What have the records of the National Debt Office to do with the census returns of 1851? The portion of the latter, referring to the ages of the population, was published in 1854, and the facts were generally commented upon at the time. Why, then, has Mr. Porter, with his strong opinion of their worthlessness, allowed them to pass for seven years unnoticed?

Mr. Porter has made a general summary of his remarks under six heads, upon each of which I will say a few words.

"1. The inordinate extent to which the ages of females between the ages of—say, 20 and 40—are understated in the census."

Could anyone, whose only information on the subject was derived from Mr. Porter's letter, suppose that the authors of the *Report of the Census for* 1851 expressed a decided opinion of the general accuracy of the returns of the ages both of males and females, and that the passage quoted (or rather misquoted) in Mr. Finlaison's Report, on which Mr. Porter lays so much stress, was actually written in support of this opinion? Such is, however, the fact; but, by the omission of eight words, the meaning of the passage is entirely perverted, and the Registrar-General and his colleagues are made to express an opinion diametrically opposite to the one they really entertain. On reference to the *Report of the Census for* 1851, part 2, vol. i., pages 23 and 24, it is stated, that—" The mean age of the females, as they are returned in England, exceeds the mean age of the males by *ten months;* so that the tendency in women to understate their ages has only operated on comparatively small numbers; and there is no doubt of their general truthfulness." Some further reasons in support of this view are then given; at the same time, we are informed that at certain ages there are some evident misstatements, which, however, admit of being corrected. Then follows the passage quoted in your *Magazine*, page 278, which is correctly given for the first portion, but which in the original proceeds as follows:—
" The extensive immigration of the Irish into Great Britain during the 10 years, 1841–51, has exercised some disturbing influence on the proportions; but, upon comparing the above numbers with those for males at the corre-

sponding ages, the conclusion appears to be inevitable that some 35,000 ladies, more or less, who have entered themselves in the second age, 20–40, really belong to the third age, 40–60; *to which the body of delinquents are transferred in Table* 7. *Millions of women have returned their ages correctly;* thousands have allowed themselves to be called 20, or some age near it, &c." The words in italics are omitted in the quotation; and the meaning intended to be conveyed is, in consequence, entirely misrepresented. The plain signification of the passage is—that there was no doubt of the general truthfulness of the returns of the ages; that there were some inaccuracies at particular periods of life, but only in the proportion of thousands to millions; that these inaccuracies admitted of correction, and had been corrected accordingly.

Mr. Porter, having thus propounded a statement which is altogether untrue (inadvertently, of course, but which a very little research would have prevented), proceeds, in language which I must be allowed to characterise as neither courteous, nor manly, nor just, to declaim against the ladies for *their* supposed want of veracity. But the indignation which began to be aroused at this charge soon passed into a smile at the remedy suggested, viz., that ladies should add to their other accomplishments a knowledge of Horace! and that their especial attention should be directed to one of the odes, which, in my time at least, *young gentlemen* in the course of their studies were usually recommended to omit.

" 2. The over-estimate of age in very advanced life, both in the census and in the registers of death."

Mr. Porter refers particularly to ages above 100, and remarks that it is a curious fact, that members of the Peerage never attain to fabulous age. Here, again, a little investigation into the real facts quite alters his view of the case. In the census of 1851, there were found 319 centenarians out of a population of 21,185,010, or only 1 in 66,411. I see no improbability in this, especially as in some researches on the families of the Peerage, recently made by Mr. Day and myself, we found 2 centenarians out of 7,473 cases, or 1 in 3,736.

" 3. The loose system adopted in ascertaining the ages at death of all classes of people, at all ages."

This is a matter within the experience of everyone; the description given by Mr. Porter is borrowed from his own imagination. The simple fact is, that medical men exercise no judgment whatever upon the ages at death; the *cause* of death is another matter, but the age inserted in the certificate is derived from the information of relations and friends. As with the ages of the living, there are probably several errors of one or two years in both directions, but there is no reason to doubt the substantial accuracy of the returns. And in applying the results to the construction of life tables, it must be remembered that the changes made in the process of graduation are so extensive, that the individual errors in the returns are comparatively unimportant and probably would not affect any pecuniary results.

" 4. The direct pecuniary loss sustained by females in the purchase from the Government, or from such Annuity Companies as adopt special scales for female lives, on the present assumption of the vitality of females being so much in excess of that of males."

This is quite new information. Mr. Hendriks, some years ago, contributed to the *Statistical Journal* an elaborate paper on the " Loss sustained by Government in granting Annuities." As regards the mortality, the Government tables since 1829 have been deduced from their own experience; and the superior vitality of females, which Mr. Porter calls an assumption, is, at all events as regards annuitants, an incontestable fact, proved by many years' experience. The established Assurance Companies, with hardly an exception, have discontinued to grant annuities, because they have found the business unprofitable; and as the majority of annuitants are females, there can be no doubt that the "direct pecuniary loss" has been on the contrary side to what Mr. Porter alleges.

"5. The advantage the female sex might derive, were Life Assurance Companies to adopt special rates of premium for female lives."

Surely Mr. Porter cannot be ignorant that this experiment has been more than once tried and abandoned! For nearly a quarter of a century the Eagle Office adopted special premiums for female lives. A careful investigation into the results of the experience of this Office was made by Mr. Jellicoe, and published in this *Magazine* (vol. iv., page 199). In the course of his remarks, Mr. Jellicoe observes—"One inference is, at all events, fully supported by these data, viz., that the insurance of female life at less rates than that of male is scarcely justifiable."

"6. The general discredit thrown upon the English Life Table, both for males and females, and the necessity that exists for Life Assurance and Annuity Companies to abstain from using it, in consequence of the incorrect returns, which, both as regards the numbers living and the numbers dying at each age, have been, no doubt, habitually made to the census-enumerators and the District Registrars of Deaths."

With a passing remark, that all Mr. Porter's objections would apply with much greater force to the Carlisle Table, against which he says nothing, I will briefly give my reasons for taking exception to this sweeping censure. The English Life Table is based, not upon the census of 1851, but upon that of 1841, where the return of ages was made in quinquennial intervals only; and the first process was to obtain the annual mortality in quinquennial periods of age. Even when treated in this manner (vide *Registrar-General's 5th Annual Report*, page 347), it was found that, as with other similar observations, the series *for both sexes* exhibited great irregularities. Corrections were made by means of Mr. Gompertz's hypothesis, that the rate of mortality may be represented by the expression aq^x; and Dr. Farr accordingly reduced the annual mortality above the age of 15 to two geometrical series with a different common ratio, one for the interval from 15–55, and the other from 55–95. The result is, a table of mortality very convenient for practical use, upon the accuracy of which we rely—both because we believe the ages both of the living and dead to have been given with sufficient accuracy for pecuniary purposes (which is quite compatible with several individual errors), and also because it is consistent with the results of other trustworthy observations. To apply a most stringent test— let us compare the English Life Table with Mr. Finlaison's tables, taking the two identical examples quoted by him to establish the inaccuracy of the former. (1) Out of 2,000 persons living at the age of 55, there survive to 75 by the English Life Table 778, and by the Government Table 901, persons. Is not this precisely what was to have been expected? The tontine nominees, upon whom the bulk of Mr. Finlaison's observations in early life are made, are recruited in old age by a very select body of lives of a different class — the annuitants, properly so called. Can, then, anyone be surprised at the Government tables indicating a more favourable mortality between the ages of 55 and 75, than that of the general population? Had a contrary result been arrived at, would there not have been much more reason to doubt the accuracy of the English Life Table? (2) At the age of 27, taking interest at 4 per cent., the annual premium to assure £1,000,000, is, by the Government Male Table, £17,447, and by the English Male Table, £16,885, *i.e.*, £1. 14*s*. 11*d*. per cent. by the former, and £1. 13*s*. 9*d*. per cent. by the latter, the difference being one shilling and two pence per cent. To my comprehension, this proves the accuracy of the English Life Table; and to confirm my impression, I referred to the prospectus of the Office from which Mr. Porter writes, and, comparing the tables with those of the Royal Exchange Office (the selection being made almost at random), made the discovery that at this identical age of 27, the rates of premium of these two Offices differed as much as two shillings and eleven pence per cent.

But, Mr. Porter remarks, the last example is derived from male lives only, and an Assurance Company could only employ a table based on a combination of both sexes. Now, as we know by experience that the proportion of the sexes amongst assured lives is about 9 males to 1 female, I think that it would be more judicious for an Assurance Company to employ a table of mortality deduced from male lives only. Otherwise, I do not think sufficient reason has been shown for distrusting the English Female Life Table; and it is curious, that neither Mr. Finlaison nor Mr. Porter should have discovered that the blow they aim at it recoils on themselves. If, as they would have us believe, between the ages of 20 and 40 the numbers living are enormously overstated, while at the same time the ages at death are more accurately given, it follows that the mortality during that period must be much greater than the table indicates. Yet, with strange inconsistency, one of their objections to the table is, that it represents the mortality of females at the younger adult ages to be so great as to be contrary to nature.

Having myself, for some years, been in the practice of using the Male English Life Table for Assurance calculations, I should have been glad to have given my reasons for thinking that, in the present state of our knowledge, this table is better adapted for the purpose than the others in common use. But having already trespassed too much on your space, I will conclude by commending the English Life Table to Mr. Porter's more careful study, requesting him, in the words of the author whom he wishes ladies to read for their moral improvement—

> " Si quid novisti rectius istis
> Candidus imperti, si non, his utere mecum."

I am, Sir,

Your obedient servant,

ARTHUR H. BAILEY.

Equity and Law Life Office,
    11*th May*, 1861.

ON THE CALCULATION OF PREMIUMS FOR ASSURANCES ON
LIVES AND SURVIVORSHIPS BY THE AID OF MR. GOM-
PERTZ'S HYPOTHESIS.

To the Editor of the Assurance Magazine.

SIR,—Your readers will have observed, with much gratification, the
announcement made by Mr. Gompertz (in his letter to Mr. Porter, published in your April Number), that, during the last two or three years,
that eminent mathematician has been engaged in adding to his important
discoveries in connection with the law of mortality and the methods of
computing the values of life contingencies; and they will most sincerely
echo the wish that his health may permit him, ere long, to lay the complete
results of his very valuable labours before the public, through the medium
of the Royal Society.

The discovery of methods for shortening the labour required in computing correctly the values of intricate cases of survivorship assurances
must be considered of the highest importance by all engaged in the business
of life assurance; and 1 trust that, notwithstanding the announcement above
referred to, which justifies the expectation of a full and comprehensive
treatment of this important subject, the following brief description of a
general formula (based upon what Mr. Gompertz has already given us),
for the solution of the more usual cases of survivorships, may not be
altogether devoid of interest.

In a paper headed "On the law of mortality and the construction of
annuity tables," published in vol. viii. of this *Magazine*, I ventured to
suggest a method which, by means of a slight modification of Mr. Gompertz's formula, appeared to me to possess some utility in facilitating the
computation of the values of *annuities* on several lives.   The alteration in
question consisted simply in introducing into the formula for the probabilities of living an additional constant, in such way that it should combine with the constant representing the interest of money in the corresponding formula for the *values of sums* depending upon those probabilities;
and, by this means, preserve an important property of Mr. Gompertz's
formula, first observed by Professor De Morgan—viz., the power of substituting an equivalent single age, easily determined, for any combination
of joint ages; with this difference, however, that, by the introduction of
the additional constant referred to, the substitution consists of an *equal
number of lives*, of a certain common age, in lieu of a *single life.**

The subjoined tables, representing the decrements and the expectation
of life at all ages, are constructed upon the principle explained in the paper
before referred to.   They are based upon the Carlisle Table; but, for the
sake of convenience in calculation, I have slightly altered the values of the
several constants yielded by that table—considering, for the object in view,
a very close adherence to any particular table unnecessary.   Nevertheless,
upon a comparison of the expectation of life, it will, I think, be found that

* My paper having been drawn up previously to the publication of Professor
De Morgan's article in the *Assurance Magazine* for July, 1859 (which is the only one I
have seen on the subject), the enunciation of the property in question is given as *new*.
In a letter to the Editor, which accompanied my manuscript a few days after the publication of the Number for July, 1859, I referred to Mr. De Morgan's prior discovery, but,
by an oversight, omitted to insert a similar reference in the paper itself.

the difference does not exceed the licence usually allowed in the adjustment of mortality tables.

The first table, which I give entire, containing the decrements and expectation of life, needs no explanation, as the values correspond exactly with those formed by the usual methods. The annuity tables, however (of which extracts only are given), are constructed by the following formula—

$$\frac{1}{B_m^\mu} \int_0^\infty \left(B_m^\mu\right)^{q^x} \left(\frac{v}{a^\mu}\right)^x . \, dx,^*$$

which denotes the value of an annuity *payable momently* during the joint existence of μ lives each aged m years, and consequently they differ in this respect from annuity tables in ordinary use. Various methods, more or less convenient, may be adopted for calculating the values of the above integral, but it is not my purpose to enter into this subject further than to state that in the tables in question the values are computed to the fourth decimal place. As, however, three decimal places are sufficient for most purposes, that number only is given in the extracts, and it will be observed that each annuity table for any given number of lives to three places could be comprised in the same amount of space as a table for single lives.

Adopting a table of annuities payable *momently* as the basis of calculation, let such an annuity on the joint existence of any given number of lives aged respectively m, n, r . . . be represented by $A_{m, n, r \ldots}$. To find the value of a similar annuity, payable t times a year, we have the following simple formula—

$$A_{m, n, r \ldots} \pm \frac{1}{2t},$$

the upper sign being taken when the payments are in advance, and the lower sign when in arrear.

By the aid of these *momently* annuities, a convenient general formula may be deduced for the *exact* solution of the following comprehensive problem in survivorship assurances.

Problem.—Required the value of £1 payable at the failure of the joint existence of μ lives aged respectively m, n . . . and r, provided that event shall happen before the failure of the joint existence of ν other lives aged respectively u, v . . . and z.

The formula by which this problem is solved is—

$$\frac{S_m}{S_m + S_u} \left\{ 1 + \left(\log_e v - \log_e a . \frac{\nu S_m - \mu S_u}{S_m} \right) A_{m, n \ldots z} \right\} \quad . \quad [1]$$

where $S_m = q^m + q^n + \ldots + q^r$, $S_u = q^u + q^v + \ldots + q^z$ and $A_{m \ldots z} =$ the value of an annuity (payable momently) during the joint existence of all the lives involved.

In the case of an absolute assurance on the joint lives m, n . . . and r, the quantities ν and S_u vanish, and the formula becomes—

$$1 + \log_e v . A_{m.n \ldots r} \quad . \quad . \quad . \quad . \quad . \quad . \quad [2]$$

For calculating the value of an assurance on a single life (m) against ν other lives (u), (v) . . . (z), the general formula [1] may be put in the following more convenient form—

* The characters correspond with those used in my paper of January, 1860.

$$\frac{q^{m-y}}{q^{m-y}+q^{u-y}+..+q^{z-y}}\left\{1+(\log_e v+C_{u-m}+C_{v-m}+..+C_{z-m})A_{m,u}...\right\}\ [3]$$

where $C_n=\log_e a\,(q^n-1)$, and y is the age of the youngest of all the lives involved.

When the problem refers to one life (m) against another life (u), the formula may be still further simplified, thus—

$$F_{u-m}+G_{u-m}.A_{m,u}\quad .\quad .\quad .\quad .\quad .\quad .\quad [4]$$

where $F_n=(1+q^n)^{-1}$ and $G_n=F_n.(\log_e v+C_n)$.

It should be observed that the preceding formulæ give the value of the reversion payable at the instant of death, from which the value of the same, payable at the expiration of any time, t, after death, may be accurately determined by multiplying by v^t; and the value of the given sum payable at the expiration of the year of death may be considered equivalent to the same payable six months after death, and determined accordingly.

In taking as the basis of calculation the values of annuities payable momently, it would, perhaps, be preferable that the annual premiums should be supposed to be payable in the same manner—*i. e.*, by momently instalments. The general formula for the annual premium payable momently during the joint existence of all the lives would be—

$$\frac{S_m}{S_m+S_u}\left\{\frac{1}{A_{m,n}...}+\left(\log_e v-\log_e a\,\frac{\nu S_m-\mu S_u}{S_m}\right)\right\},$$

and the annual premium payable by $\frac{1}{p}$thly instalments (in advance) could be deduced from the value so found by simply deducting therefrom $\frac{1}{2p}$th of a year's interest, provided that the proportion of premium paid in advance for any period beyond the time at which the death takes place be returned to the assured. For instance, if π denote the annual premium payable momently, and t the unexpired fraction of the current year at the date of death, the amount of premium returnable to the assured at the instant of death will be $\pi t\left(1-\frac{rt}{2}\right)$ (r being the yearly interest on £1), for the Office may be supposed to owe the assured the sum πt payable by momently instalments during the time t.

A practice somewhat similar to the above is, I believe, generally followed in India with regard to half-yearly and quarterly premiums; and, independently of the facilities afforded in calculation, it is, perhaps, preferable to the plan adopted in this country. However, the annual, half-yearly, and quarterly premiums, according to the usual practice, may be found from the single premium by the aid of the formula for determining the value of an annuity payable p times a year. ·

The general formula [1] is the key to the exact solution of all cases of assurances on lives, whether absolute or contingent, treated in Baily's work on annuities and assurances, with the exception of those involving the problem of determining the value of a reversion on a given life, subject to the condition of a second life surviving a third during the lifetime of the first—a satisfactory solution of which has, I believe, never yet been *published*. And the term " exact solution" is to be taken in its fullest sense;

for the hypothesis of constant decrements for single years, or any similar hypothesis, is not resorted to in deducing the formula. The remaining cases of survivorship assurances, involving the problem before referred to, are also capable of exact solution on principles similar to those by which the formula [1] is deduced; but it is not my purpose to enter here upon the subject of these interesting but unusual cases.

A very important point in Mr. Gompertz's letter is the announcement that the method which he is now engaged upon admits of the solution of cases involving combinations of lives subject to different laws of mortality. This problem also admits of solution by tables constructed on the method suggested by me, provided one uniform value of q can be adopted in the construction of the several tables of mortality. By a slight arbitrary modification of the several constants, I have ascertained that this is perfectly practicable in the case of Indian and European lives, and in all probability it would be found equally so in other cases. Mr. Gompertz's promised paper will doubtless treat this subject in a manner worthy of its great practical importance.

I will not add to the length of this communication by inserting the demonstration of the solution of the general problem, which would occupy considerable space, although no particular difficulty is involved in it. I will, therefore, conclude by a single example of the most usual case of survivorships involving three lives.

Example.—Required the value of £1 payable at the death of (28), provided (30) and (33) shall be then both living.

(m) being the youngest life, the formula becomes—

$$\frac{1}{1+q^2+q^5}\left\{1+(\log_e v + C_2 + C_5)A_{28,\,30,\,33}\right\}.$$

1·
1·1860
1·5317
———
3·7177 log. = ·57027*
log. 3 = ·47712
———
·09315 × $\left(27 = \dfrac{1}{\log. q}\right)$

2·7945
·2795
———
2·515
28·
———
30·515 = Eq. com. age.

$A_{30.30.30} = 12\cdot784$　　$-167 \times \cdot515$
　　　　　　　　　　　　———
　　　　　　　　　　　　　84
　　　　　-86　　　2
　　　　———
　　12·698

$\log_e v = \overline{1}\cdot96078$
$C_2 = \quad\cdot00132$
$C_5 = \quad\cdot00376$
　　———
　$\overline{1}\cdot96586$

$\overline{1}\cdot96586 = -\cdot03414$ log. $= \overline{2}\cdot53326$
　　　　log. $12\cdot698 = 1\cdot10374$
$-$log. $(1+q^2+q^5) = \overline{1}\cdot42973* =$ log. ·26899
　　　　　　　　　———
　　　　$\overline{1}\cdot06673$ = log. ·11661

(*Answer*) ·15238

I am, Sir,
Your very obedient servant,

5, *Lothbury.*　　　　　　　　　　　W. M. MAKEHAM.

Age.	Living.	Decrements.	Expectation of Life.	Age.	Living.	Decrements.	Expectation of Life.
15	10000·000	75·323	44·86	60	5677·788	165·630	14·57
16	9924·677	75·183	44·20	61	5512·158	171·522	13·99
17	9849·494	75·070	43·53	62	5340·636	177·472	13·42
18	9774·424	74·997	42·86	63	5163·164	183·437	12·87
19	9699·427	74·958	42·19	64	4979·727	189·345	12·32
20	9624·469	74·959	41·51	65	4790·382	195·127	11·79
21	9549·510	75·001	40·83	66	4595·255	200·701	11·27
22	9474·509	75·093	40·15	67	4394·554	205·971	10·76
23	9399·416	75·226	39·47	68	4188·583	210·838	10·27
24	9324·190	75·418	38·79	69	3977·745	215·185	9·79
25	9248·772	75·660	38·10	70	3762·560	218·888	9·32
26	9173·112	75·964	37·41	71	3543·672	221·825	8·86
27	9097·148	76·332	36·72	72	3321·847	223·856	8·42
28	9020·816	76·764	36·02	73	3097·991	224·850	7·99
29	8944·052	77·272	35·33	74	2873·141	224·677	7·58
30	8866·780	77·860	34·63	75	2648·464	223·214	7·18
31	8788·920	78·530	33·93	76	2425·250	220·351	6·79
32	8710·390	79·286	33·23	77	2204·899	216·004	6·42
33	8631·104	80·142	32·53	78	1988·895	210·115	6·06
34	8550·962	81·094	31·83	79	1778·780	202·660	5·72
35	8469·868	82·158	31·14	80	1576·120	193·657	5·39
36	8387·710	83·336	30·44	81	1382·463	183·177	5·08
37	8304·374	84·640	29·74	82	1199·286	171·341	4·78
38	8219·734	86·070	29·04	83	1027·945	158·323	4·49
39	8133·664	87·642	28·34	84	869·622	144·355	4·22
40	8046·022	89·360	27·64	85	725·267	129·714	3·95
41	7956·662	91·233	26·95	86	595·553	114·715	3·71
42	7865·429	93·274	26·25	87	480·838	99·703	3·47
43	7772·155	95·483	25·56	88	381·135	85·029	3·25
44	7676·672	97·877	24·87	89	296·106	71·029	3·04
45	7578·795	100·465	24·19	90	225·077	58·013	2·84
46	7478·330	103·248	23·51	91	167·064	46·231	2·65
47	7375·082	106·244	22·83	92	120·833	35·869	2·48
48	7268·838	109·453	22·16	93	84·964	27·029	2·31
49	7159·385	112·887	21·49	94	57·935	19·730	2·15
50	7046·498	116·546	20·82	95	38·205	13·912	2·01
51	6929·952	120·444	20·16	96	24·293	9·448	1·87
52	6809·508	124·578	19·51	97	14·845	6·156	1·74
53	6684·930	128·947	18·87	98	8·689	3·838	1·63
54	6555·983	133·556	18·23	99	4·851	2·277	1·51
55	6422·427	138·396	17·60	100	2·574	1·283	1·41
56	6284·031	143·458	16·97	101	1·291	·681	1·32
57	6140·573	148·735	16·36	102	·610	·341	1·23
58	5991·838	154·204	15·75	103	·269	·158	1·15
59	5837·634	159·846	15·16	104	·111	·069	1·07

Annuities (4 per Cent.).

Ages. (n)	One Life. (A_n)		Two Lives. $(A_{n,n})$		Three Lives. $(A_{n,n,n})$		Ages. (n)	One Life. (A_n)		Two Lives. $(A_{n,n})$		Three Lives. $(A_{n,n,n})$	
		(−)		(−)		(−)			(−)		(−)		(−)
30	17·492	·164	14·668	·171	12·784	·167	35	16·620	·191	13·758	·199	11·899	·193
31	17·328	·169	14·497	·177	12·617	·171	36	16·429	·197	13·559	·204	11·706	·197
32	17·159	·174	14·320	·182	12·446	·177	37	16·232	·202	13·355	·209	11·509	·203
33	16·985	·180	14·138	·187	12·269	·183	38	16·030	·207	13·146	·215	11·306	·208
34	16·805	·185	13·951	·193	12·080	·187	39	15·823	·214	12·931	·220	11·098	·212

n	q^n	d_n	C_n	C_{-n}	F_n	F_{-n} $(=1-F_n)$	G_n	G_{-n}
1	1·089023	·511	·0006299	$\overline{1}$·9994216	·4786927	·5213073	$\overline{1}$·9815269	$\overline{1}$·9792524
2	1·185971	1·043	·0013158	$\overline{1}$·9988905	·4574626	·5425374	$\overline{1}$·9826599	$\overline{1}$·9781194
3	1·291550	1·596	·0020628	$\overline{1}$·9984028	·4363858	·5636142	$\overline{1}$·9837848	$\overline{1}$·9769944
4	1·406527	2·170	·0028763	$\overline{1}$·9979550	·4155366	·5844634	$\overline{1}$·9848976	$\overline{1}$·9758817
5	1·531740	2·765	·0037623	$\overline{1}$·9975438	·3949853	·6050147	$\overline{1}$·9859945	$\overline{1}$·9747849

ON THE SUPERANNUATION OF EMPLOYÉS IN ASSURANCE OFFICES.

To the Editor of the Assurance Magazine.

SIR,—May I solicit the favour of your allotting a small space in your *Journal* for the insertion of a few remarks upon the subject of superannuation of *employés* in Assurance Offices, in the hope that it may engage the attention that it certainly deserves, but which, I believe, it has not hitherto received.

There may be, and, no doubt, are, systems of superannuation in connexion with some of our public institutions, but they are not, I believe, general; indeed, I am only aware of one instance in which a scheme exists in connexion with a Joint Stock Company for granting retiring pensions after certain periods of service. The Company referred to is the National Provincial Bank of England, and the main features of the scheme are, the option of retirement, after 20 years' service, on one-third of salary; after 30 years, on half salary; or, after 40 years, on two-thirds salary; and in proportion for intermediate periods of service—one of the conditions being, that the age of 60 shall be attained before retiring.

It will be seen that there is here no inducement held out to withdraw from active duties, but there is an option given of doing so, that would be esteemed a boon by very many who yet might never avail themselves of it.

Habit, we all know, has a powerful hold upon men generally—probably upon no class is its influence greater than upon those engaged in official routine—and it may reasonably be supposed that few men would, if in health, readily sacrifice two-thirds, or a half, of their income, merely for the sake of living in idleness.

It may be suggested, that a person has no claim upon the Company by which he has been employed—whatever may have been the length of his services—when incapacitated, by sickness or other infirmity, for further duty; true, he has no *legal* claim, but I am happy to think that boards of directors of Assurance Companies are not usually composed of men who take this view of things.

Assuming, then, a willingness to entertain the question of superannuation, a difficulty may arise respecting the cost; this is, however, rather imaginary than real, as I hope to show.

Waiving the consideration of retiring pensions to heads of departments, I will take the general staff of an Office, and assume that all engagements commence with a salary of £60, rising £10 annually until a maximum of £250 is reached, which would be in 20 years. Supposing, then, an individual who has attained the age of 60, and has been engaged in the service of a Company for 30 years, should feel desirous to retire from the cares and responsibilities that have for so long a period devolved upon him, what would be the pecuniary effect upon a Company, of allowing him to do so with a pension of £125 per annum, and appointing a junior to fill up the vacancy that would be caused? The cost of a Government annuity may be considered to be a fair basis upon which to form an estimate, and this, at the age of 60, would amount to about £1,300, against which has to be placed the present value of the annual differences between the salary paid to the newly appointed clerk and that which *would* have been payable had no change taken place; this will amount to about £1,500, and the result would be an actual gain to the Company of £200 in the 20 years that would elapse before the maximum salary of £250 would be attained.

True, it may be said that a person of 60 years of age is not likely to remain at his post another 20 years; but there is also the contingency of a junior not continuing 20 years. Indeed, I believe that the average length of time is likely to be as great in the former as in the latter case.

Did space permit, very many arguments might be adduced in favour of thus making provision for the retirement of officials of long standing; and I am not aware that any objections against it can be raised. However, I offer these few suggestions, trusting that an interest may be created which will lead to some more definite action in the matter.

Perhaps the Institute of Actuaries might give the subject their attention; it certainly is one that may fairly challenge discussion, involving, as it does, considerations affecting all grades in the profession, from the actuary or secretary to the junior clerk; and the adoption of some such scheme of superannuation as I have sketched out may not unfairly be considered as likely to influence the prosperity of such Companies as might be induced to entertain it; such a course being one that would appear to be eminently calculated to secure a continuance of the services of tried and efficient officers.

I am, Sir,
Your obedient servant,

London, June, 1861.   H. A.

[NOTE.—The question of superannuation allowances by public institutions is one which we should be glad to see engage attention. The scheme recently adopted by the National Provincial Bank of England appears to be a very liberal one. This bank, moreover, several years ago, with the view to encourage provident habits among their *employés*, determined to pay one-half of the yearly premiums of assurances on the lives of their officers and clerks. We may refer our readers to a letter on this subject, which will be found at page 72, vol. v., of this *Magazine*. It is there suggested by a correspondent—Mr. Porter—that those engaged in life assurance, like those employed in all other descriptions of business, should be allowed to obtain the article in which their employers deal at *cost price*.—ED. *A. M.*]

INSTITUTE OF ACTUARIES.

PROCEEDINGS OF THE INSTITUTE.

Fifth Ordinary Meeting, Session 1860-61.—Monday, 25th March, 1861.

CHARLES JELLICOE, President, in the Chair.

The minutes of the last ordinary meeting were read and confirmed.

The Secretary announced several donations to the library.

Mr. W. Newmarch, duly nominated at the last ordinary meeting, was elected a Fellow of the Institute on the recommendation of the Council.

The undermentioned gentleman, duly nominated at the last ordinary meeting, was elected a member of the Institute, viz. :—

Associate—Charles Evans Newbon.

A communication from Mr. Gompertz was read by Mr. Porter, and a paper by Mr. Sprague, "On the graduation of the series giving the expectation of life, and the nature of the corresponding curves," was also read by that gentleman.

Sixth Ordinary Meeting, Session 1860-61.—Monday, 29th April, 1861.

CHARLES JELLICOE, President, in the Chair.

The minutes of the last ordinary meeting were read and confirmed.

The Secretary announced several donations to the library.

The undermentioned gentleman, duly nominated at the last ordinary meeting, was elected a member of the Institute, viz.:—

Official Associate—John Messent.

Mr. Bailey read a paper "On the mortality amongst the families of the peerage during the nineteenth century," by Mr. A. H. Bailey and Mr. Archibald Day.

Thanks having been voted to Mr. Bailey and Mr. Day, the meeting adjourned to Monday, the 25th November next.

Fourteenth Annual General Meeting, Saturday, 1st June, 1861.

The circular convening the meeting having been read,

The minutes of the last ordinary meeting were read and confirmed.

Mr. J. Hill Williams, one of the Honorary Secretaries, then read the following Report of the Council on the progress of the Institute during the past year :—

" REPORT.

"The Council have again the pleasure of submitting to the members a Report as to the progress of the Institute during the past year.

"There has been a slight increase in the number of members, which is now 155, as compared with 147, at the date of the last Report. The numbers in each class are 46 Fellows, 21 Official Associates, and 88 Associates.

"The income of the year is £367. 4s., and the expenditure £350. 12s. 4d. The assets amount to £428. 9s. 2d., of which £192. 8s. 6d. is invested in the 3 per cent. Consols.

"The papers read during the Session have been of much interest ; they are as follows :—

"'On the theory of probabilities.' By Robert Campbell, Esq., M.A.

"'On the stability of results based on average calculations.' By the same gentleman.

"'On the rates of premium required to provide for certain periodical returns to the assured.' By Robert Tucker, Esq., V.P.

"'On Mr. Gompertz's law of mortality.' By T. B. Sprague, Esq., M.A.

"'On the graduation of series, expressing the expectation of life, and on the curves connected therewith.' By the same writer.

"'On the rate of mortality prevailing amongst the families of the peerage during the nineteenth century.' By A. H. Bailey and Archibald Day, Esqs.

"In connection with the subject of the papers which have appeared in the *Journal* of the Institute, it may be interesting to state that the historical notice by Professor De Morgan, entitled 'Account of a correspondence between Mr. George Barrett and Mr. Francis Baily,' published in Vol. IV., led to the laborious calculations which Barrett left behind him in manuscript being brought under the notice of the members of the Institute. Professor De Morgan, in that paper, states that Barrett's manuscript tables were purchased, a few years after his death, by Mr. Babbage ; and one of the members of the Council having written to Mr. Babbage, requesting to know if the papers were still in his possession, that gentleman obligingly appointed a time for an interview, and exhibited all of these documents, which he still retains. After explaining the circumstances under which he purchased them of the executors, he most kindly allowed them to be sent to the Institute for the inspection of the members. The manuscripts comprise all the tables mentioned in Professor De Morgan's historical notice ; and though it will be seen that, from the improved tables of population now in use, these volumes of calculations, the fruits of a laborious life incessantly employed upon them, are not at the present time of much practical use ; yet they are worthy of examination as the work of an author whose columnar method is now so well known, and who, by means of it, has so greatly facilitated and improved assurance calculations.

" The Council have expressed, on behalf of the Institute, the sense which they entertain of the courteous attention of Mr. Babbage ; and they are sure that every member will be fully sensible of his obliging consideration.

" The members are aware that the prize offered, in 1859, for an essay ' On the methods of distributing the surplus amongst the persons assured in a Life Assurance Company,' to be written by an Associate of the Institute, was again offered last year ; and the Council have now the satisfaction of announcing, that essays on the prescribed subject have been sent in, and that one of them has been deemed worthy of the prize, and will be shortly published in the *Journal.* The author of it is Mr. William Pollard Pattison.

" Since the last annual meeting of the Institute, the International Statistical Congress has held its fourth session in this metropolis, with what success may be estimated from the Report which has been recently issued of its proceedings. That most remarkable and able production owes its existence, as is well known, to the labours of Dr. Farr and his coadjutors ; and it is gratifying to the Council that members of the Institute have assisted so materially in the work. The papers and contributions of Mr. Samuel Brown, Mr. Hodge, Mr. Newmarch, and Mr. Hill Williams, are not surpassed in interest and importance by any to be found in the volume."

An abstract of the receipts and payments of the Institute, for the financial year ended 31st March last, was then read (*see* p. 372).

On the motion of the Chairman, the Report was unanimously adopted.

The election of a President, Vice-Presidents, and Officers, for the year ensuing, was then proceeded with.

Mr. Pattison and Mr. Strachan were appointed scrutineers.

On the result of the ballot being obtained, the following was declared to be the list, viz. :—

President.

CHARLES JELLICOE.

Vice Presidents.

SAMUEL BROWN.	WILLIAM BARWICK HODGE.
PETER HARDY, F.R.S.	ROBERT TUCKER.

Treasurer.

JOHN LAURENCE.

Honorary Secretaries.

JOHN REDDISH.	JOHN HILL WILLIAMS.

The following gentlemen were unanimously re-elected Auditors for the ensuing financial year, viz., John Coles, Edward Cutbush, and James Terry.

Mr. Jellicoe said—" I have to express the best thanks of my colleagues and myself for the compliment you have paid us, and for the confidence you have shown in our desire and ability to serve the Institute, by again electing us. It may be remembered, that at our last annual meeting I took occasion to refer to a few questions connected with our pursuits, and to remark on the desirableness of arriving at a determination of them. In the year which has since elapsed, progress has been made towards the solution of some of these. There is reason, for instance, to believe, that the Report of the American Convention, on the rate of mortality of persons whose lives are assured in the United States, is now nearly completed, and will shortly be made public, and we may hope, therefore, soon to have the means of deciding the very interesting question, whether there really is, or is not, any material difference in the duration of human life thus circumstanced in the two countries; and it must, I think, be looked upon as fortunate, that all the materials for this investigation have been collected before the events which are calculated to create so great a disturbance in the results, and which are so much to be lamented, have had their origin.

"Another question to which I briefly alluded, has also, I am happy to say, received considerable illustration at the hands of one of our members. I refer to the inquiry as to the most accurate mode of dividing surplus in Assurance Companies. Mr. Pattison's essay on this subject cannot but be read with advantage by all who are desirous of forming correct ideas upon it; nor will the reader fail to appreciate the amount of useful information brought together, as to the various methods hitherto pursued in this process by those having the management of these important institutions.

"The members of the Institute have not addressed themselves during the past Session to the discussion of the vexed question of direct taxation; but the suggestions made at the last annual meeting have received remarkable confirmation in the measures of the Government with reference to the abolition of the duties on the manufacture and importation of paper, and in the efforts made by Mr. Hubbard to modify the system under which the income tax is at present levied. I cannot but regard the measures of the one and the efforts of the other as essentially in the right direction, and, having no other desire than the growth and prevalence of right principles, I am happy to think that a similar opinion is entertained by a large majority of those whom I have now the honour to address.

"Turning, however, to the consideration of what has been accomplished in the Institute during the Session, a source of much satisfaction arises from the contemplation of the many able and useful communications which have been laid before it. Apart from the importance of the practical results given in Mr. Campbell's last paper, the method therein suggested of eliminating quantities of small value, and thus reducing expressions, unmanageable from their intricacy and diffuseness, to a form easily tabulated, is likely very much to facilitate calculations in probabilities; the labour being by such means brought within reasonable compass, whilst the results are quite sufficiently accurate for all ordinary purposes.

"The mode, too, of investigation, which has of late been pursued, with a view to determine the relation between the premiums charged by Assurance Companies and the returns made by them, is useful and instructive. Heretofore, writers have sought to show what returns the premiums charged would justify. Of late, the object has been to determine what premiums are necessary to provide for such returns as are usually made; and as this latter process brings into forcible contrast the rates which are necessary and those ordinarily adopted, attention is more likely to be drawn to the absolute necessity of bringing the two into relations consistent with safety. Mr. Tucker has given formulæ with this object applicable to almost every case. I believe there is no system of division which he has not brought under investigation in this way.

"To Professor De Morgan and to Mr. Sprague we are indebted for a most thorough and complete exposition of the Gompertzian theory. (I do not speak here of Dr. Farr's masterly treatise.) The extent to which the logarithmic expression which arises out of it may be usefully applied has been clearly shown and well defined, and enough has transpired to prove that the expression itself must be handled with much delicacy and judgment, if we wish to adhere rigorously to the records of experience. As is the case with most other expressions for concatenated series, the series yielded by this is dependent in every part upon the whole; like a cane compressed into a curve and confined at the centre, we cannot give a direction to the one portion without influencing the other, and thus our instrument is not quite of that degree of ductility which we require; whether it will better adapt itself to other kinds of series in use with us remains to be seen. Mr. Sprague seems to be directing his attention to this branch of our inquiries, and you will all admit that it cannot be in better hands.

" Prior to the investigation made by Mr. Bailey and Mr. Day into the rate of mortality prevailing amongst the families of the peerage, I believe the impression has been very general (derived from previous writings on the subject) that the duration of life in that class is considerably below the average. At all events, a great deal of doubt has existed as to the real facts of the case. The labours of Messrs. Bailey and Day have, however, set the question at rest. They have proved, beyond all controversy, that, so far as regards the experience of the present century, the male portion of the class in question enjoys a degree of longevity rarely known; and they have also shown, and are, I believe, the first writers who have done so, that the longevity of the female portion is at all ages equal to the greatest on our modern records, and at some ages even greater.

" These results are such as we might reasonably look for when the condition of the class in question is taken into account, and when we know that the great means it generally possesses are no longer dissipated, as they once may have been, in over-indulgence and dangerous excesses. A very satisfactory consideration also arises out of the deductions thus so conclusively arrived at by Mr. Bailey and his coadjutor, and one which I remember Mr. Newbatt ably adverted to at the reading of the paper; it is, that if the probabilities of living are so great amongst the highest classes, they cannot be very inferior amongst the classes immediately succeeding, since the habits and modes of life in all are much the same, making some allowance for the wear and tear of professional life.

" Hence there is reason to hope that the experience of our Assurance Companies will, at the least, be as favourable in future as it has hitherto been, since the names on the registers of those institutions are, for the most part, those of persons in the middle and upper ranks of the population. Whether the evidence which Messrs. Bailey and Day have brought forward will warrant the Offices in dispensing with the existing restrictions on foreign travel and residence I will not now inquire. The suggestion is, however, well worthy of consideration, and points to the adoption of an arrangement which, though at first somewhat startling, is but a trifle in advance of the practice almost universal with the Companies in the north, and not very unfrequent with those of this metropolis.

" Such, gentlemen, is a brief review of the subjects which have, for the most part, engaged our attention within the Institute during the last Session, whilst externally, during this period, an extraordinary call on the activity and intelligence of some of our colleagues has arisen from the advent hither of the International Statistical Congress. Amongst the statements and reports submitted to that body, organized with such extraordinary skill and ability by Dr. Farr and those around him, will be found Mr. Samuel Brown's programme on the statistical units of money, weights, and measures, and his report on the proceedings and progress of the Institute; Mr. Hodge's programme on military statistics; Mr. Newmarch's on prices and wages, and his report on the proceedings and progress of the Statistical Society; and Mr. Hill Williams' programme on the subdivision, transfer, and burdens of real property.

" How well these gentlemen have executed their several tasks—undertaken, I believe, in each case, at a very short notice—will be seen by the most cursory reference to them. Taken in connexion with those of which I have already spoken, they will at least serve to show that the labours of the gentlemen connected with this Institute are by no means light, and that no little consideration is due to them for devoting so freely to the common advantage, the time and ability already so fully absorbed by the duties of an arduous and engrossing occupation."

A vote of thanks to the President, moved by Mr. Lodge and seconded by Mr. Pinckard, was passed unanimously, and the proceedings terminated.

INSTITUTE OF ACTUARIES.

Abstract of Receipts and Payments for the Year ending 31st March, 1861.

Dr.

RECEIPTS.

		£	s.	d.
April 1, 1859. To balance brought forward		220	9	11
Subscriptions for 1859–60 (arrears)		7	7	0
Subscriptions due for 1861, from—				
47 Fellows .... ..35 Town ..at £3 3 0	£110 5 0			
12 Country ,, 2 2 0 ..	25 4 0			
21 Official Associates..19 Town ,, 3 3 0..	59 17 0			
2 Country ,, 2 2 0..	4 4 0			
87 Associates........69 Town ,, 2 2 0..	144 18 0			
18 ..ry ,, 1 1 0..	18 18 0			
		363	6	0
Subscriptions of Members not paid, viz.—				
1 Fellow, Town .....at £3 3 0	£3 3 0			
4 Associates, Town ,, 2 2 0	8 8 0			
1 ,, Country ,, 1 1 0..	1 1 0	12	12	0
		350	14	0
Examination Fee		5	5	0
Sundries		3	18	1
		£587	14	0
March 31, 1861. To Balance brought down		£236	0	8

Cr.

PAYMENTS.

	£	s.	d.	£	s.	d.
1860.						
By Rent	75	0	0			
Sales	99	5	10			
Journal	88	0	0			
Library	10	14	5			
Stationery and Printing (2 years)...	29	12	9			
Postage and Receipt Stamps	3	17	6			
Lighting	6	7	3			
Ordinary Meetings	15	3	0			
Advertising Examinations ...	3	9	6			
Statistical Congress expenses ...	5	9	0			
Miscellaneous	13	13	1			
Total of general expenses				350	12	4
March 31, 1861.						
Subscription partly repaid				1	1	0
Balance carried forward				236	0	8
				£587	14	0

Examined and approved :—

JOHN COLES,
EDWARD CUTBUSH, } Auditors.
JAMES TERRY,

Note.—The Assets of the Institute, on the 31st March, 1861, consisted of £198. 16s. 2d. Consols, say 183 19 4
Cash 236 0 8
Books in Library, say 300 0 0
Total........... ....£720 0 0

12, ST. JAMES'S SQUARE, LONDON,

INDEX TO VOLUME IX.

2 D

END OF VOL. IX.

Printed by C. & E. Layton, 150, Fleet Street.

EAGLE INSURANCE COMPANY.

REPORT OF THE DIRECTORS FOR THE YEAR ENDING 30TH JUNE, 1860.

THE Directors have again the pleasure to make their Annual Report to the Proprietors—the Fifty-third since the commencement of the Company's operations, and the Third since the last Quinquennial distribution of surplus.

The Income and Outgoings of the year ending on the 30th June last, will appear in the following abstract from the Surplus Fund Account, as shown by the Company's Books:—

SURPLUS FUND ACCOUNT.

INCOME OF THE YEAR ENDING JUNE 30TH, 1860.	£ s. d.	£ s. d.	CHARGE OF THE YEAR.	£ s. d.	£ s. d.
Balance of Account, June 30th, 1859	659,013 17 2		Dividend to Proprietors		10,343 8 6
Ditto of a small Assurance Company	39,264 0 10		Claims on decease of Lives Assured	238,552 12 7	
		698,277 18 0	Additions to those under Participating Policies.. ..	21,167 18 6	
Premiums on New Assurances	19,588 17 6		Policies surrendered	9,733 7 2	
Ditto on Renewed ditto ..	283,250 19 11		Reassurances, New	1,838 6 5	
	302,839 17 5		Ditto, Old ..	30,124 6 3	
Interest from Investments ..	81,203 1 11			301,416 10 11	
		384,042 19 4	Commission	10,722 14 1	
			Medical Fees	1,071 16 3	
			Income Tax	3,603 3 1	
			Expenses of Management ..	11,044 4 10	
					327,858 9 2
			Balance of Account, June 30th, 1860, as below		744,118 19 8
		£1,082,320 17 4			£1,082,320 17 4

Examined and found to be correct,

(Signed) THOMAS ALLEN,
WILLIAM HENRY SMITH, Jun., } *Auditors.*

The Proprietors will observe that another small Assurance Company has merged into the Eagle during the year, and that it has contributed about £39,000 to the Surplus Fund.

The Premiums on new Assurances amount to £19,588. 17s. 6d., and the total Income from Premiums and interest to £384,042. 19s. 4d. This is short by about £6,000 of the actual Income, in consequence of the junction above mentioned not taking place at the commencement of the financial year.

Deducting the sums immediately payable, the realized Assets of the Company on the 30th June, 1859, were, in round numbers, £1,789,900; and, since the interest received during the year amounts, as above shown, to £81,203. 1s. 11d., it follows that the Company's funds of that date, productive and unproductive, have been accumulating in the interval at rather more than the average rate of 4½ per cent.

The claims on decease of Lives Assured and the general expenses are, as it is reasonable to expect they would be, somewhat more than they were the previous year. It will be observed that the total expenses, including commissions, but excluding income tax, are not quite six per cent. of the income.

The Company's Liabilities and Assets on the 30th June last, stated with as much accuracy as they can be in the absence of a re-valuation, will be seen in the following Balance Sheet:—

BALANCE SHEET.

LIABILITIES.	£ s. d.	£ s. d.	ASSETS.	£ s. d.
Interest due to Proprietors, not claimed ..	6,555 12 9		Amount invested in Fixed Mortgages ..	1,195,493 16 3
Claims on decease of Lives Assured and additions thereto unpaid	88,494 2 4		Ditto ditto decreasing Mortgages ..	154,783 10 3
Cash Bonus due to Policy-holders	12,811 10 4		Ditto ditto Reversions	77,846 1 11
Sundry Accounts..	12,541 7 10		Ditto ditto Funded Securities	257,708 2 1
Value (1857) of Sums Assured, Annuities, &c.	4,387,426 2 11		Ditto ditto temporary Securities ..	61,402 14 10
Proprietors' Fund £203,743 10 3			Current Interest on the above Investments..	26,636 3 11
Surplus Fund, as above .. 744,118 19 8			Cash and Bills	33,973 17 3
	947,862 9 11		Advanced on Security of the Company's Policies, &c. ..	89,784 7 11
			Agents' Balances	26,965 14 1
			Sundry Accounts	12,723 2 6
			Value (1857) of Assurance Premiums ..	3,518,373 15 1
	£5,455,691 6 1			£5,455,691 6 1

Examined and found to be correct,

(Signed) THOMAS ALLEN,
WILLIAM HENRY SMITH, Jun., } *Auditors.*

From this it appears that the realized Assets amount to £1,937,317. 11s., and that those to be realized are estimated at £3,518,373. 15s. 1d. (about 11¼ years' purchase), the two together being not far from Five Millions and a Half in amount.

The Surplus Fund has increased during the year from £659,013. 17s. 2d. to £744,118. 19s. 8d., the increase being £85,105. 2s. 6d.

The Proprietors will thus observe that the Income of the Company still exceeds the Outgoings, and that its funds are still on the increase from year to year. But it may be well to point out that, although this state of things may yet continue for some years, a time must arrive when it will be reversed, and when the

Outgoings will, first be equal to, and then for some years exceed the Income, as is the case with many of the older Companies at the present day.

This course is one which must be followed by all Life Assurance Institutions, without exception, and has nothing in it indicative, as persons not conversant with their nature are apt to suppose, of loss or disadvantage; on the contrary, it not unfrequently happens that Societies of this description become relatively more wealthy, or accumulate a larger divisable surplus, as their funds decrease.

In a well-regulated Company, however, the surplus fund should always be maintained in its due proportion, let the fluctuations in the General Fund be what they may, and it will be for the Directors to see that, as regards the Eagle, this principle is carefully carried out, and that every participating Policyholder has his full and proper share of the divisible surplus accruing throughout the period of his connection with the Company, whether the particular phase under which it may then present itself be increasing, decreasing, or stationary.

The Proprietors' Fund, and the Income arising out of it, are of course exempt from the fluctuations here spoken of.

Equity and Law Life Assurance Society,

18, LINCOLN'S INN FIELDS, LONDON, W.C.

CAPITAL — ONE MILLION, in £10,000 SHARES of £100 EACH.

REDUCTION OF PREMIUM.—Parties effecting assurances within Six Months of their last Birthday are allowed a proportionate diminution in the Premium.

FOREIGN RESIDENCE.—Persons whose lives are assured are allowed, without licence or extra charge, in time of peace, to proceed to and reside in any part of the World distant more than thirty-three degrees from the Equator; and to reside within the prohibited degrees upon payment of an extra premium.

SECURITY TO THIRD PARTIES.—Policies do not become void by the lives assured going beyond the prescribed limits,—so far as regards the interest of Third Parties, provided they pay the additional Premium so soon as the fact comes to their knowledge.

BONUS.—NINE-TENTHS of the Profits are divided at the end of every five years among the assured. The additions made to Policies have averaged very nearly *Two per Cent. per Annum,* on the sums assured. Policies becoming Claims between the periods of Division are entitled to a Bonus, in addition to that previously declared.

PUBLICATION OF ACCOUNTS.—The Annual Reports and accounts are printed periodically. Copies may be had, with Forms of Proposal and every requisite information, upon written or personal application to the Office.

Gresham Life Assurance Society,

37, OLD JEWRY, LONDON, E.C.

DIRECTORS.
WILLIAM TABOR, Esq., *Chairman.*
JOHN BEADNELL, Esq., *Deputy-Chairman.*

J. LYNE HANCOCK, Esq.	EDWARD SOLLY, F.R.S.
GEORGE LOWE, F.R.S.	W. H. THORNTHWAITE, Esq.
ALFRED SMEE, F.R.S.	GEORGE TYLER, Esq.

JOSEPH WILLIAMS, Esq.

Policies effected, without loss of time, every day from 10 to 4; Saturdays, 10 to 2; Medical Officer, daily, at 11. The Board assembles on Thursdays, at half-past 12.

Loans may be obtained in connexion with Policies effected with the Company. There has been advanced in this respect upwards of a Quarter of a Million since July, 1848.

Annual Reports, Prospectuses, and other Forms on application.

EDWIN JAMES FARREN, *Actuary & Secretary.*

Guardian
FIRE AND LIFE ASSURANCE COMPANY,
No. 11, LOMBARD STREET, LONDON, E.C.
ESTABLISHED 1821.

DIRECTORS.
HENRY VIGNE, Esq., *Chairman.*   Sir MINTO T. FARQUHAR, Bt., M.P., *Deputy-Chairman.*

HENRY HULSE BERENS, Esq.	JOHN HARVEY, Esq.	JAMES MORRIS, Esq.
CHAS. WM. CURTIS, Esq.	JOHN G. HUBBARD, Esq., M.P.	HENRY NORMAN, Esq.
CHARLES F. DEVAS, Esq.	JOHN LABOUCHERE, Esq.	HENRY R. REYNOLDS, Esq.
FRANCIS HART DYKE, Esq.	STEWART MARJORIBANKS, Esq.	Sir GODFREY J. THOMAS, Bt.
Sir WALTER R. FARQUHAR, Bart.	JOHN MARTIN, Esq.	JOHN THORNTON, Esq.
THOMSON HANKEY, Esq., M.P.	ROWLAND MITCHELL, Esq.	JAMES TULLOCH, Esq.

AUDITORS.
LEWIS LOYD, Esq.   HENRY SYKES THORNTON, Esq.
JOHN HENRY SMITH, Esq.   CORNELIUS PAINE, Jun., Esq.

THOS. TALLEMACH, Esq., *Secretary.*—SAMUEL BROWN, Esq., *Actuary.*

LIFE DEPARTMENT.—UNDER THE PROVISIONS OF AN ACT OF PARLIAMENT, this Company now offers to new Insurers **Eighty per Cent. of the Profits, at Quinquennial Divisions, or a Low Rate of Premium** without participation of Profits.

Since the establishment of the Company in 1821, the amount of Profits allotted to the Assured has exceeded in cash value £660,000, which represents equivalent Reversionary Bonuses of £1,058,000.

After the Division of Profits at Christmas, 1859, the Life Assurances in force, with existing Bonuses thereon, amounted to upwards of £4,730,000 ; the Income from the Life Branch, £207,000 per annum ; and the Life Assurance Fund exceeded £1,618,000.

LOCAL MILITIA & VOLUNTEER CORPS.—No extra Premium is required for service therein.

INVALID LIVES assured at corresponding extra Premiums.

LOANS granted on Life Policies to the extent of their values, if such value be not less than £50.

ASSIGNMENTS OF POLICIES.—Written Notices of, received and registered.

MEDICAL FEES paid by the Company, and no charge for Policy Stamps.

Notice is hereby given, That Fire Policies which expire at Midsummer must be renewed within fifteen days at this Office; or with Mr. SAMS, No. 1, St. James's Street, corner of Pall Mall; or with the Company's Agents throughout the Kingdom; otherwise they become void.

Losses caused by Explosion of Gas are admitted by this Company.

The London Assurance,

INCORPORATED A.D. 1720,

FOR LIFE, FIRE, AND MARINE ASSURANCES.

HEAD OFFICE—No. 7, ROYAL EXCHANGE, CORNHILL.

JOHN ALVES ARBUTHNOT, Esq., *Governor.*
JOHN ALEX. HANKEY, Esq., *Sub-Governor.*
BONAMY DOBREE, Jun., Esq., *Deputy-Governor.*

DIRECTORS.

NATHANL. ALEXANDER, Esq.	F. G. DALGETY, Esq.	LOUIS HUTH, Esq.
RICHARD BAGGALLAY, Esq.	JOHN ENTWISLE, Esq.	CHARLES LYALL, Esq.
HENRY BONHAM BAX, Esq.	ROBT. GILLESPIE, Jun., Esq.	JOHN ORD, Esq.
JAMES BLYTH, Esq.	HARRY GEO. GORDON, Esq.	CAPT. R. W. PELLY, R.N.
EDWARD BUDD, Esq.	EDWIN GOWER, Esq.	DAVID POWELL, Esq.
EDWARD BURMESTER, Esq.	SAMUEL GREGSON, Esq., M.P.	P. F. ROBERTSON, Esq.
CHARLES CRAWLEY, Esq.	A. C. GUTHRIE, Esq.	ALEXANDER TROTTER, Esq.
SIR FREDK. CURRIE, Bart.	EDWARD HARNAGE, Esq.	LESTOCK P. WILSON, Esq.

WEST END OFFICE—No. 7, PALL MALL.

COMMITTEE.

TWO MEMBERS OF THE COURT in rotation, and

HENRY KINGSCOTE, Esq. AND JOHN TIDD PRATT, Esq.

Superintendent.—PHILIP SCOONES, Esq.

LIFE DEPARTMENT.

Actuary.—PETER HARDY, Esq., F.R.S.

THIS CORPORATION has granted Assurances on Lives for a **period exceeding One Hundred and Thirty Years,** having issued its first Policy on the 7th June, 1721.

Two-thirds, or **66 per cent.,** of the entire Profits are given to the Assured.

Policies may be opened under any of the following plans, viz. :—

At a low rate of Premium, without participation in Profits, or at a somewhat higher rate, entitling the Assured, either after the first five years, to an annual abatement of Premium for the remainder of Life, or, after payment of the first Premium, to a participation in the ensuing Quinquennial Bonus.

The high character which this ancient Corporation has maintained during **nearly a Century and a Half,** secures to the public a full and faithful declaration of Profits.

The Corporation bears the whole EXPENSES OF MANAGEMENT, thus giving to the Assured, conjoined with the protection afforded by its **Corporate Fund,** advantages equal to those of any system of Mutual Assurance.

All Policies are issued Free from Stamp Duty, or from charge of any description whatever, beyond the Premium.

The Fees of **Medical Referees** are paid by the Corporation.

Annuities are granted by the Corporation, payable Half-Yearly.

FIRE DEPARTMENT.

Manager.—THOS. B. BATEMAN, Esq.

Common Assurances, One Shilling and Sixpence per Cent.
Hazardous Assurances, Two Shillings and Sixpence per Cent.
Doubly Hazardous Assurances, Four Shillings and Sixpence per Cent.
Foreign and Special Assurances accepted at moderate Rates.

Prospectuses and all other Information may be obtained by either a written or personal application to the Actuary, the Manager of the Fire Department, or to the Superintendent of the West End Office.

JOHN LAURENCE, *Secretary.*

Pelican

LIFE INSURANCE COMPANY,

ESTABLISHED IN 1797,

70, LOMBARD STREET, CITY;

AND

57, CHARING CROSS, WESTMINSTER.

This Company offers

COMPLETE SECURITY.

Moderate Rates of Premium, with Participation in Four-fifths, or Eighty per Cent., of the Profits.

Low Rates, without Participation in Profits.

LOANS

in connection with Life Assurance, on approved Security, in Sums of not less than £500.

ANNUAL PREMIUM

required for the Assurance of £100 for the Whole Term of Life:—

Age.	Without Profits.			With Profits.			Age.	Without Profits.			With Profits.		
	£.	s.	d.	£.	s.	d.		£.	s.	d.	£.	s.	d.
15	1	11	0	1	15	0	40	2	18	10	3	6	5
20	1	13	10	1	19	3	50	4	0	9	4	10	7
30	2	4	0	2	10	4	60	6	1	0	6	7	4

ROBERT TUCKER, *Actuary & Secretary.*

The following Divisions of Mr. Scratchley's "TREATISE ON ASSOCIATIONS FOR PROVIDENT INVESTMENT" *may be had separately :—*

DIVISION I. **MANUAL TREATISE ON SAVINGS BANKS;** containing a Review
420 *pp.* of their Past History and Present Condition, and of Legislation on the Sub-
14s. ject; together with much Legal, Statistical, and Financial Information for the
  use of Trustees, Managers, and Actuaries. (*Longmans'.*)

DIVISION II. **BENEFIT BUILDING SOCIETIES, TONTINES, & EMIGRATION**
Third Edition. **SOCIETIES.**
310 *pp.*
7s. 6d.

Continuation of **CHURCH LEASES. — ADVOWSONS, NEXT PRESENTATIONS,**
DIVISION II. **HERIOTS, FINES, TITHES,** &c. Numerous New Tables, with Instruc-
Fourth Edition. tions for the use of Clergymen, Solicitors, and Estate Agents desirous of
192 *pp.* knowing the Values of the above, are given in the COPYHOLD & CHURCH
3s. 6d. LEASE ENFRANCHISEMENT MANUAL.

DIVISION III. **LIFE ASSURANCE SOCIETIES** and **FRIENDLY SOCIETIES;**
Tenth Edition. with an Exposition of the **TRUE LAW OF SICKNESS,** Instructions for
316 *pp.* Valuing Post Obits and Reversions, and for Investigating the Affairs of
7s. 6d. Assurance Societies, &c.

Each Division contains a Set of Model Rules, with numerous Tables, and the substance of the Acts of Parliament.

LONDON: CHARLES & EDWIN LAYTON, 150, FLEET STREET, E.C.

July 1st, price 3s. 6d.,

THE

MEDICAL CRITIC

AND

PSYCHOLOGICAL JOURNAL.

No. III.

EDITED BY FORBES WINSLOW, M.D., D.C.L. OXON.

A New Series of the "Psychological Journal," enlarged to 200 pages; one hundred devoted to PSYCHOLOGICAL Subjects, and the remaining portion to Articles upon general MEDICAL POLITICS, LITERATURE, and SCIENCE.

The MEDICAL CRITIC consists of Essays, illustrative of the PRESENT and PROSPECTIVE condition of the MEDICAL PROFESSION in its MORAL, SOCIAL, POLITICAL, LITERARY, and SCIENTIFIC relations.

It will be the object of the Editor and his Literary Staff to discuss the important Questions of the day and Books that come under review in a fair, impartial, and liberal spirit, apart from feelings of a personal, private, or party character.

Contents of Nos. I. and II.

To be continued Quarterly, price 3s. 6d. each Number.

Just Published, Second Edition, Revised, 8vo. cloth, price 16s.

ON OBSCURE

DISEASES OF THE BRAIN,

AND

DISORDERS OF THE MIND.

BY FORBES WINSLOW, M.D., D.C.L. OXON.

LONDON: JOHN W. DAVIES, 54, PRINCES STREET, LEICESTER SQUARE, W.

Lightning Source UK Ltd.
Milton Keynes UK
UKHW021326250219
337978UK00013B/1586/P